MASCULINITY/FEMININITY

THE KINSEY INSTITUTE SERIES

June Machover Reinisch, General Editor

Volume I
MASCULINITY/FEMININITY: *Basic Perspectives*
Edited by
June Machover Reinisch, Leonard A. Rosenblum,
and Stephanie A. Sanders

MASCULINITY/ FEMININITY

Basic Perspectives

Edited by

June Machover Reinisch, Ph.D.

Leonard A. Rosenblum, Ph.D.

Stephanie A. Sanders, Ph.D.

New York Oxford

OXFORD UNIVERSITY PRESS

1987

OXFORD UNIVERSITY PRESS

Oxford New York Toronto
Delhi Bombay Calcutta Madras Karachi
Petaling Jaya Singapore Hong Kong Tokyo
Nairobi Dar es Salaam Cape Town
Melbourne Auckland

and associated companies in
Beirut Berlin Ibadan Nicosia

Published by Oxford University Press, Inc.,
200 Madison Avenue, New York, New York 10016

Library of Congress Cataloging-in-Publication Data
Masculinity/femininity.
(The Kinsey Institute series, vol. I)
Revised papers from a conference held in 1984.
Includes bibliographies and index.
1. Sex differences (Psychology)—Congresses.
2. Masculinity (Psychology)—Congresses. 3. Femininity (Psychology)—Congresses. 4. Identity (Psychology)—Congresses. 5. Sex—Cause and determination—Congresses.
6. Psychology, Physiological—Congresses. I. Reinisch, June Machover. II. Rosenblum, Leonard A. III. Sanders, Stephanie A. IV. Series. [DNLM: 1. Gender Identity—congresses. 2. Sex Characteristics—congresses.
3. Sex Differentiation—congresses. BF 692.2.M395]
BF692.2.M36 1986 155.3'3 86-5252
ISBN 0-19-504105-4

Printing (last digit): 9 8 7 6 5 4 3 2 1

Printed in the United States of America on acid-free paper

To Frank A. Beach whose insights and incisive research have immeasurably illuminated our modern views of the sources of masculinity and femininity and whose deeply human dedication to science has inspired two generations of students and colleagues.

Preface

Masculinity/Femininity: Basic Perspectives is the inaugural volume in *The Kinsey Institute Series*. In each volume, researchers from a wide range of academic disciplines draw on their own data and on the viewpoints of their own area of expertise to address the central issues in a specific arena of discourse. The chapters of each volume are written after the contributors participate in a *Kinsey Symposium* on the topic. As a result, they reflect the diverse perspectives that emerge during sustained discussions among colleagues from many different fields. The editors of each volume provide an introduction based on the full range of discussions at the Symposium and the contents of the final contributions. This overview highlights the central themes and research findings of the volume as well as the major issues for future consideration.

We would like to acknowledge the support of Indiana University and in particular Dr. Morton Lowengrub, Dean of Research and Graduate Development, and Dr. Kenneth R. R. Gros Louis, Vice President, Bloomington. Special thanks go to Ruth Beasley, Kathryn Fisher, Janet Watkins, and the other members of The Kinsey Institute staff for their assistance in planning and conducting the First Kinsey Symposium and in the preparation of this volume. We are deeply indebted to Lillian Reinisch for her exemplary efforts at the formidable task of developing and compiling the encyclopedic index to this volume.

Bloomington, Indiana
January 1987

J.M.R.
L.A.R.
S.A.S.

Contents

PART VI Psychosocial Perspectives

PART VII Cultural Perspectives

Contributors

Frank A. Beach, Ph.D.
Professor Emeritus, Department of Psychology
University of California
Berkeley, California

Sandra Lipsitz Bem, Ph.D.
Director, Women's Studies Program
Professor, Psychology and Women's Studies
Cornell University
Ithaca, New York

David Crews, Ph.D.
Associate Professor of Zoology and Psychology
The University of Texas at Austin
Austin, Texas

Kay Deaux, Ph.D.
Professor of Psychological Sciences
Purdue University
West Lafayette, Indiana

Jacquelynne Eccles, Ph.D.
Professor of Psychology
Research Scientist, Institute for Social Research
University of Michigan
Ann Arbor, Michigan

Anke A. Ehrhardt, Ph.D.
Professor of Clinical Psychology
Department of Psychiatry
College of Physicians & Surgeons
Columbia University
and
Research Scientist
New York State Psychiatric Institute
New York, New York

David A. Goldfoot, Ph.D.
Associate Scientist
Wisconsin Regional Primate Research Center
University of Wisconsin
Madison, Wisconsin

Roger A. Gorski, Ph.D.
Professor and Chairman
Department of Anatomy
Center for the Health Sciences
UCLA School of Medicine
Los Angeles, California

Robert W. Goy, Ph.D.
Director
Wisconsin Regional Primate Research Center
University of Wisconsin
Madison, Wisconsin

Michael Lewis, Ph.D.
Professor of Pediatrics
UMDNJ-Rutgers Medical School
New Brunswick, New Jersey

Eleanor E. Maccoby, Ph.D.
Professor, Department of Psychology
Stanford University
Stanford, California

Bruce S. McEwen, Ph.D.
Professor, Laboratory of Neuroendocrinology
Rockefeller University
New York, New York

John Money, Ph.D.
Professor of Medical Psychology
Associate Professor of Pediatrics
Department of Psychiatry and Behavioral Sciences
and Department of Pediatrics
The Johns Hopkins University and Hospital
Baltimore, Maryland

Ronald D. Nadler, Ph.D.
Research Professor
Yerkes Regional Primate Research Center
Emory University
Atlanta, Georgia

Deborah A. Neff, Ph.D.
Associate Scientist
Wisconsin Regional Primate Research Center
University of Wisconsin
Madison, Wisconsin

Milos Novotny, Ph.D.
Professor, Department of Chemistry
Indiana University
Bloomington, Indiana

June Machover Reinisch, Ph.D.
Director, The Kinsey Institute for Research
in Sex, Gender, and Reproduction
Professor of Psychology
Indiana University
Bloomington, Indiana

Leonard A. Rosenblum, Ph.D.
Professor, Department of Psychiatry
State University of New York
Health Sciences Center at Brooklyn
Brooklyn, New York

Anya Peterson Royce, Ph.D.
Dean of the Faculties
Professor, Department of Anthropology
Indiana University
Bloomington, Indiana

Stephanie A. Sanders, Ph.D.
Assistant Scientist
Science Assistant to the Director
The Kinsey Institute for Research
in Sex, Gender, and Reproduction
Indiana University
Bloomington, Indiana

Lionel Tiger, Ph.D.
Professor of Anthropology
Rutgers, The State University of New Jersey
New Brunswick, New Jersey

Steven G. Vandenberg, Ph.D.
Professor of Behavioral Genetics and Psychology
Institute for Behavioral Genetics
University of Colorado, Boulder
Boulder, Colorado

James R. Wilson, Ph.D.
Associate Professor of Behavioral Genetics and Psychology
Institute for Behavioral Genetics
University of Colorado, Boulder
Boulder, Colorado

MASCULINITY/FEMININITY

College JAM
27 West 20 street

1

Masculinity/Femininity: An Introduction

June Machover Reinisch, Leonard A. Rosenblum, and Stephanie A. Sanders

> *Feminine:* Soft, delicate, gentle, tender, docile, submissive, amenable, deferential. . . .
> *Masculine:* Robust, strong, lusty, energetic, potent, brave, bold, fearless. . . .
>
> Sample of synonyms from *The Synonym Finder* by J. O. Rodale, 1978

Categorization of the world around them is perhaps intrinsic to the very nature of higher organisms; humans, of course, are the most adept at this almost reflexive cognitive activity. As the authors of the current volume reflect in so many ways, no dichotomous classification is as pervasive and sometimes as pernicious in its ramifications as the efforts toward division of our species into the two, often mutually exclusive, categories, classes, subsets, groups, types, partners, and adversaries as those embodied in the terms *masculine* and *feminine.* Whereas other divisions of humankind might have more or less clarity, utility, or validity, no other formulation is as ubiquitous, is imbued with as much emotional conviction, or is underwritten with as systematic and energetic a societal and developmental commitment as the assignment of genders and their presumed characteristics.

As we reviewed the varied materials that fill this volume, we were reminded of the story of the three baseball umpires: The first umpire says, "I call 'em the way I *see* 'em!" The second responds, "Not me, I call 'em the way they *are!*" To these confident views, the third umpire, presumably older and wiser, proclaims, "They aren't anything until I *call* them!" In a sense, this parable captures much of the concern and controversy in our effort to elucidate and understand the origins of those features most often associated with males and females. Problems of interpretation remain, even though we now approach these issues in a more scientific manner than was characteristic of the anecdotal approach of the past.

Many of the chapters that follow reflect the efforts of a multidisciplinary group of scientists to delineate, at a variety of levels in nature and in contemporary society, the features common to each of the sexes and those that presumably distinguish between them. Other authors utilize approaches that seek to reveal the biological dimorphisms that may subserve the appearance of differences and similarities between the sexes. The view that emerges from the totality of material before us is that our judgments about masculinity and femininity are a function of the questions we ask, how we seek to answer them, and how we interpret those answers—all undoubtedly reflecting, in their own right, the limitations of the particular cultural context within which we live. As we have learned in so many domains of inquiry, the question is not a matter of simple dichotomies (Chapters 19, Bem; 2 and 3, Money and Beach; 17, Ehrhardt)—nature or nurture? are there true differences or aren't there? are the differences good or bad?—for all, "they aren't anything until *we* call 'em."

The scientific method alone is clearly not a surefire means of unscrambling the complex intermixing of biological, individual, contextual, and cultural inputs that structure our views of the world around us. As Gould (1981) points out, "scientific" efforts at establishing "known truths" produce the type of error that is often the most insidious and difficult to discern. Consider the following two nineteenth-century quotations from Gould's *The Mismeasurement of Man*. One reflects an evolutionary and one a developmental perspective on the "recognized" distinctions between men and women. Pondering the apparently growing differences in brain sizes of men and women over evolutionary time, P. Topinard (1888), a prominent neuroanatomist said:

> The man who fights for two or more in the struggle for existence, who has all the responsibility and the cares of tomorrow, who is constantly active in combatting the environment and human rivals, needs more brain than the woman whom he must protect and nourish, than the sedentary woman, lacking any interior occupations, whose role is to raise children, love, and be passive. (Gould, 1981, p. 104)

Similarly, from a peculiar developmental perspective, reflecting the significance of the contemporary recapitulation views, E. D. Cope, a distinguished biologist of the same period, noted that the characteristics of women are:

> very similar in essential nature to those which men exhibit at an early stage of development. . . . The gentler sex is characterized by a greater impressibility; . . . warmth of emotion, submission to its influence rather than that of logic; timidity and irregularity of action in the outer world. All these qualities belong to the male sex, as a general rule, at some period of life . . . some early period of their lives when the emotional nature predominated. . . . Perhaps all men can recall a period of youth when they were hero-worshippers—when they felt the need of a stronger arm, and loved to look up to the powerful friend who could sympathize with and aid them. This is the "woman stage" of character. (Gould, 1981, p. 118)

The questions, the data, and the resulting interpretations fit well together, and conform as well with the presumptions and expectations of both society and the scientists of the era. When the pieces fit together in such a seeming coherent and culturally satisfying whole, does it matter that it is all incorrect? As the current volume reveals, it matters very much indeed!

Much progress has obviously been made in our understanding of these issues during the hundred years since perspectives such as those presented were the cherished views of both the scientific and lay communities. While bearing in mind the usual caveat that much remains to be done, and the usual conclusion that perhaps we have found more new questions than we have answers, readers will see that this volume demonstrates that in many ways the nature of these new questions is clearer than ever before. In addition, the contributions to the volume begin to provide concrete suggestions as to how best to proceed in answering the questions. That is not to say that this generation of scientists will not be making its own errors of judgment. Surely we are no better than our predecessors at "seeing the lens through which we look." Nonetheless, as the twenty-one chapters in this volume testify, we may for the first time be reaching an understanding of the kind of errors to be avoided.

Consider the hoary, always dying, but never dead, pseudoquestion of the role of nature and/or nurture in determining the most commonly observed masculine and feminine characteristics. Consensus has grown regarding the important contributions of biological substrata to the ultimate expression of sexually dimorphic behavior observed in the organism as it functions within its complex and changing environment. (Chapters 6, Crews; 7, Novotny; 4, Gorski; 5, McEwen; 10, Vandenberg; 11, Wilson). Culture may seek to diminish, exaggerate, ignore, or even reverse the impact of these biological factors (Chapters 20, Royce; 21, Tiger; 19, Bem). Nevertheless, although the pathways from genes to behavior may be extremely complex in all organisms (Chapter 17, Ehrhardt), culture and experience act on a given constellation of capacities and propensities present at the start of the whole process, even when that interactive process is seen as beginning prior to conception.

As a number of authors point out (Chapters 2, Money; 15, Maccoby; 18, Deaux), to consider the classic nature/nurture question correctly, we must first ask what is meant by *commonly observed* masculine and feminine characteristics. Whether we consider relatively simple forms (Chapter 6, Crews), nonhuman primates (Chapters 12, Goldfoot and Neff; 13, Goy; 8, Nadler; 9, Rosenblum), human children (Chapters 14, Lewis; 16, Eccles), or adults (Chapter 18, Deaux), the constellation of patterns that we might consider masculine and feminine are truly "fuzzy sets." Contradictions are apparent within the sets (e.g., what is sexy? what is masculinity/femininity?—Chapter 15, Maccoby), and some have difficulty and disagreement as to set boundaries, but all agree that no matter how precise the definitions (removing the fuzz from the sets, as it were), masculine and feminine characteristics are found in greatly overlapping distributions within any given species.

Even if we can agree on what is meant by *commonly observed,* the next question concerns whether the features we agree to assess represent the innate capacities of the organisms in question or whether these are "merely" the features that have been learned or chosen for expression. Regardless of which elements we *believe* we are including or excluding from our focus of inquiry, we still must decide within which setting we will examine the expression of these characteristics. This contextual factor often either is ignored or its effects go unestimated, and yet it appears to be a factor of inestimable importance (Chapters 12, Goldfoot and Neff; 13, Goy). Whether we attempt to compare across species, ages, societies, or settings, individual expressions of behavior can vary as a function of the interplay of salient stimuli and the cognitive interpretation of these stimuli at any given time. Generalizations from laboratory to field or from laboratory to home or school may be distorted because of differences in demand characteristics of each setting (Chapters 8, Nadler; 9, Rosenblum). Thus, when different species or individuals are observed in varied settings, they may appear to be quite different from one another, not solely because of intrinsic distinctions among them, but in part as a result of the disparate characteristics of the diverse settings. So it's all an interaction effect—where have we heard that before? We are still left with the original question one step removed. Is the tendency to adopt certain expectancies or to alter patterns in a particular direction within different situations a reflection of nature or . . . ?

The seductive quality of simple dichotomies like those of male/female or of nature/nurture can also be seen in the scientific efforts to avoid the pitfalls of cultural or contextual biases by examining masculinity/femininity questions from biological (biochemical, anatomical, physiological), comparative (interspecific), or cross-cultural perspectives. Unfortunately, our procrustean efforts notwithstanding, the universe in which we live can rarely be split into "either/or" categories. For example, whereas significant differences between male and female brains in various species have been well documented, and particular brain structures may differentiate as the result of sex-specific hormonal events, the parameters of the relationships among physiology, anatomy, biochemistry, and behavior are not yet clear, and the forces that mediate effects across these levels are generally unknown. Even a crude isometry between singular biological features and particular behavioral outcomes does not appear to exist. Nonetheless, few would deny the intrinsic importance of the biological substrate in providing the boundaries within which behaviors come to be expressed. As this volume amply reflects, we have made considerable strides in the crucial process of identifying the types of structures and related biological events that are at least consistently correlated with certain masculine and feminine characteristics in a number of settings.

The task of dissecting the role of these biological forces will not be simple, however. Correlations between hormonal and behavioral differences may be bidirectional, are far from perfect within and between subjects, and are not always consistent across species. Complicating the task even further, certain behaviors may be the correlates of antecedent instead of contemporaneous physiological processes, or they may be the consequence of prior behavioral factors that were themselves the sequelae of even earlier biological events. If, as seems to be suggested in the material of this volume, the same individual does not show the same set of behaviors in each different setting, our understanding of the role of particular biological factors must also be viewed from an extremely complex contextual perspective.

We confront the same configuration of exciting progress and significant problems as we attempt to look past our cultural and contextual biases by examining sexually differentiated patterns in other species. In the effort to seek constancies in the midst of diversity, other species (presumably devoid of the types of cultural impact found in humans) can provide opportunities for untangling certain aspects of the biological and experiential predeterminants of sexually dimorphic patterns.

All our problems are not circumvented by the comparative approach. When, for example, several different species show similar patterns of behavioral sex differences, we may have more confidence that an evolutionarily conservative pattern may be present and that such stable distinctions may be sought or expected elsewhere (Chapters 4, Gorski; 3, Beach). As a number of authors (Chapters 5, McEwen; 12, Goldfoot and Neff; 13, Goy) point out, however, we must remain acutely aware of the crucial distinction between analogous and homologous comparisons, when looking at different societies or when leaping from branch to branch in the phylogenetic tree!

We see repeated instances in nature of similar environmental demands eliciting apparently similar outcomes in very different animals, each building the common outcome on a different substrate and with different developmental and contemporaneous mechanisms. Whether we are examining the relationship of particular early events to later adult performance (Chapters 16, Eccles; 14, Lewis), or the role of current hormonal status on behavioral expression (Chapters 5, McEwen; 6, Crews; 7, Novotny), we must remember that each species has its own evolutionary history and hence distinctive genetic structure (Chapters 10, Vandenberg; 11, Wilson), each is found in its own niche (Chapters 8, Nadler; 9, Rosenblum), each may have unexamined differences in sensory capacities (Chapter 7, Novotny) resulting in responses to different aspects of the environment, and each has passed through a different chain of developmental events. The acceptance of the multidetermined nature of femininity and masculinity, however defined, provides both the excitement and the cautionary view with which we must proceed in its study. Common outcomes do not necessarily imply common causes!

A mixture of causative factors can also be seen when we seek to determine the origin of masculinity and femininity across cultures. The authors of this volume agree that culture acts on a biological substrate that presents to the world organisms already differentiated in a number of ways prior to birth, and with perhaps a differential readiness to enlarge on those initial distinctions in a variety of ways. We may reasonably conclude that subsequent experience acts to enhance, diminish, or allow the relatively unaffected expression of these genetically or prenatally determined differences. Most distinctions fall into largely overlapping distributions across the sexes, however. Perhaps because of these similarities between the sexes, societies expend considerable energy to delineate many of the differential properties and to blur or counteract others with such vigor and conviction (Chapters 16, Eccles; 19, Bem; 18, Deaux; 21, Tiger). The persistent struggle of societies to articulate the boundaries between the sexes is perhaps most dramatically portrayed on the narrower stage of theater and art, where, through the varied pigments of paint and technique, masculinity and femininity are exemplified, highlighted through contradiction, or discarded altogether (Chapter 20, Royce).

Bearing all these perspectives in mind (and such a feat of mental cohesion is by no means easy), we see that current research on the development of masculine and feminine characteristics in our own culture reflects our growing sophistication regarding the nature of our questions and the answers to them. Mixing metaphors, as we mix our data sets, all authors agree that biology stands as the gatekeeper to expressions of gender-identity/role (Chapters 2, Money; 3, Beach; 10, Vandenberg; 11, Wilson), but that even before birth environmental factors enter the fray to begin their constant transaction with the biological potentials (Chapters 4, Gorski; 17, Ehrhardt). Once the differentiated organisms emerges, the forces of the parental dyad (Chapters 14, Lewis; 9, Rosenblum) and then the child and adolescent subcultures (Chapter 15, Maccoby), as well as the wider culture that surrounds them (Chapters 20, Royce; 21, Tiger) all come vigorously into play. Through it all, the biological machinery—sometimes apparently robotically, more often not—affects and is affected by the changing organism and the diversifying environments through which it moves.

Are the early events or features (whether morphological, hormonal, or experiential) predictive of the later ones? Will relatively simply linear models of developmental processes account for the changes and the constancies over time? If two organisms appear to begin at the same point or status and appear to end with the same outcome, are the intervening processes the same? The current answers to these questions are both yes and no! Clearly, no one rule of thumb, theoretical approach, methodological sophistication, or system of measurement can yet extricate all aspects of the development of masculinity and femininity from the complex web of factors that we now know acts to produce the constellation of features we seek to define as masculine or feminine.

Like all of science, the study of masculinity and femininity must be seen as a process, not as a product. We see the contents of this volume as reflecting both the accomplishments of the road already traveled and the directions of the road ahead. The contributions to this volume include chapters rich in empirical details and their immediate and theoretical importance; others provide less bounded interpretations of the array of materials that lie before us and consider in more varied fashion the vigilance we must maintain as we proceed. The chapters are apportioned to seven interrelated and overlapping perspectives with which this complex problem area may be viewed. The chapters are ordered into seven broad sections: Psychobiological Perspectives, Neuroscience Perspectives, Evolutionary Perspectives, Behavioral Genetics Perspectives, Developmental Perspectives, Psychosocial Perspectives, and Cultural Perspectives. The reader should be aware, however, that the heuristic device of applying varied labels to these perspectives is not an attempt to maintain disciplinary boundaries among them. Even though we have attempted to delineate sections, the goal should be an integration and not a disjunction of these related approaches. Each contributes to a fuller understanding of the other; in more than just principle, each contributes to a more comprehensive and comprehensible appreciation of the scientific concept of masculinity/femininity.

Reference

Gould, S. J. (1981). *The Mismeasurement of Man.* NY: W. W. Norton & Company.

I

PSYCHOBIOLOGICAL PERSPECTIVES

2

Propaedeutics of Diecious G-I/R: Theoretical Foundations for Understanding Dimorphic Gender-Identity/Role

John Money

Of the Flesh and of the Spirit: Nature and Nurture

The difference between male and female is something that everybody knows and nobody knows. Everybody knows it, proverbially, as an eternal verity. Nobody knows it, scientifically, as an absolute entity for, as day and night merge under the glare of the midnight sun, male and female merge under the scrutiny of empirical inquiry.

In the technical vocabulary of science, a monecious species of animal or plant is one in which male and female are accommodated in only one house (Gr. *oikos,* "house"). In a diecious species, there are two houses, one for the male and one for the female. The single house has been cut or divided—etymologically, *sex* derives from L. *secare,* "to cut or divide." Thus, according to the wisdom of etymology, a male and female are defined, respectively, on the criterion of the morphology of the body in which they are housed. Beyond etymology, the great interest of science is in the live building materials that differentiate not only the construction of the two houses, but also the totality of their function.

In the prescientific thought of the era that preceded modern science, the constituents of sex were divided between the flesh and the spirit. The flesh was responsible for the animal nature of the sexes, and for their carnal desires that needed the purification of the spirit. The flesh represented the sin of passion, and the spirit the purity of reason. Women were creatures of weakness and passion, men of power and rationality. These stereotypes have changed but have not become extinct.

The religious doctrine of the sin of the flesh versus the righteousness of the spirit underwent a scientific metamorphosis in the nineteenth century. It became the secular doctrine of nature versus nurture. In contemporary scholarship regarding sex difference, nature is equated with biology, and nurture with what is taught and learned. Biology is of the flesh and immutable. Learning, if not of the spirit, is of the mind and mutable—ostensibly, what is learned can be unlearned, and unlearning takes place under the auspices of behavior modification or psychotherapy.

The faultiness of this line of reasoning is that there is a biology of learning and remembering. It is part of neuroscience. Although the biology of learning is only at the beginning of its development, its very existence is sufficient to demonstrate that the converse of biology is not social learning and memory. Logically, the converse of biology should be spiritualism and the astral body.

A New Paradigm: Nature/Critical-Period/Nurture

The biology of learning includes learning that becomes imprinted and immutable. Thus, a new paradigm is needed to replace nature/nurture with a three-term principle: nature/critical-period/nurture. The interaction of nature and nurture at a critical period may produce a permanent sequela that, in turn, may react at another critical period with a new facet of nurture. This principle has long been known in embryology. It is only slowly being applied to postnatal development.

Even nature is not inevitably immutable. The genes that govern heredity, the very epitome of nature, can be altered at a critical period by the intervention of nurture in the guise of gene splicing. The alteration then becomes permanent.

There is something quaintly archaic in the application of a resurgent nature/nurture ideology to the science of sex difference. The hidden agenda of this archaism is political. Formulated in terms of reductionistic biology, nature is a political strategy of those committed to the status quo of sex differences. They use reductionist biology to maintain the biological inevitability of sex differences and to exclude the possibility of their being historically stereotyped. Conversely, neglect of biological determinism is the political strategy of the advocates of the political and social liberation of the sexes from historical stereotypes. I can attest to these strategies with authority born of experience, for I have been attacked by nurturists for representing the absolutism of prenatal hormonal determinism, and by naturists for representing the relativism of social determinism.

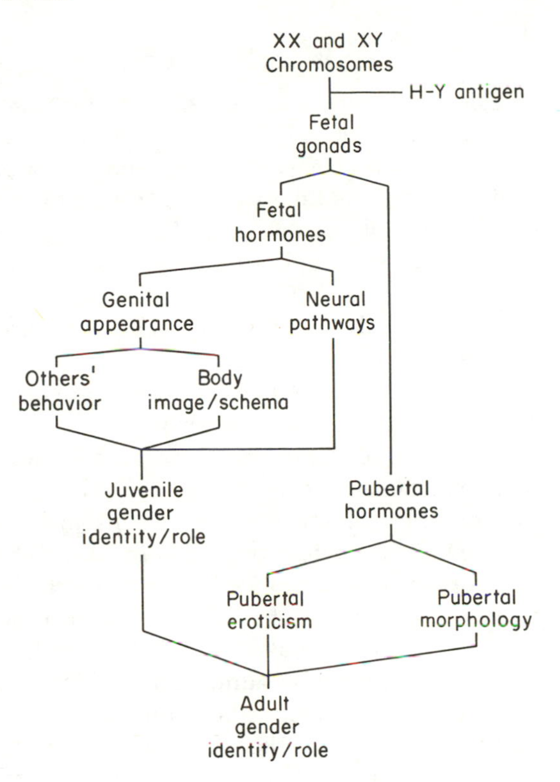

Figure 2–1. The multivariate and sequential determinants of G-I/R. When H-Y antigen was first discovered, it was believed to be invariably associated with differentiation of the gonadal anlagen as testes. Exceptions to this rule have subsequently been discovered.

Unification of the nature/nurture polarity can be found in the paradigm of native language. At birth, a baby must be equipped with a healthy human brain in order to develop a native language. But no genes program a native language. It must be programmed into the brain, socially, from the example of others who use it, and the programming must be done at a critical period of development.

As applied to sex difference, this paradigm means that some of the program of sex difference is phyletically designed and shared by most, if not all, members of the species, both male and female. This phyletically shared segment is not the whole program, however. The remainder is the ontogenetic program, designed by personal and sociocultural history, specific to time and place. The great scientific challenge is to identify empirically the variety and number of phyletic and ontogenetic determinants, regardless of whether they are conventionally attributed to nature or nurture, and to uncover the mechanisms by which they underlie, overlie, and influence one another developmentally. This requires not a reductionist theory, but one that is both multivariate and sequential (Fig. 2-1).

Multivariate Sequential Determinism

I designed Figure 2-1 originally for the *American Handbook of Psychiatry* (Money, 1974); it first appeared in print in Money and Ehrhardt (1972) and has since been used in several textbooks. Its purpose is to show that the adult status of gender-identity/role (G-I/R) is the culmination of a sequential and multivariate process. The multivariables range from prenatal to pubertal hormones, from body morphology to body image, and from prenatal hormonalization of the brain to postnatal assimilation of cultural stereotypes of male and female.

In ordinary, healthy people, the multiple variables of sex, both male and female, correlate perfectly with one another, so that their potential independence of one another is not self-evident. This perfect correlation does not exist in syndromes of hermaphroditism. By reason of the discrepancies that exist in these syndromes, I could specify several of the variables of sex that are potentially independent of one another (Money, 1952; 1955; Money, Hampson, & Hampson, 1955).

In order of appearance, these variables were, as specified in 1955: chromosomal sex, gonadal sex, fetal hormonal sex, internal morphologic sex, and external morphologic sex. In postnatal life, the variables were, successively: assigned sex, sex of rearing (including clinical habilitation), and pubertal hormonal sex. Together these variables were held responsible for the differentiation of gender role and orientation (identity) as male, female, or androgynous in the course of growing up. Since 1965, beginning with the 24th edition, *Dorland's Illustrated Medical Dictionary* has incorporated these variables into its definition of sex.

In 1955, the relationship of fetal sex hormones to the masculinization or feminization of brain and behavior was a branch of science that would not make its debut until the end of the decade.

Gender Role and Identity

The term *gender role* first entered the literature and was defined in the previously cited articles (Money, 1955; Money et al., 1955). I found it not only expedient but also obligatory to coin this new term because, as a parallel to chromosomal sex or gonadal sex, a term such as *sex role sex* would make little sense. Moreover, because of the birth defect of the sex organs in hermaphroditism, the term *sex role* is needed to apply literally to the person's male or female capability with the sex organs in a sexual partnership. In traditional sex role psychology, the term *sex role* refers to everything that differentiates male and female, except the use of the sex organs.

As I originally defined it, *gender role* meant everything (literally everything, the erotosexual included) that a person says or does to indicate to the self or others his/her status as male, female, or ambiguous (androgynous). In any given case, this definition reciprocates an introspective, private self-definition, given to the self alone, with an observational and public definition arrived at by others on the basis of what they see, hear, touch, smell, or taste, with respect to the person concerned.

This reciprocal definition did not take root. It split. The introspective part was renamed *gender identity*. In its most simplistic form, the term *gender identity* is defined by some who use it as the declaration, "I am male" or "I am female."

The term for the observational part of the definition did not change, but the definition did: *gender role* became a social script or social stereotype to which an individual either did or did not conform.

In obedience to the nature/nurture cleavage, *gender* thus became allocated to nurture, and set off against *sex*, which became the property of nature. In this way, the vested interest of social science was left unperturbed. Social scientists claimed *gender role* for themselves. Media people and the general public followed suit, as in the political term *gender gap*. Thus the obligation to relate *gender role* to biology as well as to history and sociology was circumvented.

Biologists themselves have adopted the fashion of using the term *gender* outside of its former grammatical domain. But they use it chiefly as a harmless synonym for *sex*—as in referring to the gender of an animal. Going a step further, one biologist claimed to have found a hormonal determinant (testosterone) of gender identity at puberty (Imperato-McGinley, Guerrero, Gautier, & Peterson, 1974). And so the nature/nurture pincer tightens its grip!

The people most caught in the squeeze are those who deal with the gender-identity syndrome of transexualism, renamed gender dysphoria. Whereas, for the most part, they do not favor a social science etiology for gender dysphoria, their hormonal data do not support the hypothesis that the testosterone level correlates with gender identity.

G-I/R: Heterosexual/Bisexual/Homosexual

My own present gender terminology forces the union of gender identity and gender role as two sides of the same coin, by using the acronym G-I/R for gender-identity/role. G-I/R has different components, one of which is erotosexual. Another is occupational, another recreational, and so on.

Erotosexually, a G-I/R may be heterosexual, bisexual (androgynous), or homosexual, or it may be transexual or transvestic. It also may or may not be paraphilic ("kinky"). It may or may not be congruous with the occupational component of G-I/R. Thus, one may, for example, say of a person that he has a homosexual G-I/R erotosexually, but not occupationally or recreationally. This avoids the conceptual diffuseness and contradiction inherent in saying that such a person has a male gender identity and a male gender role, but a homosexual preference or object choice.

Being homosexual—or bisexual or heterosexual—is not a preference or choice. It is a status—the status of having a G-I/R erotosexually concordant or discordant with the reproductive status of the genitalia, or an androgynous combination of both possibilities. The ultimate criterion of homosexual, bisexual, or heterosexual is not simply the sex of the partner with whom one's own sex organs are shared, but the sex of the partner toward whom one undergoes the experience of being love smitten.

The person who falls in love with someone of the same genital morphology is erotosexually homosexual. He or she may procreate with a partner of the opposite sex, and may or may not appear to have an incongruous G-I/R at work and in other nonerotic, nonsexual contexts.

Sex-Irreducible G-I/R

Four grades of sexual dimorphism are found in G-I/R: sex irreducible, sex derivative, sex adjunctive, and sex arbitrary.

Among primates, the sex-irreducible difference is that males impregnate and females menstruate (which also usually means ovulate), gestate, and lactate. This division of reproductive labor holds throughout the diecious mammalian kingdom, but it does not invariably apply to diecious fishes. In many diecious fish species, an individual spends part of its lifetime breeding as a male and part breeding as a female (Chan, 1977), or vice versa. Some may alternate, more than once, their capability to produce both eggs and sperms. With each alternation is a concurrent capability of alternating the mating behavior of female and male. Both behavioral schemata are, by inference, on call in the brain and nervous system.

The concept of a brain that has encoded in it both the male and female schemata of behavior is especially well illustrated in a parthenogenetic species of whiptail lizard, *Cnemidophorus uniparens,* studied by Crews (1982). Being parthenogenetic, this species has no sex; all members of the species are clones. The clones hatch from eggs. When a lizard is in an ovulatory phase, it matches up with a nonovulatory clone, and together they simulate the behavior of mating seen in closely related diecious whiptail species. At a later phase, their roles may be reversed.

Once science uncovers the secret of the hermaphroditic versatility of sex-changing fish and parthenogenetic lizards, then on the criterion that today's science fiction becomes tomorrow's science, it will undoubtedly be applied to mammals. Thus, one can envisage a future when the sex-irreducible G-I/R will no longer be fixed and irreducible, but, by a process equivalent to reverse embryogenesis, it will be sex reversible.

Sex-Derivative G-I/R

A male or female difference that is not sex irreducible may nonetheless be derived from the difference that is irreducible. That is, it may have its origin in the same hormones that in prenatal life program the differentiation of the genitalia.

Fetal morphologic differentiation as male or female follows what may, on the basis of current knowledge, be known as the Adam/Eve principle. This principle signifies that regardless of its chromosomal sex, the primary destiny of the fetus is to become Eve, not Adam. Adam requires that something be added—namely, MIH (Müllerian inhibiting hormone) from the embryonic testes, to vestigiate the Müllerian ducts and prevent the growth of a uterus and fallopian tubes; and also testosterone, again from the testes, to induce development of the male organs.

If the testes fail to develop, a fetus will develop the genital morphology of a female, minus ovaries. An obvious sex-derivative by-product of such a deandrogenized development in the human fetus is that the baby will grow up to urinate in the sitting position. The reverse may also occur, so that an excessively androgenized 46,XX fetus with ovaries is born with a fully formed penis and empty scrotum.

When the latter condition is experimentally produced in animals, there is a species difference regarding the extent to which subsequent mating behavior will also be masculinized. The sheep has proved to be preprogrammed to be what I have called a hormonal robot. That is, the complete repertoire of its adult mating behavior is preprogrammed prenatally by a sex-hormonal effect on the brain. A film from Edinburgh by Short and Clarke (no date) shows ewes impersonating rams at the first mating season. Their blood hormone levels at the time were normally female. They had been altered only in fetal life by the injection of testosterone into the pregnant mother. The timing was precisely calculated so that it was too late for the external genital morphology to be masculinized, but not too late for the sexual brain to be masculinized. Thus, the animal was preprogrammed to behave like a ram when old enough for first mating—and to be responded to as a male rival by other rams.

This same hormonal-robot effect is not seen in primates. It is also not encountered in human beings with clinical syndromes that induce prenatal masculinization. In its place is a greatly attenuated effect in the nature of a tendency or predisposition that will become incorporated into either a masculine or feminine G-I/R, in accordance with the vicissitudes of rearing as a boy or girl, respectively. The terminology I have used for these sex-derivative aspects in human beings is that they are sex shared but threshold dimorphic. They are not dimorphic in the absolute sense of either male or female, but in the relative sense of being manifested with greater facility in one sex than the other. Parentalism is an example. Parentalism is sex shared insofar as it is exhibited by the father as well as the mother, but is threshold dimorphic insofar as an infant or child evokes it more readily and more frequently in the mother than the father. For example, the sleeping mother is typically more sensitive to the stirrings of the neonate than is the sleeping father.

Nine Sex-Shared Threshold Dimorphisms

To date, nine phyletically basic behavioral dispositions have been isolated. Although they might appear as sexually dimorphic, they are actually sex shared and dimorphic only in either threshold or frequency of manifestation. With the advent of new research, the list may well need to be revised.

First is kinetic energy expenditure, which, in its more vigorous, outdoor, athletic manifestations, is typically more readily elicited and prevalent in males than in females, even before males reach the postpubertal stage of being, on the average, taller, heavier, and more lean and muscular than females.

Second is roaming and becoming familiar with or marking the boundaries of the roaming range. Whereas pheromonal (odoriferous) marking is characteristic of some small animals, in primates including humans vision takes the place of smell. The secretion of marker pheromones is largely under the regulation of male sex hormone and thus is more readily elicited in males than females. The extent of a sex difference in the threshold for visual marking in primates is still conjectural.

Third is competitive rivalry and assertiveness for a position in the dominance hierarchy of childhood, which is more readily elicited in boys than in girls. A position of dominance may be accorded an individual without fighting or after a victory. Whereas fighting and aggressiveness per se are not sexually dimorphic, despite a widespread scientific assumption that they are, sensitivity to eliciting stimuli may or may not be. An example of the latter is retaliation against a deserter or rival in love or friendship, which is not sex specific.

Fourth is fighting off predators in defense of the troop and its territory, which, among primates, is typically more readily elicited in males than females.

Fifth is fighting in defense of the young, which is more readily elicited in females than males. Females are more fiercely alert and responsive to threats to their infants than, in general, are males.

Sixth is a provision of a nest or safe place for the delivery, care, suckling, and retrieving of the young. This variable may be associated with a greater prevalence of domestic neatness in girls than in boys, as compared with the disarray that is the product of, among other things, vigorous kinetic energy expenditure.

Seventh is parentalism, exclusive of delivery and suckling. Retrieving, protecting, cuddling, rocking, and clinging to the young are more prevalent in girls' rehearsal play with dolls and/or playmates.

Eighth is sexual rehearsal play. Evidence from monkeys is that juvenile males elicit presentation responses in females, and juvenile females elicit mounting responses in males more readily than vice versa. The taboo on human juvenile sexual rehearsal play and on its scientific investigation prohibits a definitive generalization regarding boys and girls at the present time.

Ninth is the possibility that the visual erotic image more readily elicits an initiating erotic response in males than in females, whereas the tactile stimulus more readily elicits a response in females. Here again, no definitive generalization can yet be made because of the effects of the erotic taboo and erotic stereotyping in our society.

Sex-derivative roles are, by definition, not sex exclusive, but are sex shared or interchangeable. Statistically there is overlap, so that extremely masculinized people of either sex resemble one another more closely than do men at either end of the scale of masculinization—and, conversely, women. In fact, men find it a severe hardship to be forced by cultural tradition into an ultramacho sex-derivative role to a degree beyond which they are, by disposition, ill prepared.

Sex-Adjunctive G-I/R

Whereas sex-derivative is secondary to sex-irreducible G-I/R, sex-adjunctive G-I/R is tertiary. To illustrate: fetal androgen has a sex-irreducible effect in forming the male genitalia. It has also a sex-derivative effect in setting thresholds for certain constituents of behavior that are manifested more prevalently, or with more facility, in the play of boys than in the play of girls. With puberty, androgen reasserts its sex-irreducible effect, although it can be attenuated or, by castration, interrupted. It governs the maturation of the male reproductive system, and also the growth of bone, fatty tissue, and muscle in such a way that males—on the average, and despite much overlap with females—are more muscularly powerful than females, on the basis of being taller and leaner. Thus, in all the millennia prior to the modern age of industrial, farming, and domestic laborsaving devices, it made sense for societies to have a sex-based division of labor.

This division of labor was, in ancient times, based on the greater mobility of the male as compared with the restricted mobility of the female while pregnant or breast-feeding. Women moved in closer proximity to the home base than did men, and they were chiefly responsible for feeding not only their babies but all members of their troop. The men ranged more widely in hunting, fighting, and trading. This ancient system of the division of labor survived the partial equalization of the sexes made possible first by the domestication of animals of transport. It is surviving less well the mechanization of transport by land, sea, and air, the invention of automated and computerized laborsaving devices, and the discovery of infant feeding formulas, prepared baby foods, and ready-to-eat meals for the family.

The age-old division of labor has also been rendered anachronistic by the invention of contraception, which permits an abbreviated breeding period. Fewer births are needed, because infant mortality is down, and a world already exploding with population needs a smaller next generation.

Added to these population dynamics is the dramatic increase in longevity from an average age of 35 in 1776 to around 70 or older today. Adulthood begins earlier, with the age of puberty having lowered by 4 months per decade for a century and a half, or longer. These life-span changes mean that men and women have longer adult lives in which not to be enchained to the obligation of a sex-stereotyped division of labor in order to ensure that they will successfully launch a family in their formerly short life span. They have more years in which to do different types of work.

These tides of change that have set in motion the destereotyping of sex-adjunctive roles have also provoked a backlash. Most of the current feuding and political debate about equal rights for men and women rage around social, legal, and economic rights and restrictions of sex-adjunctive work roles and play roles.

Science has been drawn into this dispute. Funding for research on the psychology of sex differences, on closer scrutiny, turns out to be funding for research in sex differences in work or play. The current fashion is to research these differences in terms of aggression, delinquency, emotionality, learning disability, space-form perception, verbal ability, mathematical ability, and homosexuality in relation to sex differences in hormone levels, age of pubertal onset, and brain lateralization. All too often, the hidden assumption behind such research is that if a biological (physical or organic) correlate can be found, then the difference is one of real nature, and not of trivial nurture—an assumption that requires a blind adulation of biology. It is totally oblivious to the critical-period hypothesis and to the possibility that in some instances that which is postnatally learned becomes permanently encoded in the brain.

Sex-Arbitrary G-I/R

To a visitor from another planet, political and scientific feuding about sex-derivative G-I/R could well seem incomprehensible, but its incomprehensibility would be minor in comparison with that engendered by militancy over sex-arbitrary roles. Sex-arbitrary roles pertain to issues of sex-divergent body language, ornamentation, grooming, clothing, and etiquette. Often some sort of connection can be conjectured between sex-arbitrary and sex-adjunctive roles. For example, the former Polynesian custom of restricting the amount of tattoo a woman might have on her face as compared with that of a man ostensibly reflected her lesser bravery as a warrior. In fact, it signified her role as a lesser warrior who might fight only in defense of the tribal home territory when it was under attack by the enemy, whereas the male traveled far to maraud and initiate attack.

The ancient and widespread custom of the greater mobility of men has also been reflected in the footwear of women. Until the 1949 Communist revolution in China, families who aspired to wealth and prestige deformed the feet of their young daughters by binding them. These daughters were a living testament to their fathers' wealth, and also to that of their prospective husbands, because their deformed feet rendered them incapable of working competitively with men; henceforth they could work only as courtesans or prostitutes. Today in our own society, we have an adumbration of this custom of deformed feet in the fashion of high-heeled shoes for women that hobble their gait and enforce a method of locomotion that men may interpret as needing their support.

The contemporary American customs of feminine decoration can be traced to an earlier era in which aristocratic women were idle display models, exhibiting the wealth of their fathers or husbands. Now that women work and have their own wealth, women may opt to abandon cosmetic and jewelry decoration, or their male partners may opt to be decorated similarly. The furor in the 1960s regarding long hairstyles for males as well as females alienated many fathers from their own sons as effectively as if long hair were a badge of homosexual effeminacy—even though the new style also dictated a macho moustache and beard as a badge of masculinity. Here is certain evidence of the pervasive extent to which the superficialities of the male or female permeate the average person's perception—or misperception—of irreducible masculinity and femininity. *Misperception* is the correct term; the average person, including the average physician or scientist, is not accustomed to differentiating the constituents of G-I/R that are not erotosexual from those that are.

Identification/Complementation

The constituents of G-I/R that are sex arbitrary, and likewise those that are sex adjunctive, become incorporated into the G-I/R totality by a joint process of assimilation and learning—namely, identification and complementation. Identification is a familiar concept in psychology to signify learning that takes shape by copying, imitating, or replicating an example or model. In learning the sex-arbitrary conventions of couple dancing, for example, a girl identifies with a female dancer and a boy with a male dancer. The girl then complements her dancing to that of a male partner, and the boy to that of a female.

Complementation is not yet a familiar concept in psychology. I first became aware of the necessity of the concept on the basis of an account given to me by a father whose hermaphroditic child was sex reassigned at the age of 18 months. When she was 3 years old, the child made a boisterous game of copying her older brother's rock-and-roll dancing. Her father spent time with his two children when he arrived home from work. His impulse was to have his daughter dance with him as a partner, more sedately. Initially she resisted, in favor of copying her big-shot brother. After a few evenings, when her brother became a rival for the father's attention, she was won over. The father's lesson to his son was that little boys copy their fathers, but dance with their mothers, and little girls copy their mothers, but dance with their fathers. He was aware of the role he was playing as a partial architect of G-I/R. The mother did not actually have to be there in person; she could have been symbolically represented as if present.

Complementation or identification each builds up its own representation or schema in the brain, and each becomes incorporated with its sex-derivative counterpart—but a sex-irreducible counterpart exists for only one of the pair—namely, the identification schema coded "this is me." The complementation schema is coded "this is other," the reciprocal of me. Complementation and identification schemas exist not only for gender, but also for age, social class, hierarchical authority, and so on.

The two gender schemata, male and female, exist compatibly in the brains of most people, one dominant over the other, their dual presence scarcely recognized. To encounter the duality, one needs a special syndrome—for example, transvestism of the type that is characterized by two names, two wardrobes, and two personalities, one male and one female (Money, 1974).

In some instances, this syndrome evolves to the point at which the complementation and identification schemata become permanently transposed (although the sex-irreducible constituent remains unchanged). Then the syndrome becomes one of the forms of transexualism. Eventually, the patient is likely to make an application for sex reassignment.

The transposition phenomena include also the little-known syndrome of the woman with a penis, for which I have coined the term *gynemimesis.* This syndrome is close to transexualism, whereby the person lives and passes as a woman and takes female hormones, but does not apply for surgical sex reassignment. In street argot, the syndrome is subsumed under the rubric *drag queen,* which recognizes its relationship to extreme homosexual effeminacy of the G-I/R.

The counterpart argot for females is *butch dyke,* for which the formal syndrome name is andromimesis. Although the andromimetic may apply for mastectomy, sex-reassignment genital surgery is not sought, and rarely is hormonal virilization.

The more partial degrees of G-I/R transposition manifest themselves as the varied phenomenology of homosexual and bisexual G-I/R. These phenomena are not designated as syndromes. They are now officially classified as typological distinctions, analogous to left-handedness and ambidextrousness.

Masculinization and Defeminization Are Not Synonymous

The coexistence of identification and complementation schemata in the brain parallels the biphasic alternation of masculine and feminine in sex-changing fish and in parthenogenetic lizards, as previously mentioned. Another parallel is found in mammalian prenatal experiments demonstrating that masculinization of the fetal brain is not automatically synonymous with defeminization, nor is feminization with demasculinization (Ward, 1972; 1977; Ward & Weisz, 1980; Baum, 1979; Baum, Gallagher, Martin, & Damassa, 1982).

Applied to behavior, these experiments mean that prepotency of sexual mounting in an animal with a masculinized brain is compatible with a lesser potency of sexual crouching or presenting. Whether the one or the other will manifest itself will depend partly on the strength, insistence, timing, or context of the external evoking stimulus, and partly on the internal status of the arousal threshold.

Erotosexual Rehearsal Play

Juvenile erotosexual rehearsal play is one of the sex-shared/threshold dimorphic constituents of G-I/R in the sex-derivative category. It is subject also to some sex-adjunctive overlay, insofar as it can be affected by the sex ratio of the age-mate play group.

The chief source of empirical data on juvenile erotosexual rehearsal play is the Wisconsin Regional Primate Center where juvenile rhesus monkeys have been studied. Isolation-reared monkeys are deprived of play and grow up unable to perform their part in the strategy of mating (Goldfoot, 1977). Consequently, they do not reproduce their species. Males are more vulnerable to impairment as a consequence of play deprivation than are females.

Impairment is partly offset by as little as a half hour daily of age-mate play. Given this opportunity, about one third of the young ones became competent in foot-clasp mounting and in presenting, as male and female, respectively. Whereas the juveniles not deprived of play achieved such competency by the age of 6 to 9 months, those permitted only a half hour of play a day did not achieve the same competency until as late as 18 to 24 months of age. Despite their success, the retardation had a still later effect on their breeding success, for they had an abnormally low birth rate. The two thirds of their unsuccessful age-mates continued to be unsuccessful as adults. They produced no offspring; they either did not mate or were incompetent when they attempted to do so.

Juvenile sexual rehearsal play and adult sexual interaction are also related to the sex ratio of playmates (Goldfoot, Wallen, Neff, McBrair, & Goy, 1984). In sex-segregated experiments, baby monkeys were reared with their mothers and with age-mates of only their own sex. Males in the all-male group showed more rear-end presenting and less foot-clasp mounting than did those in the male/female group. Conversely, females in the all-female group showed a significant excess of foot-clasp mounting and a decrease of rear-end presenting. The conspicuous effect of segregated rearing on the aforesaid changes in sexual play rehearsal was an increase in heterotypical sexual play: males did more presenting and females more mounting than did their counterparts in the mixed-sex group.

The relationship of juvenile sexual rehearsal play to adult G-I/R in human beings has a history of insufficient research. It may well play an extremely influential role as a critical-period phenomenon wherein nature and nurture merge to establish future erotosexual health, male and female. Heterosexual rehearsal play, without question, establishes the primacy of the sex organs and their use as the definitive criterion of the difference between male and female. Children who grow up secure and confident in the immutability of this primary difference can be equally secure and confident in the mutability of other sex differences that are secondary and more or less arbitrary products of social and historical stereotyping. For them, adherence or nonadherence to these stereotypes is arbitrary or optional. Thus, they have the option of sex sharing, free from the threat that male/female equality will somehow rob them of their quintessential identity and status as girl or boy, man or woman.

Acknowledgments

Supported by USPHS Grant HD00325 and funds from the William T. Grant Foundation.

References

Baum, M. J. (1979). Differentiation of coital behavior in mammals: A comparative analysis. *Neuroscience and Biobehavioral Reviews, 3,* 265–284.

Baum, M. J., Gallagher, C. A., Martin, J. T., & Damassa, D. A. (1982). Effects of testosterone, dihydrotestosterone, or estradiol administered neonatally on sexual behavior of female ferrets. *Endocrinology, 111,* 773–780.

Chan, S. T. H. (1977). Spontaneous sex reversal fishes. In J. Money & H. Musaph (Eds.), *Handbook of sexology,* Amsterdam/New York: Excerpta Medica.

Crews, D. (1982). On the origin of sexual behavior. *Psychoneuroendocrinology, 7,* 259–270.

Dorland's illustrated medical dictionary (26th ed.). (1981). Philadelphia: Saunders.

Goldfoot, D. A. (1977). Sociosexual behaviors of nonhuman primates during development and maturity: Social and hormonal relationships. In A. M. Schrier (Ed.), *Behavioral primatology, Advances in research and theory* (Vol. 1). Hillsdale, NJ: Lawrence Erlbaum.

Goldfoot, D. A., Wallen, K., Neff, D. A., McBrair, M. D., & Goy, R. W. (1984). Social influences upon the display of sexually dimorphic behavior in Rhesus monkeys: Isosexual rearing. *Archives of Sexual Behavior, 13,* 395–412.

Imperato-McGinley, J., Guerrero, L., Gautier, T., & Peterson, R. E. (1974). Steroid 5α-reductase deficiency in man: An inherited form of male pseudohermaphroditism. *Science, 186,* 1213–1215.

Money, J. (1952). *Hermaphroditism: An inquiry into the nature of a human paradox.* Cambridge: Harvard University Library (Ann Arbor: University Microfilms, 1967).

Money, J. (1955). Hermaphroditism, gender, and precocity in hyperadrenocorticism: Psychologic findings. *Bulletin of The Johns Hopkins Hospital, 96,* 253–264.

Money, J. (1974). Intersexual and transexual behavior and syndromes. In S. Arieti & E. B. Brady (Eds.), *American handbook of psychiatry,* Vol. III, 2nd ed. rev. New York: Basic Books.

Money, J. (1974). Two names, two wardrobes, two personalities. *Journal of Homosexuality, 1,* 65–70.

Money, J., & Ehrhardt, A. A. (1972). *Man and woman, boy and girl: The differen-*

tiation and dimorphism of gender identity from conception to maturity. Baltimore: Johns Hopkins Press.

Money, J., Hampson, J. G., & Hampson, J. L. (1955). An examination of some basic sexual concepts: The evidence of human hermaphroditism. *Bulletin of the Johns Hopkins Hospital, 97,* 301–319.

Short, R. V., & Clarke, I. J. (no date). *Masculinization of the female sheep.* Distributed by R. V. Short, MRC Reproductive Biology Unit, 2 Forrest Road, Edinburgh EH1 2QW, Scotland, U.K.

Ward, I. L. (1972). Prenatal stress feminizes and demasculinizes the behavior of males. *Science, 175,* 82–84.

Ward, I. L. (1977). Exogenous androgen activates female behavior in noncopulating, prenatally stressed male rats. *Journal of Comparative and Physiological Psychology, 91,* 465–471.

Ward, I. L., & Weisz, J. (1980). Maternal stress alters plasma testosterone in fetal males. *Science, 207,* 328–329.

3

Alternative Interpretations of the Development of G-I/R

Frank A. Beach

John Money has presented us with so many original and stimulating ideas in Chapter 2 that they cannot possibly be dealt with in a brief discussion. For this reason, I will concentrate on one or two that I find particularly provocative. First, I will consider development of gender identity and gender role.

Ontogeny of G-I/R

Money gave birth to the concept of gender identity in 1955. Presumably the process was parthenogenetic. He viewed the infant as having two components. One consisted of an introspective self-definition, and the other was a public definition arrived at by others on the basis of the individual's behavior. These two parts were simply opposites of the same coin.

Shortly after parturition, attendants at the birth dismembered the helpless child. They named the introspective component *gender identity* and defined *gender role* as a social script to which the neonate eventually would or would not conform. The outcome of this process is described by Money as follows.

> In obedience to the nature/nurture cleavage, *gender* thus became alloted to nurture, and set off against *sex*, which became the property of nature. In this way, the vested interest of social science was left unperturbed. Social scientists claimed *gender role* for

> themselves. . . . Thus the obligation to relate *gender role* to biology as well as to history and sociology was circumvented.

I have read this passage over and over but still am not sure I fully comprehend it. Somewhere in the argument the distinction between gender role and gender identity gets lost. I understand that sociologists consider gender role as a script imposed on the individual by society. But what happens to gender identity? Is it relegated to Immanual Kant's category of innate ideas? This seems improbable, but what is the difference between factors responsible for development of gender role and gender identity. Are both assignable to nurture?

My suspicion is that the apparent confusion can be resolved by approaching the issue from a particular ontogenetic point of view which Money has used so effectively in other situations. Instead of speculating about contributions of nature and nurture to development of characters that identify the individual, it may be more illuminating to identify organismic and environmental variables that jointly determine the process of development. If we happen to be concerned with morphological or behavioral sex-related characters, my modification of Money's figure provides a very sketchy outline of some of the interactions involved.

Under the general heading of organismic factors, we begin with sex differences in the chromosomes that in turn determine the type of gonad that will differentiate. If this is a testis it will secrete Müllerian inhibiting factor and, later, testosterone. Testosterone or one of its metabolites will in turn influence differentiation of urogenital sinus tissue into a penis and scrotum, and also will induce certain irreversible modifications in morphology and functional capacity of hypothalamic mechanisms in the brain.

During this same period of prenatal development, specific factors in the uterine environment will affect ontogeny of the individual. Hormones secreted by the placenta that influence development of the fetus include pregnenolone, progesterone, various corticoids, estradiol, and androsterone (Vernadakis & Timiras, 1982). In mice and rats a female fetus located between two male fetuses is exposed to testis hormone, which passes through the amniotic fluid to induce moderate degrees of genital and neural masculinization (Clemens, Gladue, & Coniglio, 1978).

As soon as a human infant is born, its genital morphology elicits special reactions from adults in the social environment that result in sex assignment and thus determine the sex role to which the neonate eventually will be trained. With development of language as a means of communication and a cognitive tool, the child gradually acquires the ability to internalize elements of the childhood sex role and to compare it with her or his own behavior. If a high degree of concordance be-

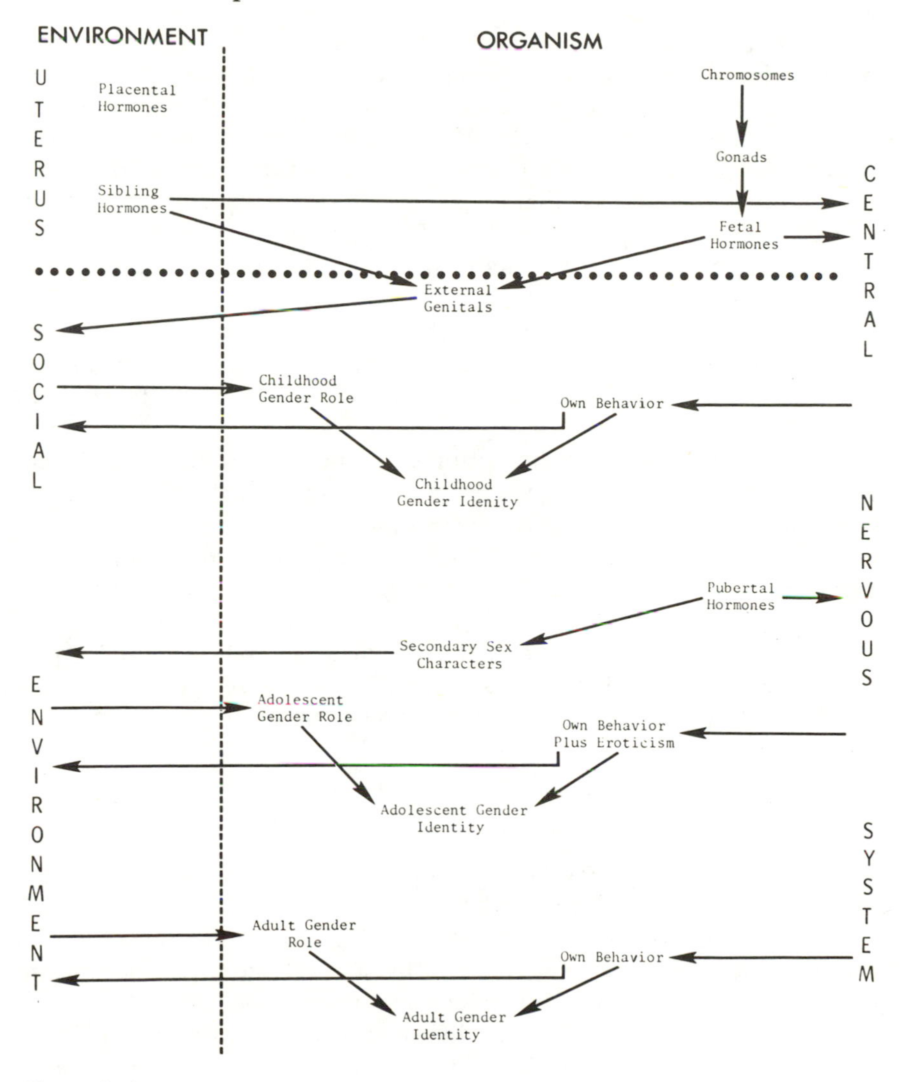

Figure 3–1.

tween the two exists, a strong and unambiguous gender identity will result. The child should easily learn the "appropriate" behavior and enjoy performing it. Sex differences in the brain may influence the facility with which the individual learns and experiences positive affect involved in fulfilling the socially defined gender role. A child who finds difficulty in learning behavior appropriate to its assigned sex role, or who does not enjoy performing it, may not develop a clearly defined gender identity.

The advent of puberty and entrance into adolescence are marked by both organismic and environmental changes. Secretion of gonadal hormones plus numerous other physiological alterations modifies many behavioral proclivities; at the same time altered perception on the part of others places new demands and expectations on the individual. The socially defined gender role changes from one fitted to childhood but does not yet become one that is appropriate for adults. Money's addition of an erotic component to the G-I/R is an essential innovation, and the concept of love attachments is an especially important element as it influences new forms of heterosexual interaction.

Development of adolescent gender identity, or introspective feelings of masculinity or femininity, depends on the individual's success and pleasure in matching her or his own behavior with the internalized sex role.

The final step in establishment of an adult G-I/R involves internalization of an adult gender role and making adjustments whereby overt behavior and gender role become consonant.

I can thus conceive of gender identity and gender role as being two sides of the same coin just as Money regards them. Both are products of learning, and as such depend on modification in brain mechanisms. But learning does not occur in the absence of environmental input; much that the individual learns about G-I/R is embedded in the propaedeutical program of the society in which she or he grows up.

Sex-Derivative Characters

The other aspect of Money's paper upon which I wish to comment is his treatment of "sex-derivative" aspects of the G-I/R. Although I am totally in sympathy with his underlying logic, I cannot help but feel that development of the thesis would be strengthened by more careful attention to differences both between and within species.

My reference to within-species differences is a subterfuge that allows me to correct a minor error in Money's paper and at the same time to record for posterity a hitherto unobserved scientific verity.

In his discussion of human males who develop without androgenic stimulation, Money states, "An obvious sex-derivative by-product of such deandrogenized development of the human fetus is that the baby will grow up to urinate in the sitting position." Fourteen years ago, when I was conducting an experiment on sex differences in urinary behavior of dogs, I wondered whether sex-typical postures assumed by men and women in European societies were species specific or culturally variable. I asked C. S. Ford, Professor of Anthropology at Yale University, to consult the Cross Cultural Areas Files to see what information was available regarding preliterate societies. In a cursory sur-

Table 3-1
Urinary Postures in 15 Preliterate Societies

	Percent of Total
Dimorphic Postures	47
Men stand, women squat (N=7)	
Monomorphic Postures	53
Both sexes squat (N=4)	
Both sexes stand (N=3)	
Both sexes "on all fours" (N=1)	

vey, he was able to obtain information for 15 societies. The results are summarized in Table 3-1. This should be a lesson to all of us. Never assume the significance of *p* until we have collected an adequate sample.

Now, to return to a more serious consideration of Money's major thesis, I agree that many types of behavior can occur in both sexes but are "threshold-dimorphic," which means they are more easily and completely evoked in one sex than in the other. I agree also that this is prima facie evidence that the brains of both sexes have encoded within them both male and female schemata for such behavior.

When Money passes from the general to the particular, however, and lists nine sex-shared threshold-dimorphic types of behavior, I experience a degree of uncomfortable dubiety. Because his list includes both human and animal behavior, he seems to imply shared mechanisms, but in several cases no such implication appears justified.

One example, designated as "parentalism," is illustrated in human beings by noting that girls indulge in rehearsal play with dolls, and female animals retrieve, lick, and hover over their young. The comparison would be more convincing if it involved behavior of prepubescent female animals or adult human females. Another difference with respect to animals is that females provide a safe place or nest for delivery and care of young. This is true, of course, and in most species the choice is made solely by the female, but comparable behavior is socially arranged in human beings.

Another difference is described as roaming and becoming familiar with or marking the boundaries of the roaming range. This accurately describes the behavior of adult male wolves, and of adult *female* spotted hyenas. I do not know precisely what the human counterpart might be, unless Money is referring to the relative freedom of movement accorded little boys in many societies.

My criticisms may be unfair and unfounded, but even if Money does not intend to imply common mechanisms mediating analogous behavior in different species, certainly many of his readers will make such an inference on their own.

As we all know, outwardly similar patterns of behavior in different species can have quite different developmental histories and be mediated by quite dissimilar mechanisms. When it comes to sex-related patterns of behavior, the underlying brain mechanisms for male and female responses may very well be represented in both sexes, but these mechanisms need not be the same in different species.

For example, it is conceivable that diencephalic circuits involved in male and female copulatory behavior are similar in many mammalian species; but that there are marked interspecific differences in the degree to which these circuits are regulated by telencephalic mechanisms that are more highly developed in primates and especially in human beings.

If they existed, such differences in organismic determinants of sex-related behavior might be associated with equally important differences in the degree to which or the manner in which environmental determinants help to shape the final product.

References

Clemens, L. G., Gladue, B. A., & Coniglio, L. P. (1978). Prenatal endogenous androgenic influences on masculine sexual behavior and genital morphology in male and female rats. *Hormones and Behavior, 10,* 40–53.

Vernadakis, A. & Timiras, P. S. (Eds.) (1982). *Hormones in Development and Aging.* New York: Spectrum Publications Medical and Scientific Books.

II

NEUROSCIENCE PERSPECTIVES

4

Sex Differences in the Rodent Brain: Their Nature and Origin

Roger A. Gorski

Any concept of masculinity and femininity must obviously differ depending on the behavioral repertoire of a given species. From my perspective, developed over 25 years of experimentation with the laboratory rat, I am convinced of the existence of clear, sometimes dramatic, sex differences in the functional potential, microscopic, and macroscopic structure of the rat brain, and of the process of sexual differentiation through which these sex differences arise. Although the concept of the sexual differentiation of the brain can be illustrated with an anthropomorphic cartoon (see Fig. 4-1), my objective in this discussion is not to define masculinity and femininity as applied to the laboratory rat, let alone to the wild species. Instead, I will illustrate some of the sex differences in the rat brain and present my current views on their development. I will do so in an attempt to show that in the rat, and perhaps in most mammals, the concept of sexual differentiation of the nongonadal components of the reproductive system (i.e., the internal reproductive organs and external genitalia) also applies to the brain, which is, in a real sense, another important reproductive organ. Because human beings are also mammals, the following may represent a review of one possible basis of our "biological" masculinity and femininity, with the obvious caveat that environmental and cultural factors may also shape "social" masculinity and femininity.

Although I will review the evidence for sexual differentiation of both brain function and structure, I want to emphasize several general points to put this evidence in proper perspective. When we consider func-

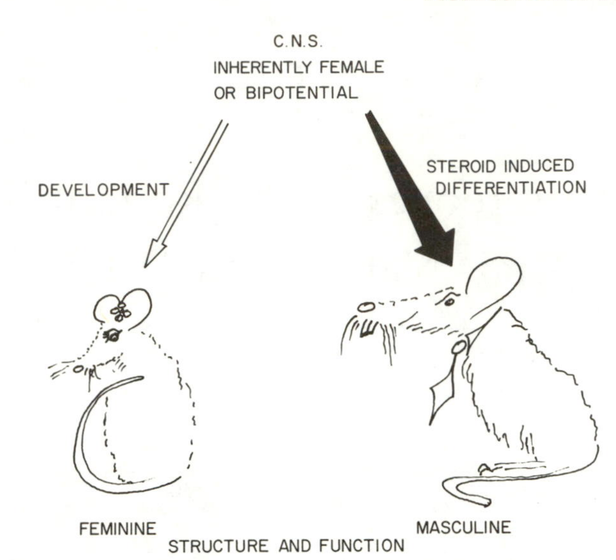

Figure 4–1. Schematic representation of the concept of the sexual differentiation of the rat brain. (Reprinted with permission from Gorski, 1985a.)

tional sex differences in the brain, we must remember that we actually do not know their neurochemical or morphological bases. Scientists merely look at some output of brain function—hormone release or a behavior—and infer a sex difference in brain function, often in the abstract. I will point out in this chapter two examples of the inherent weakness in this descriptive approach. On the other hand, when we consider structural sex differences in the brain, which I believe constitute the most significant recent development in this area of research, we may not know their functional significance. This is certainly true in the system I will discuss in detail. Important advances in our understanding await the correlation of structure and function.

Although we know that steroid hormones act on neurons presumably via intracytoplasmic receptor mechanisms, direct membrane effects of these hormones may also exist (McEwen et al., 1982). Moreover, the precise mechanism(s) by which a steroid hormone-steroid receptor-induced modification of genomic activity alters neuronal function remains unknown.

The fact that gonadal hormones alter neuronal function, has obvious relevance to my specific comments because the hormonal milieus in the adult male and female differ. Thus, one could well argue that the most meaningful concept of masculinity or femininity should address the intact animal—the male with his testes, the female with her ovaries. The focus of my comments, however, is not to define masculinity and femininity in rats; instead, I focus on fundamental sex differences in structure and function of the brain and how they come about.

To identify inherent sex differences in the brain, we have essentially two basic experimental approaches: to study animals in the absence of their gonads, or to study animals in the presence of the same hormonal environment as far as can be determined (i.e., the male and female exposed to the same ovarian or testicular hormones). Clearly, studying the actions of the sex-specific hormone milieu (that is, testicular hormones in the male, ovarian hormones in the female) is important, but the first two approaches will be emphasized here.

The Concept of the Sexual Differentiation of the Brain

A convenient way to illustrate our approach is to consider steroid hormone action on the central nervous system in conceptual terms: activational versus organizational (originally proposed by Phoenix, Goy, Gerall, & Young, 1959). *Activational* refers to the transient changes in brain function induced by steroid hormones presumably via the receptor mechanisms described earlier. *Organizational* refers to the permanent changes in the central nervous system that may or may not involve the same mechanisms. We have a critical point here, particularly in terms of how brain function is tested. The concept of the sexual differentiation of the brain (Fig. 4-1) proposes that functional sex differences are imposed on the developing brain by the organizational action of steroids, yet many functions that are sexually dimorphic—certainly the two I choose to emphasize—are dependent on the activational action of steroids. Thus, the hormonal environment is modified perinatally in order to study the organization of the brain, but the activational effects of steroids are utilized to measure these organizational events. The two functional systems I choose to illustrate are the pattern of the secretion of luteinizing hormone (LH) and reproductive behavior.

LH Secretion

In the female rat, rising titers of plasma estradiol act on a cyclic control system within the hypothalamus to facilitate the release of LH-Releasing Hormone, which in turn triggers the ovulatory surge of LH (see Gorski, 1984a; Harlan, Gordon & Gorski, 1979). The intact male does not exhibit this dramatic cyclic pattern of LH secretion in adulthood, but, as indicated earlier, the male is exposed to a different hormonal environment (testicular versus ovarian hormones). Fortunately, the same or at least a similar surge of LH can be induced in the ovariectomized female by the injection of estradiol alone or followed by progesterone

(Taleisnik, Caligaris, & Astrada, 1971). This effect is called facilitatory feedback. Note that in the male rat, orchidectomized as an adult, estrogen and progesterone do not exert facilitatory feedback; the genetic male does not appear to have the neural substrate to activate the surge of release of LH-Releasing Hormone.

Sexual Receptivity

Along with the cyclic pattern of LH secretion, the intact female rat exhibits a similar and correlated cyclic change in sexual receptivity. When the female rat is ovariectomized, however, she no longer exhibits sexual receptivity unless she is exposed to exogenous estradiol alone, or (even more physiologically) to estradiol followed by progesterone (see Gorski, 1974, 1976). Particularly in the latter case, sexual receptivity is high. The intact male, as well as the male that is gonadectomized as an adult and treated with these ovarian hormones, will not exhibit the lordosis reflex, an important behavioral component of sexual receptivity in the rat. More precisely, the gonadectomized male primed with ovarian hormones will not display lordosis at the same frequency displayed by the female. Thus, the genetic male rat will occasionally exhibit lordosis; the neural substrate for this reflex must be present, at least in part, but appears to be more difficult to activate by hormones in order to achieve female levels of lordosis responding. In the male the threshold of hormonal activation of the lordosis neural substrate appears to be much higher than that of the female.

Thus, the capacity to secrete LH cyclically and to display lordosis behavior are two clear functional sex differences in the rat brain. Such differences could well be determined by the neural genome during development. This, however, is not the case (Gorski, 1983, 1985a; Goy & McEwen, 1980; Harlan et al., 1979; McEwen, 1983). It is firmly established that exposure to exogenous testosterone or to testicular grafts during the perinatal period permanently blocks the ability of the female rat to exhibit facilitatory feedback or lordosis behavior (defeminization) and may actually exhibit levels of masculine behavior more typical of the male (masculinization). In contrast, if the genetic male is gonadectomized within the first 2–3 days of postnatal life, as an adult he will display high (femalelike) levels of lordosis responding and facilitatory feedback (feminization), but a reduction in masculine sexual behavior (demasculinization).

Using the terms *masculinization* or *demasculinization* to refer to the presence or absence, respectively, of functional traits characteristic of the normal male is common practice, as is using *feminization* or *defeminization* to refer to the presence or absence, respectively, of functional traits characteristic of the normal female. One value in the use of these terms is to emphasize the possibility that the mechanisms for each may differ, and/or that these are independent processes (see Goy & McEwen, 1980). I want to point out, however, that these rubrics may actually promote an unjustified bias for common mechanisms. A study of the comparative aspects of sexual differentiation, among either different species or different functional processes within the same species, actually argues against the concept of a "common" mechanism for defeminization or even masculinization.

In terms of sexual differentiation of the neural regulation of LH secretion, testosterone exposure has been reported to block ovulation in the following species: hamster, mouse, rat, guinea pig, pig, sheep (see Gorski, 1983). Note, however that a temporal difference is found in these species when the process of sexual differentiation occurs. In the first three species, the sexual differentiation of LH secretory regulation occurs postnatally, presumably because these animals are born in a relatively immature state, whereas in the other species, sexual differentiation occurs prenatally. Similarly, the testosterone exposure of the female will suppress female sexual behavior in the following species: prenatal exposure—dog, ferret, guinea pig, rhesus monkey, sheep; postnatal exposure—hamster, mouse, pig, rat (see Gorski, 1983). In addition to temporal differences between species, note that species specificity also exists in terms of functional processes. The evidence available to date suggests that in the primate the phenomenon of facilitatory feedback does not differ between the sexes (Hodges & Hearn, 1978; Karsch, Dierschke, & Knobil, 1973), whereas this has been one of the intensely studied sex dimorphisms in the laboratory rat.

The species specificity in the functional sexual dimorphisms is further evident from the following list of functional processes that have been reported to be sexually dimorphic and undergo the process of sexual differentiation: the pattern of prolactin release in rats; aggressive behavior in rats, primates, and mice; social and play behavior in primates and rats; territorial marking in gerbils; urination posturing in the beagle; and the regulation of food intake and body weight in rats (Gorski, 1983, 1985a; Goy & McEwen, 1980; Harlan et al., 1979). Thus, we must emphasize that the concepts of functional masculinization or defeminization may be highly species specific.

Moreover, even within one species such as the rat, it is likely that individual functional processes exhibit differences with respect to hormonal sensitivity in terms of the actual hormonal molecule, the time of highest sensitivity to exogenous or endogenous steroids, and the site and perhaps actual mechanisms of hormone action (Arnold & Gorski, 1984; Christenson & Gorski, 1978; Harlan et al., 1979). For example, when the neonatal female rat is given increasing doses of exogenous testosterone, defeminization occurs in the order indicated among these functional processes LH secretion, lordosis behavior, and food intake. (Tarttelin, Shryne, & Gorski, 1975).

The Relevance of Testing Conditions

At this point I want to digress from the main theme to present two examples of weaknesses in what can be called the descriptive studies of sexual differentiation of the brain using functional parameters. Both relate to the manner in which sexual behavior is usually tested. In the laboratory situation, rodent sexual behavior is commonly tested in a relatively restricted testing arena. In such a situation the female appears to play a minor role; the receptive female may display soliciting or proceptive behavior (Beach, 1976), but the male actually dominates the testing situation. Thus, the routine testing procedure provides one concept of "femininity" in the laboratory rat.

This concept may be far from correct. When rat sexual behavior is tested in a situation where the female can escape but the male cannot, the female actually paces the mating behavior (see Gilman & Hitt, 1978). Thus, the constraints of the testing situation clearly have an impact on the investigator's view of the "femininity" of the female rat.

The second example relates more directly to the concept of the sexual differentiation of the brain. I pointed out earlier that exposure of the neonatal female to testosterone permanently suppresses her ability to display lordosis behavior. A number of years ago we studied the influence on sexual receptivity of a low dose of testosterone propionate (TP; 10 μg) on different days postnatally. In these studies (Clemens, Hiroi, & Gorski, 1969) we confirmed that 10 μg TP given on day 4 or 6 of postnatal life effectively suppressed adult lordosis responsiveness; according to classical dogma, this was safely considered to be a permanent change in the brain of the female. At that point, however, we did some reasoning. It is well known that the male rat's sexual performance is inhibited by a strange or novel environment. Therefore, one usually allows the male to adapt to the testing arena, or one actually may test reproductive behavior in the home cage of the male. The female, at least the ovariectomized animal primed with EB and progesterone, clearly does not necessarily have to adapt to the testing arena in order to display high rates of lordosis in response to the stimulus of mounting. In fact, in most situations the female is introduced into the testing arena and the behavioral test begins immediately.

We reasoned as follows: If males must adapt to a novel environment in order to perform sexually, and the androgen-exposed female has a more masculine brain, shouldn't we allow this female to adapt to the testing arena before beginning any behavioral test? In spite of the need for male rats to adapt to a testing arena, this was possible because we used trained stud males who do not require this period of adaptation. Thus, we allowed these same neonatally androgen-exposed female rats to adapt to the testing arena for a period of 2 hours before the male was introduced to begin the test. The rate of lordosis responding in these females under these conditions was significantly increased (Clemens et al., 1969). Thus, adaptation to the testing arena overcame what we had assumed was a permanent change in the functional capacity of the brain! Interestingly, adaptation to the testing arena does not facilitate lordosis behavior in the male (Clemens, Shryne, & Gorski, 1970).

It is beyond the scope of this chapter to discuss this further. Suffice it to say that the normal ovariectomized estrogen-treated female will exhibit this adaptation facilitation of lordosis behavior when she is primed with a less than optimal dosage of progesterone (Clemens et al., 1969). We view the adaptation phenomenon, therefore, as a possible reflection of the action of progesterone on the mesencephalic reticular formation and the suppression of a cortical function that is inhibitory to lordosis behavior (Gorski, 1976).

The point of these examples is straightforward: Even in the case of the rat, which may be considered a relatively simple animal in social terms, the precise nature of the testing conditions of a particular functional parameter can dramatically influence the data obtained and the resultant interpretations.

Structural Sex Differences in the Brain

Fortunately, as already indicated, studies of the sexual differentiation of the brain have recently been advanced beyond the descriptive level of functional analysis by the discovery of structural sex differences in the brain. Beginning with early reports of sex differences in the organization of various regions of the cortex, the now impressive list of structural sex differences in the central nervous system continues to grow (Gorski, 1979, 1983; McEwen, 1983; Tobet, Gallagher, Zahmiser, Cohen, & Baum, 1983; Yahr & Commins, 1983). In 1973, Raisman and Field published a very influential study. They demonstrated that there were statistically significant sex differences in the pattern of synaptic connectivity in the preoptic area of the rat hypothalamus as determined at the level of the electron microscope. Moreover, they demonstrated that this structural sex difference was modified by manipulating the gonadal hormone environment perinatally in a manner consistent with the concept of the androgen-dependent sexual differentiation of the brain as determined from functional studies. Subsequently, Greenough, Carter, Steerman, and DeVoogd (1977) demonstrated a sex dif-

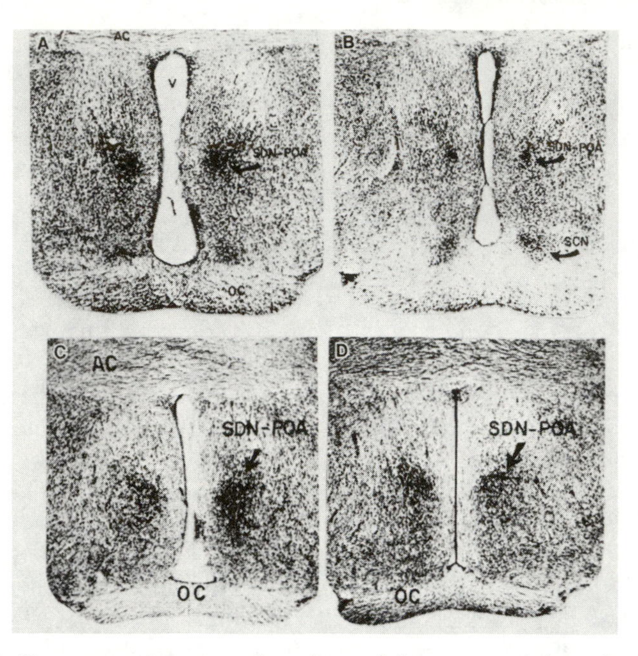

Figure 4–2. Representative coronal sections of the young adult rat brain through the center of the Sexually Dimorphic Nucleus of the Preoptic Area (SDN-POA) of the male (A) and female (B). The lower two figures illustrate the SDN-POA in young adult genetic females exposed from day 16 of gestation through postnatal day 10 to daily injections of testosterone propionate (C) or diethylstilbestrol (D). (All at same magnification. Modified from Döhler, Hines, Shryne, & Gorski, 1984. Abbreviations: AC, anterior commissure; OC, optic chiasm; V, third ventricle.)

ference in the dendritic organization of neurons of the preoptic area of the hamster, and several additional structural sex differences at the ultrastructural level have been reported (Matsumoto & Arai, 1980; Nishizuka & Arai, 1981). Perhaps the most influential discovery was that of Nottebohm and Arnold (1976): the existence of a series of relatively gross sex differences in the brain of two species of songbird, the canary and zebra finch. This observation strongly rekindled the concept that there might be relatively major structural sex differences in the brain. This has indeed been confirmed with the discovery of the Sexually Dimorphic Nucleus of the Preoptic Area [SDN-POA; see Fig. 4-2 (A, B)] in the rat (Gorski, Gordon, Shryne, & Southam, 1978; Gorski, Harlan, Jacobson, Shryne, & Southam, 1980), and subsequently of a similarly marked structural sex difference in the rat spinal cord at the level of the motor neurons that innervate the perineal musculature, the Spinal Nucleus of the Bulbocavernosus (SNB; Breedlove & Arnold, 1980). Most recently, Loy and Milner (1980) have demonstrated a surprising structural sex difference in the rat hippocampus. Destructive of the medial septum removes a cholinergic input to the hippocampus; subsequent to this denervation, noradrenergic fibers from overlying blood vessels sprout into the hippocampus. Loy and Milner (1980) have shown a sex difference in this plastic response to denervation in that noradrenergic reinnervation of the hippocampus is much greater in the female both in quantitative and qualitative terms. Moreover, once again, this sex difference in the adult brain is sensitive to the perinatal hormonal environment (Milner & Loy, 1982).

Recently, Swaab and Fliers (1985) have reported the existence of a structural sex difference in the volume of a nucleus in the human hypothalamus which they have labeled the sexually dimorphic nucleus (SDN). In their study the SDN was significantly larger in volume in males. In an independent study we have evaluated the volume of four small nuclei, which we have labeled Interstitial Nucleus of the Anterior Hypothalamus (INAH) 1–4 (Allen, Hines, Shryne, & Gorski, 1986). Two of these were found to be sexually dimorphic in age-matched pairs; again, nuclear volume is greater in males. We believe that the nucleus we call INAH-1 is equivalent to the SDN of Swaab and Fliers (1985), but in our sample of human brains, this nucleus was not sexually dimorphic. Nevertheless, structural sex differences in the nuclear organization of the human brain may be assumed to exist, but it will be difficult to establish homologies among different species and to evaluate the influence of gonadal hormones on the human brain during either development or adulthood.

In the case of the bird song system and the SNB in the rat, the function of the sexually dimorphic neural circuitry appears to be known. My laboratory is interested in the SDN-POA, which is a marked structural sex difference but its function is totally unknown. Even though we do not know the function of the neurons of the SDN-POA, we believe this nucleus represents a valuable model system for the study of basic mechanisms of hormone action during sexual differentiation. In fact, as I will demonstrate here, the study of the formation of the SDN-POA suggests that fundamental processes of developmental neurobiology may be modified by the hormone environment.

The Sexually Dimorphic Nucleus of the Preoptic Area

Figure 4-2 (A, B) illustrates the structural sex difference we have discovered in the rat brain. The SDN-POA is approximately five times larger in volume in the adult gonadectomized male than it is in the ovariectomized adult female (Gorski et al., 1978). We labeled the SDN-POA a nucleus because of our observation that the density of neurons within this region is significantly higher than that in the surround (Gorski et al., 1980; Jacobson & Gorski, 1981). Importantly, however, the den-

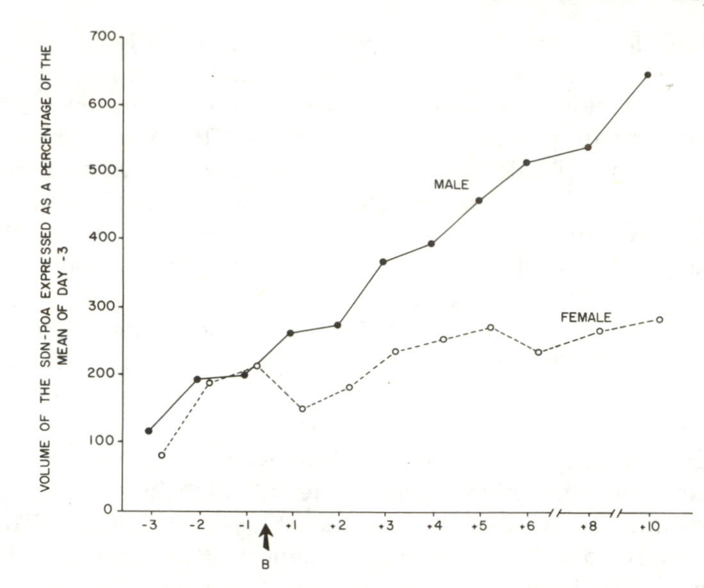

Figure 4–3. Schematic illustration of the relative change in the volume of the SDN-POA related to the day of birth (B). Data are expressed as a percentage of the mean volume observed in males and females sacrificed at 20 days post-fertilization [3 days (−3) before expected birth]. The day of birth (23 days post-fertilization) is called day +1. (Data from Jacobson et al., 1980. Reprinted with permission from Gorski, 1984a.)

sity of neurons within the SDN-POA is not different between males and females. Because the overall volume of this area of increased neuronal density is much larger in the male than in the female, the SDN-POA of the male is comprised of more neurons. Obviously, the discovery of a structural sex difference, even one as marked as that of the SDN-POA, does not establish that it undergoes sexual differentiation. Indeed, we must know when the sex dimorphism in the SDN-POA develops, and whether or not SDN-POA volume of the adult animal is determined or influenced by the hormonal environment perinatally.

With respect to the development of the SDN-POA, the data we have obtained fit remarkably well with the concept of the functional sexual differentiation of the brain (Jacobson, Shryne, Shapiro, & Gorski, 1980). Thus, we first detected a statistically significant sex difference in the volume of the SDN-POA on the day of birth; essentially that sex difference is maintained and increases in magnitude over the first 10 days of life (Fig. 4-3). In fact, there is a statistically significant increase in the volume of the SDN-POA of the male over this time, but there is no significant change with age in the female. The structural difference in the SDN-POA therefore develops and in fact reaches its apparent maximum during the precise period of the functional sexual differentiation of the brain. Interestingly, although there is a dramatic overall growth of the brain during this period, in the parameters we have analyzed, the only significant sex differences were observed in SDN-POA volume (Jacobson et al., 1980).

We have approached the question of the possible influence of the hormonal environment on the development of the SDN-POA in several different ways. First, we have determined that the cells of the SDN-POA do take up and retain radiolabeled steroids in adulthood (Jacobson, Arnold, & Gorski, 1982). In fact, the region of the SDN-POA appears to be an area where such neurons are especially concentrated. (We have also initiated a study of the ontogeny of the ability of the cells of the SDN-POA to take up and retain labeled steroids, but this will be described later.) A most critical approach is to determine whether manipulations of the hormonal environment perinatally alter the volume of the SDN-POA attained by the adult. As summarized in Figure 4-4, this is indeed true. The administration of a single injection of 1.25 mg testosterone propionate (TP) or 100 μg estradiol benzoate (EB) to the female rat within the first 5 days of postnatal life causes a statistically significant increase in SDN-POA volume in the adult. Similarly, castration of the newborn male rat produces approximately a 50% decrease in nuclear volume attained in adulthood. Although the reduction in nuclear volume induced in the male by neonatal castration can be completely prevented by the administration of exogenous TP, the nuclear volume in the neonatally castrated male rat is still significantly larger than that of the normal female. In addition, the SDN-POA of the steroid exposed female, although larger than that of the female, does not attain the volume of the nucleus in the male. This inability to sex reverse SDN-POA volume completely raises several questions: Do neuronal genomic factors play a role in the development of the SDN-POA? Is the hormonal environment prenatally important for the development of the SDN-POA? Must the developing brain be exposed to gonadal steroids for an extended period in order to sex reverse the SDN-POA? Finally, does the surge of testosterone that has been reported to occur in the male rat during or shortly after parturition (Corbier, Kerdelhue, Picon, & Roffi, 1978), and leads to an increase in hypothalamic estrogen (Rhoda, Corbier, Roffi, Castanier, & Scholler, 1983), play a role in development of this nucleus?

In an attempt to answer the last question, we collaborated with Corbier and demonstrated that castration of the male rat just before normal parturition, or 6, 12, or 24 hours after parturition, produces approximately a 50% decrease in SDN-POA volume attained by the adult (Handa, Corbier, Shryne, Schoonmaker, & Gorski, 1985). Thus, we see

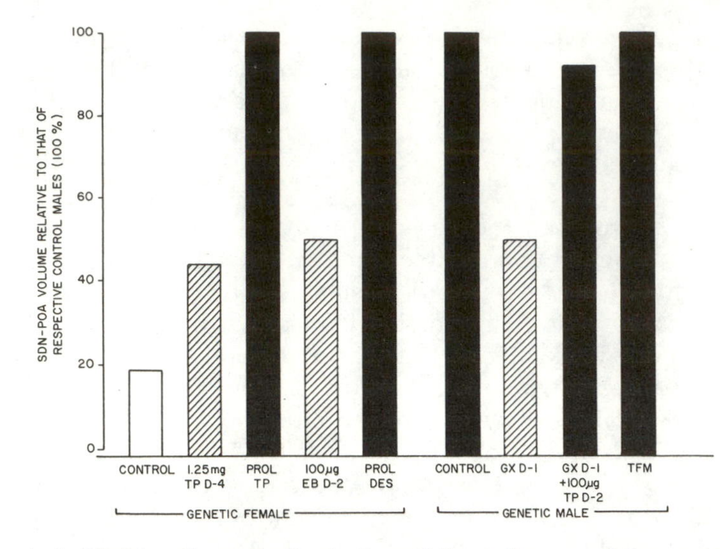

Figure 4–4. Highly schematic illustration of the influence of the gonadal hormonal environment perinatally on the volume of the SDN-POA expressed as a percentage of the volume of this nucleus in control males from several independent studies. Genetic females received a single injection of testosterone propionate (TP) on postnatal day 4 (TP D-4; Gorski et al., 1978), or estradiol benzoate (EB) on postnatal day 2 (EB D-2, Jacobson & Gorski, unpublished observations), or prolonged daily exposure (day 16 postconception through postnatal day 10) of TP (PROL TP; Döhler, Coquelin, Davis, Hines, Shryne, & Gorski, 1982) or diethylstilbestrol (PROL DES; Döhler, Hines, Coquelin, Davis, Shryne, & Gorski, 1982). Genetic males were gonadectomized on postnatal day 1 (Gx-D-1; Gorski et al., 1978), some were injected with TP on postnatal day 2 (TP D-2; Jacobson, Csernus, Shryne, & Gorski, 1981); some were animals with the testicular feminizing mutation (Tfm; Gorski, Csernus, & Jacobson, 1981). Solid, shaded, and open bars significantly different from each other are based on the analysis of the original data from the several studies represented here. (Reprinted with permission from Gorski, 1984b.)

no indication that the surge of testosterone at parturition plays a role in the development of the SDN-POA, even though there are behavioral (Corbier, Roffi, Rhoda, & Valens, 1981) and hormonal (Handa, et al., 1985) sequelae to such early castration.

In an attempt to determine whether the steroid hormone environment alone could lead to the full development of the masculine SDN-POA, we administered exogenous TP for a prolonged period during perinatal development (Döhler, Coquelin, Davis, Hines, Shryne, & Gorski, 1982). Thus, pregnant females were injected daily beginning on day 16 postconception with 2 mg TP. At birth, all pups continued to receive 100 μg TP daily for the first 10 days of postnatal life. It was necessary to inject all pups because the prenatal treatment masculinized the genitalia so that it was impossible to sex the pups. As illustrated in Figures 4-2C and 4-4, this prolonged treatment with exogenous TP did completely masculinize the SDN-POA volume of female animals. It did not influence the volume of the SDN-POA in males, however (Döhler, Coquelin, Davis, Hines, Shryne, & Gorski, 1984). (There would appear to be a maximal SDN-POA volume.) Obviously, the complete sex reversal of the SDN-POA produced by such prolonged exposure to exogenous TP does not rule out a possible genomic factor in normal development, but it clearly indicates that gonadal hormones alone can indeed determine the volume of, and neuronal numbers that comprise, the SDN-POA. The minimal exposure (in terms of either amount of hormone or duration) necessary to sex reverse SDN-POA volume is unknown.

We also treated pregnant females and their offspring for the same period with the nonsteroidal estrogen diethylstilbestrol (Döhler, Hines, Coquelin, Davis, Shryne, & Gorski, 1982). As illustrated in Figures 4-2D and 4-4, perinatal exposure to this estrogen also completely sex reverses SDN-POA volume. This observation, as well as the fact that the administration of a single injection of 100 μg EB can also increase SDN-POA volume in female rats (Fig. 4-4; Jacobson & Gorski, unpublished observations), brings up an interesting and still unresolved conflict that may very well have relevance to concepts of masculinity and femininity and of sexual differentiation of the brain itself. These data raise two questions. What is the hormonal species that is actually involved in the sexual differentiation of the brain? Is the brain actually inherently feminine, or is it neuter requiring exposure to some steroid in order to develop typically feminine characteristics?

The balance of evidence available today suggests that at least in the rat, the molecular species active in sexual differentiation of the brain is actually estradiol (Lieberburg, Wallach, & McEwen, 1977). First, as is well known, testosterone can be metabolized, either aromatized to estrogens or 5-α reduced to dihydrotestosterone. Estradiol clearly appears to be more potent than testosterone in defeminizing the brain (Gorski, 1966); the enzyme aromatase is present in the rat brain during development (Naftolin et al., 1975; Selmanoff, Brodkin, Weiner, & Siiteri, 1977); nonaromatizable androgens are relatively ineffective in defeminizing the brain (Korenbrot, Paup, & Gorski, 1975; Whalen & Rezek, 1974); treatment with either antiestrogen or an aromatase inhibitor blocks spontaneous defeminization in male rats (Booth;, 1977); Doughty & McDonald, 1974; McEwen, Lieberburg, Chaptal, & Krey, 1977). Initially, these observations, which suggest that it is estrogen that defeminizes the brain, appeared to be a metabolic peculiarity of nature. Many believed, on the basis of the apparent ineffectiveness of neonatal ovariectomy on subsequent neuroendocrine function (Gorski & Wagner, 1965), that the ovaries in the female were essentially inactive.

The observation (Ojeda, Kalra, & McCann, 1975; Weisz & Gunsalus, 1973) that estrogen levels are actually very high in the plasma of both the neonatal male and female rat (although the precise source is uncertain) came as a surprise and initially appeared to challenge basic concepts of the sexual differentiation of the brain. If estrogen is the hormone active in masculine sexual differentiation, and both the female and male have very high plasma levels during sexual differentiation, how could normal females develop? The explanation for this dilemma appears to be the existence perinatally of a special estrogen-binding protein, which has been shown to be alpha-fetoprotein in the rat (Nunez, Savu, Engelmann, Bennassayag, Crepy, & Jayle, 1971; Plapinger, McEwen, & Clemens, 1973; Raynaud, Mercier-Bodard, & Baulieu, 1971). Thus, as illustrated in Figure 4-5A, the high plasma titers of estrogen in the neonatal rat of either sex may be functionally sequestered by alpha-fetoprotein. (In fact, the presence of this estrogen-binding protein could actually be the cause of the very high plasma titers.) In the male, however, the testes secrete testosterone, which is not bound by alpha-fetoprotein. Thus, testosterone can enter neurons where it is aromatized to estradiol before acting presumably on the neuronal genome. (Note that if this concept is true, feminization is a natural property of the rodent brain. In this context, therefore, neonatal castration of the male rat does not actively feminize the brain; instead, it permits the development of the inherent femininity of the genetic male brain.)

Another view, however, has two main proponents. In in vitro studies of explant cultures of the mouse hypothalamus, Toran-Allerand has published evidence that estrogen stimulates (Toran-Allerand, 1976) and may actually be necessary (Toran-Allerand, 1980a) for neurite outgrowth from steroid-sensitive (Toran-Allerand, Gerlach, & McEwen, 1980) neurons in such explants. Also, she has identified immunohistochemically alpha-fetoprotein within neurons, and puts forth the hypothesis that perinatal alpha-fetoprotein might actually serve to deliver estrogen into neurons (Toran-Allerand, 1980b). In this regard, we note that regions of the brain that appear to be rich in steroid-uptake ability show a paucity of intraneuronal alpha-fetoprotein (Toran-Allerand, 1982). Perhaps, therefore, alpha-fetoprotein has a direct role during development.

Döhler and his colleagues have studied the influence of treatment with the antiestrogen tamoxifen on the sexual differentiation of the brain. Low doses of tamoxifen, which apparently are not estrogenic, defeminize (that is, block ovulation and lordosis behavior) without masculinizing the brain of these females (Hancke & Döhler, 1980). In a collaborative study, we showed that treatment postnatally with tamoxifen reduces SDN-POA volume in the male rat, as might be expected, but

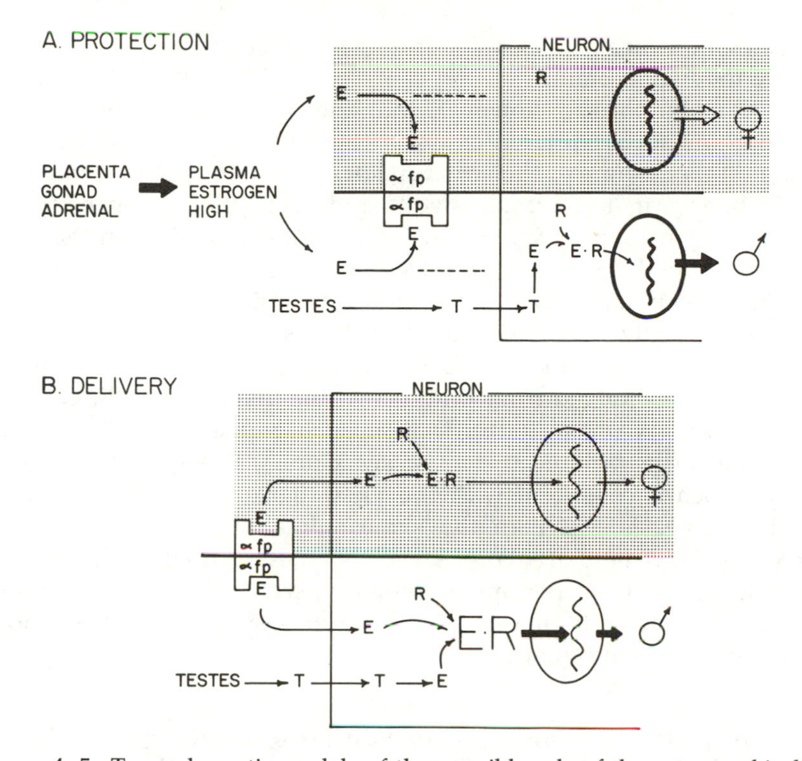

Figure 4–5. Two schematic models of the possible role of the estrogen-binding protein in the perinatal rat, alpha-fetoprotein, in the process of the sexual differentiation of the rat brain. In A, alpha-fetoprotein "protects" neurons of both the female (shaded) and male (clear) from exposure to plasma estrogen (E). In B, alpha-fetoprotein "delivers" the E to the brain of the female that is required for normal development. In both models, E derived from the intraneuronal aromatization of testicular testosterone (T) is necessary for the normal differentiation of the male brain. (Abbreviations: αfp, alpha-fetoprotein; E-R, estrogen-receptor complex; R, estrogen receptor.) (Reprinted with permission from Gorski, 1983.)

this also occurred in the female (Döhler, Srivastava, Shryne, Jarzab, Sipos, & Gorski, 1984). Thus, as illustrated in Figure 4-5B, Döhler and Hancke (1978) have proposed what I have called a delivery model for the role of alpha-fetoprotein in sexual differentiation. They argue that in species with a short gestation period, the species in which alpha-fetoprotein may be an estrogen-binding protein, it serves to prolong the prenatal hormonal environment of high levels of estrogen beyond parturition. According to this concept, the brain is essentially neuter, and even in the female the brain must have some exposure to gonadal steroids to develop or differentiate as feminine. In the case of the male, the estradiol derived from the intraneuronal aromatization of testicular testosterone is again responsible for the further masculine differentiation of the male's brain.

We must emphasize, however, that androgen per se may well have a role in the sexual differentiation of the central nervous system even in the rat. For example, the structural sex difference in the spinal cord, the SNB, appears to be responsive to androgens, not estrogens (Breedlove, Jacobson, Gorski, & Arnold, 1982). In this regard, both the spinal cord and hypothalamus of the rat with the testicular feminizing mutation have been analyzed. In rats with this mutation, androgen receptors are markedly deficient in number (Naess, Haug, Attramadal, Aakvaag, Hansson, & French, 1976), although estrogen receptors appear to be normal as judged from evidence in the mouse (Attardi, Geller, & Ohno, 1976). We have found that in such animals the SDN-POA is masculine in its volume (Fig. 4-4; see Gorski, Csernus, & Jacobson, 1981), which is consistent with the view that estrogen determines its adult volume. The SNB in the same animals is markedly feminine (Breedlove & Arnold, 1981), however, suggesting that in this nucleus an androgen is the important hormonal molecule. In the same animal, different components of the central nervous system may respond to different hormonal signals.

Possible Mechanisms by Which Gonadal Steroids Influence the Sexual Differentiation of the Brain

Whether the rat brain is inherently female or neuter, and whether the active hormonal species is estradiol or another metabolite of testosterone, we still face the fact that both endogenous and exogenous gonadal steroids increase SDN-POA volume and the number of neurons that comprise this nucleus. We view the SDN-POA as one morphological signature of the organizational action of gonadal steroids during neuronal development, and as such it offers a unique model to study basic mechanisms of steroid action. Thus, the basic question is how gonadal steroids determine the number of neurons that ultimately comprise the SDN-POA. Possible mechanisms are illustrated schematically in Figure 4-6. Gonadal hormones may influence or modulate neurogenesis itself, the migration of neurons from their origin in the ependymal lining of the ventricular system, neuronal survival during migration, the aggregation of the neurons into the structure we recognize as the SDN-POA, the survival of these neurons during the period connectivity is established, or the functional neurochemical specification and maturation of the neurons of the SDN-POA.

The last possible mechanism seems unlikely to apply to the SDN-POA because the nucleus is surrounded by a region of lower neuronal density that is equal in both sexes (Gorski et al., 1980; Jacobson & Gor-

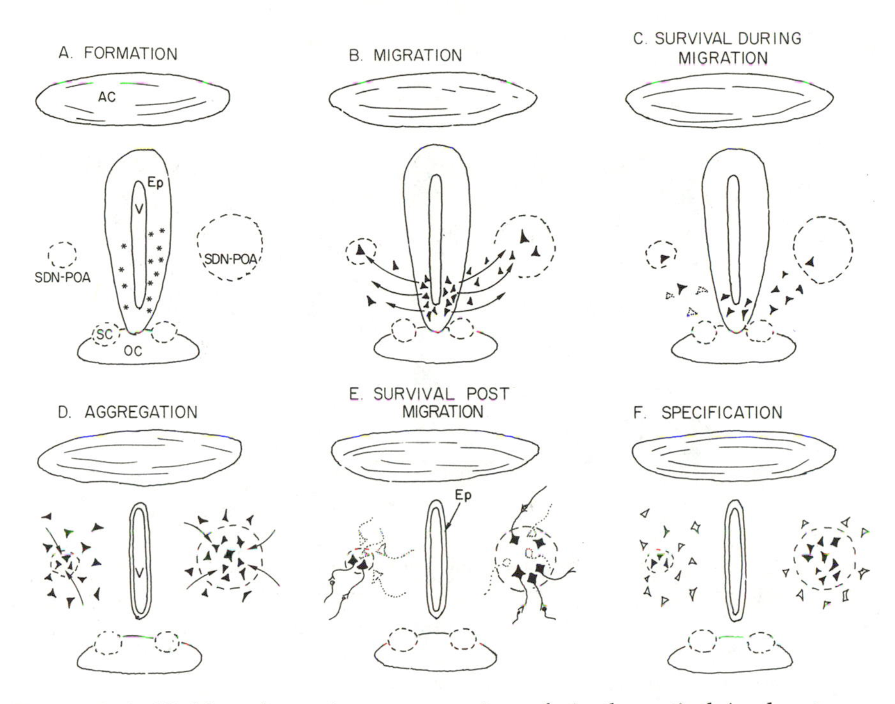

Figure 4–6. Highly schematic representation of six theoretical (and not mutually exclusive) mechanisms by which gonadal steroids (right side of each panel) might influence the number of neurons that ultimately comprise the SDN-POA. [Abbreviations: AC, anterior commissure; Ep, ependymal lining of the third ventricle (V); OC, optic chiasm.]

ski, 1981). In our laboratory we have been attempting to document the possible role of gonadal steroids in several of the remaining theoretical processes. First, we attempted to define the developmental period during which the neurons of the SDN-POA were formed. Although reports from the literature had indicated that neurogenesis of the medial preoptic area is essentially complete by about day 16 of gestation (Altman & Bayer, 1978; Anderson, 1978; Ifft, 1972) and therefore would play very little role in the postnatal influence of gonadal steroids on SDN-POA development, we attempted to confirm these observations while focusing specifically on the neurons of the SDN-POA and a control region just lateral to it. Although we confirmed the observations in the literature that the neurons of the medial preoptic area in general stop DNA synthesis by approximately day 16 of prenatal life (Fig. 4-7), we observed a dramatic prolongation of apparent neurogenesis within those neurons specifically destined to form SDN-POA (Jacobson & Gorski, 1981). (We must speak of apparent neurogenesis, because in

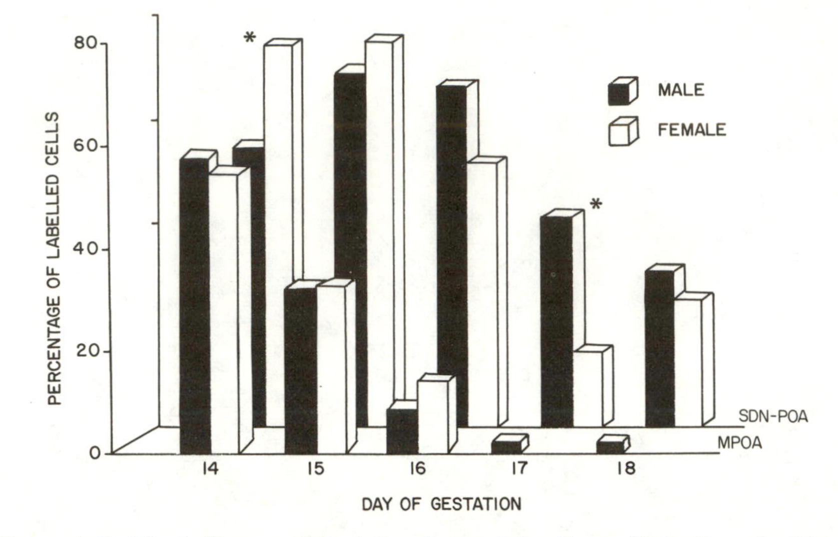

Figure 4–7. The influence of gestational age at the time of injection of tritiated thymidine to pregnant rats on the mean percentage of labeled neurons in the SDN-POA and a region of the medial preoptic area (MPOA) just lateral to the SDN-POA for their offspring sacrificed at postnatal day 30. (Asterisk indicates a significant sex difference. Data from Jacobson & Gorski, 1981.)

this study animals were not sacrificed until they were 30 postnatal days old. Thus, any or all of the other potential mechanisms including migration and cell death would also be expressed in these data.) As illustrated in Figure 4-7, we detected a significant effect of sex on the percent of labeled neurons in the SDN-POA following prenatal exposure to tritiated thymidine. On day 14 of gestation, apparent DNA synthesis is significantly greater in the female, but on day 17 this pattern is reversed. Although the increase in apparent neurogenesis in the male on day 17 of gestation could be related to the activity of the testes because of the reports of a surge of plasma levels of testosterone on day 18 of gestation (Weisz & Ward, 1980), there is currently no hormonal explanation for the observation of the reverse sex difference in apparent neurogenesis on day 14 of gestation.

DNA synthesis is clearly prolonged in a component of the neurons that comprise the SDN-POA, which means that we have a specific and permanent marker of these neurons (compare Figs. 4-8A and 4-8B). One distinct advantage of this fact is that we can now study the process of migration of these late-arising neurons and look for possible sex or hormonal differences. As shown in Figure 4-8C, when animals are exposed to tritiated thymidine on day 18 of gestation and sacrificed just 2 hours later, labeled cells are restricted to the ependymal lining of the third ventricle (Jacobson, Davis, & Gorski, 1985). By administering tri-

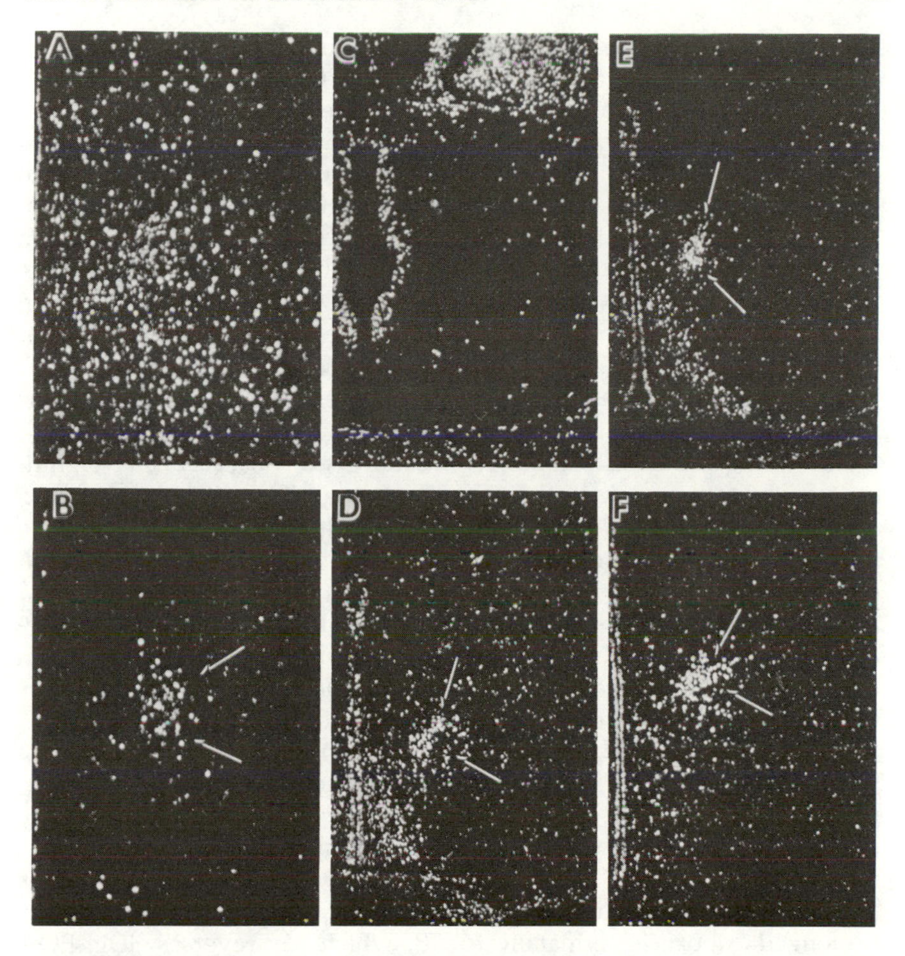

Figure 4–8. Dark-field photomicrographs of coronal sections through the SDN-POA (arrows). A and B (both at the same magnification) demonstrate the specific prolongation of apparent neurogenesis of the SDN-POA. Pregnant rats were injected with tritiated thymidine on day 15 (A) or 18 (B) postconception, and the pups were sacrificed at postnatal day 30. (Modified from Jacobson & Gorski, 1981.) C–F (all at the same magnification) illustrate the change in position of labeled cells during the perinatal period. All rat embryos were exposed to tritiated thymidine on day 18 of gestation via an injection to their mothers. These exposed rats were sacrificed at 18 days plus 2 hours (C), 22 days (D), 26 days (E), or 32 days (F) postfertilization. (Modified from Jacobson, Davis, & Gorski, 1985. Reprinted with permission from Gorski, 1985b.)

tiated thymidine to pregnant mothers on day 18 of gestation, and sacrificing their pups at intervals thereafter, we have been able to trace the migratory pathway of the neurons of the SDN-POA. As shown in Figure 4-8C–F, the labeled cells appear to migrate to the base of the third ventricle and the surrounding neural tissue, then move upward and laterally toward the region of the SDN-POA, and may coalesce to form this nucleus. Statistical analyses of these data have revealed a statistically significant movement of labeled neurons in terms of a grid system as well as a gradual increase in the number of labeled cells in the medial preoptic area and in the region of the SDN-POA until approximately day 24, which is then followed by a significant decline in cell number by day 32 postfertilization (Jacobson et al., 1985). At the present time, it is not possible to determine whether this decrease in the number of labeled cells is because of their loss through death, or their migration onto sections of the brain not analyzed in this study. Another significant observation is that cell size increases during this entire period of perinatal development.

Interestingly but unfortunately, we have not observed any significant sex effects on any of these parameters and no effect of gonadectomy of the male on postnatal day 2. Currently, we have three possible interpretations of these negative data. First, neuronal migration and cell death play no role in the development of the SDN-POA. A second and similar interpretation is that migration and cell death of the late-arising neurons (i.e., those neurons labeled with tritiated thymidine on day 18 of gestation) play no role in the development of the sex differences within the SDN-POA. Finally, the methods of analysis used in this study (Jacobson et al., 1985) were actually inadequate to detect the subtle differences that may lead to the sex difference in SDN-POA volume. At this time, I favor the third interpretation, but my bias is clearly personal; future studies will be required to clarify the precise mechanism of steroid action. In this regard, the results of preliminary studies (R. Dodson, unpublished) from this laboratory show that following the prolonged exposure perinatally to TP, which sex reverses SDN-POA volume in females, a marked increase is seen in the number of neurons labeled with tritiated thymidine on embryonic day 18 that survive. Thus, these late-arising neurons of the SDN-POA clearly do appear to be sensitive to the hormonal environment.

The mechanism that is generally accepted as pertaining to the development of the SDN-POA is the influence of gonadal steroids on the process of cell death during the period neuronal connectivity is established. As indicated, Toran-Allerand (1966, 1980a) has evidence that at least in vitro, estradiol may be required for the outgrowth of neuritic processes from steroid-responsive neurons. Moreover, during development cell death occurs to a remarkable extent (Cowan, 1978; Hamburger & Oppenheim, 1982; Silver, 1978). It is generally accepted, although not proved, that growing neurons reach a source of an unknown neuronotrophic substance that promotes their survival (Hamburger & Oppenheim, 1982). Thus, in the presence of high titers of gonadal hormones, as occurs in the male rat, we suggest that neurons that have arrived at the location of the SDN-POA are stimulated to establish connections giving them access to this unknown neuronotrophic substance that promotes their survival. In the female, however, without this hormonal stimulus, fewer neurons make their appropriate connections, and thus many more die, resulting in the marked sex difference within this nucleus.

The observation of a significant decrease over time in the number of labeled neurons following exposure to tritiated thymidine on day 18 of gestation (Jacobson et al., 1985) suggests yet another potential mechanism. Neurons that have migrated to their adult location can either establish connections and survive or fail to do so and die (histogenic cell death), but neuronal death can also occur during the process of migration (morphogenic cell death). Thus, gonadal steroids may act directly on migrating neurons to promote their survival during morphogenesis. Note that our data on the migration of tritiated thymidine-labeled neurons of the SDN-POA do not currently support this hypothesis; again, this may be because of inadequate methods of analysis.

At this point, we might consider briefly several studies that are still ongoing in our laboratory. If, as we expect, steroids have an important action on fundamental processes during early neuronal development, it becomes important to evaluate the ontogeny of the ability of neurons either in or destined to form the SDN-POA to take up steroids. At the present time we have found that at postnatal day 2, the cells of the SDN-POA do not take up radiolabeled methyltrienolone, a nonaromatizable androgen, but do take up and retain moxestrol, a nonsteroidal estrogen not bound by α-fetoprotein (Schoonmaker, Breedlove, Arnold, & Gorski, 1983). Moreover, on postnatal day 2, the SDN-POA does not appear to be homogeneous in its ability to take up and retain labeled moxestrol. By postnatal day 10, however, the neurons of the SDN-POA are much more homogeneous in their ability to take up and retain labeled moxestrol (Schoonmaker & Gorski, 1984). We would clearly be interested to know whether in younger animals the neuroblasts that form the neurons of the SDN-POA, or migrating neurons, are capable of taking up and retaining tritiated moxestrol.

In other studies we are examining immunohistochemically the connectivity and neurochemical content of the neurons of the SDN-POA. To date we have seen that the region of the SDN-POA, which on the basis of cytoarchitectonics and immunohistochemistry may be considered the central component of the medial preoptic nucleus (MPN; Fig. 4-9), is essentially devoid of terminals labeled with antibodies against serotonin (Simerly, Swanson, & Gorski, 1984). In fact, the distribution of serotonin immunoreactivity in each of the three components of the MPN is sexually dimorphic and, importantly, sex reversed by the same

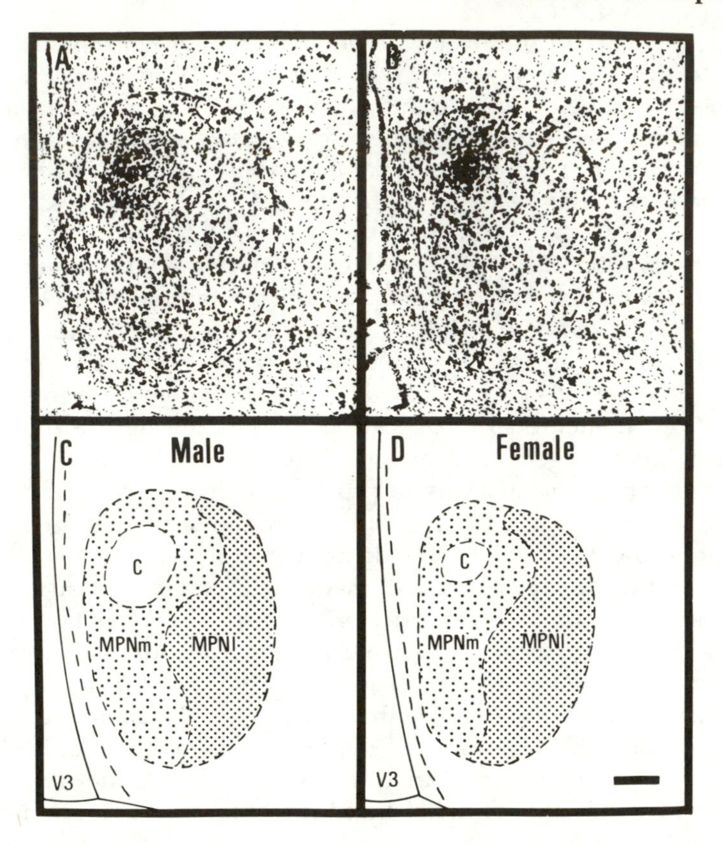

Figure 4–9. A and B: photomicrographs (at same magnification) of thionin-stained sections through the medial preoptic nucleus (MPN) in the male (A) and female (B) rat on which the boundaries of the three subdivisions of the MPN according to Simerly, Swanson, and Gorski (1984) are outlined. C and D: the same subdivisions of the MPN are reproduced schematically, and the relative number of immunoreactive serotonin fibers for each division are indicated by the intensity of stippling. (Abbreviations: C, central component of the MPN that appears comparable to the SDN-POA; MPN_l, the lateral component of the MPN; MPN_m, the medial component of the MPN; V3, third ventricle.) (Reproduced with permission from Gorski, 1984b.)

prolonged exposure to TP that sex reverses SDN-POA volume (Simerly, Swanson, & Gorski, 1985).

On the other hand, we have clear evidence that in the SDN-POA is a relative concentration of terminals that stain immunologically for cholecystokinin (Larriva-Sahd, Gorski, & Micevych, 1985; Simerly, Gorski, & Swanson, 1986). Hammer (1984) has also reported that the region of the SDN-POA is very rich in opiate receptors and that these develop during the first postnatal week.

We are just beginning to characterize the ultrastructure of the neurons and neuropil of the SDN-POA (Larrive-Sahd & Gorski, 1984), although in agreement with one published abstract (Dieter, Dellman, & Jacobson, 1983), obvious sex differences do not appear to exist in ultrastructure. It may be necessary to apply quantitative morphometrics to such analyses.

Transplantation of Neural Tissue

I would like to describe briefly an experiment that may be of considerable relevance to the subject of this volume. This experiment addresses the following question: Can "masculinity" be transplanted from the male to the female rat? The region of medial preoptic area containing the SDN-POA was punched out of newborn male rats and implanted stereotaxically into the region of the medial preoptic area of littermate females. In addition, amygdala and caudate punches were used for control purposes. When these animals were adult, we evaluated their reproductive behavior (Arendash & Gorski, 1982).

Before describing the results, we should explain one component of this experiment. Because we believe that the development of the SDN-POA is dependent at least to a degree on the action of testicular androgen postnatally, we felt that grafts from the male might be unlikely to exert any influence on the developing female because she would lack the hormonal environment thought to be necessary for the full development of the SDN-POA. Therefore, we injected half of the female recipients with a small dose of TP (8 μg). As illustrated in Figure 4-10, transplants of the male medial preoptic area into females significantly increased the masculine behavior demonstrated by these animals when adult, whether or not they had been treated with exogenous TP. Thus, we had in fact succeeded in transplanting "masculinity" from the male to the female. This interpretation is clearly premature, because feminine behavior is also increased by such transplants, especially in those animals also treated with TP (Fig. 4-10).

The mechanisms by which transplants of male brain tissue into females alter the behavior of the recipients are unknown. The approach of brain transplants can ameliorate deficits in both antidiuretic hormone (Gash, Sladek, & Sladek, 1980) and LH-releasing hormone secretion (Krieger et al., 1982), but the mechanism by which overall animal behavior is altered by such grafts is totally unknown.

Interestingly, we observed that graft volume in those females also given 8 μg TP was larger than graft volume in animals not exposed to this hormone. Therefore, we designed an experiment to test the influence of exogenous hormone treatment on graft volume. In this study, medial preoptic area and caudate grafts were implanted into the medial

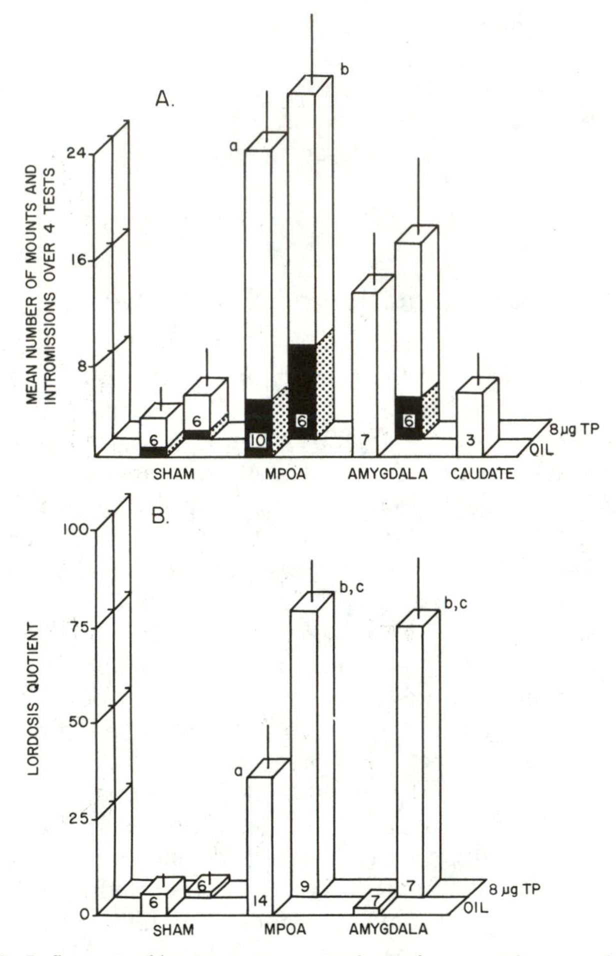

Figure 4–10. Influence of brain tissue transplants from newborn male rats placed into the medial preoptic area (MPOA) of littermate females upon their copulatory behavior when adult. A: Masculine reproductive behavior (mounts, open bars; intromissions, solid bars) in adult females treated with testosterone propionate (TP) and receiving sham, MPOA, amygdala, or caudate transplants neonatally followed by an injection of oil or 8 μg TP. B: Feminine reproductive behavior displayed by the same animals (minus the caudate group) following treatment with 2 μg estradiol benzoate daily for 3 days. Numbers at the base of each bar indicate the number of animals in each group. a: significantly different from all groups given oil after surgery; b: significantly different from comparable sham group; c: significantly different from same experimental group given oil after surgery. (Data from Arendash & Gorski, 1982. Reproduced with permission from Gorski, 1985b.)

preoptic area of female animals who were then exposed to 200 μg TP daily for the next 5 days. These animals were sacrificed on day 30 before any behavioral measurements were obtained. Such treatment with testosterone markedly increased graft volume (Arendash & Gorski, 1983). Thus, steroid hormones appear to be neuronotrophic even in this transplant paradigm. This approach may offer a new avenue to evaluate the influence of steroid hormones on neuronal development.

A final topic returns us to the beginning of this discussion—the effects of gonadal hormones in the adult. Although we have focused on the influence of gonadal steroids during perinatal development, a true representation of masculinity and femininity may require the study of the adult animal in the presence of its sex-specific gonad and, therefore, sex-specific hormonal environment. A morphological effect of gonadal steroids on the adult nervous system following either hypothalamic deafferentation (Arai, Matsumoto, & Nishizuka, 1978) or transsection of the hypoglossal nerve (Yu & Yu, 1983) has been reported. In the songbird system, a hormone-dependent seasonal fluctuation in the morphological structure of the brain has been reported (DeVoogd & Nottebohm, 1981; Nottebohm, 1981). From studies of the sexual differentiation of the developing brain, it is clear that gonadal hormones can have morphological effects. These recent findings, however, suggest that morphological changes in the brain induced by hormones may not stop at the end of the so-called critical period for sexual differentiation. Clearly, during development the rat exhibits the greatest sensitivity to steroid hormones, but we must remember that the brain even of the adult is constantly under the influence of gonadal hormones. Clear functional changes are found in the brain of the adult, and we must be aware that structural changes may also occur. Therefore, and in conclusion, it seems to me, from the perspective gained from the study of the laboratory rat, that any concept of masculinity and femininity must take into account the potential actions of gonadal hormones on both the developing and the adult brain. Gonadal hormones determine the structure and functional potential of the brain during development and in the adult modulate brain function, and perhaps brain structure.

References

Allen, L. S., Hines, M., Shryne, J. E., & Gorski, R. A. (1986). Two sexually dimorphic cell groups in the human brain. *Endocrinology, Suppl., 118,* 633.

Altman, J., & Bayer, S. A. (1978). Development of the diencephalon in the rat. I. Autoradiographic study of the time of origin and settling patterns of neurons of the hypothalamus. *J. Comp. Neurol., 182,* 945–972.

Anderson, C. H. (1978). Time of neuron origin in the anterior hypothalamus of the rat. *Brain Res., 154,* 119–122.

Arai, Y., Matsumoto, A., & Nishizuka, M. (1978). Synaptogenic action of estrogen on the hypothalamic arcuate nucleus (ARCN) of the developing brain and of the deafferented adult brain in female rats. In G. Dorner & M. Kawakami (Eds.), *Hormones and brain development* (pp. 43–48). Amsterdam: Elsevier/North Holland Biomedical Press.

Arendash, G. W., & Gorski, R. A. (1982). Enhancement of sexual behavior in female rats by neonatal transplantation of brain tissue from males. *Science, 217,* 1276–1278.

Arendash, G. W., & Gorski, R. A. (1983). Testosterone-induced enhancement of male medial preoptic tissue transplant volumes in female recipients: A "neuronotrophic" action of testosterone. *Soc. Neurosci. Abstr., 9,* 307.

Arnold, A. P., & Gorski, G. A. (1984). Gonadal steroid induction of structural sex differences in the CNS. In W. M. Cowan (Ed.), *Annual rev. neuroscience* (Vol. 7, pp. 413–422). Palo Alto: Annual Rev. Inc.

Attardi, B., Geller, L. N., & Ohno, S. (1976). Androgen and estrogen receptors in brain cytosol from male, female and Tfm mice. *Endocrinology, 98,* 864–874.

Beach, F. A. (1976). Sexual attractivity, proceptivity and receptivity in female mammals. *Horm. Behav., 7,* 105–138.

Booth, J. E. (1977). Sexual behavior of male rats injected with the antiestrogen MER-25 during infancy. *Physiol. Behav., 19,* 35–39.

Breedlove, S. M., & Arnold, A. P. (1980). Hormone accumulation in a sexually dimorphic motor nucleus of the rat spinal cord. *Science, 210,* 564–566.

Breedlove, S. M. & Arnold, A. P. (1981). Sexually dimorphic motor nucleus in the rat lumbar spinal cord: response to hormone manipulation, absence in androgen-insensitive rats. *Brain Res., 225,* 297–307.

Breedlove, S. M., Jacobson, C. D., Gorski, R. A., & Arnold, A. P. (1982). Masculination of the female rat spinal cord following a single neonatal injection of testosterone propionate but not estradiol benzoate. *Brain Res., 237,* 173–181.

Christensen, L. W., & Gorski, R. A. (1978). Independent masculinization of neuroendocrine systems by intracerebral implants of testosterone or estradiol in neonatal female rat. *Brain Res., 146,* 325–340.

Clemens, L. G., Hiroi, M., & Gorski, R. A. (1969). Induction and facilitation of female mating behavior in rats treated neonatally with low doses of testosterone propionate. *Endocrinology, 84,* 1430–1438.

Clemens, L. G., Shryne, J., & Gorski, R. A. (1970). Androgen and development of progesterone responsiveness in male and female rats. *Physiol. Behav., 5,* 673–678.

Corbier, P., Kerdelhue, B., Picon, R., and Roffi, J. (1978). Changes before, during and after birth in the perinatal rat. *Endocrinology, 103,* 1985–1991.

Corbier, P., Roffi, J., Rhoda, J., & Valens, M. (1981). Comportement sexuel femelle des rats mâles castrés à la naissance et soumis aux hormones sexuelles femelles: Effects de l'heure de la castration. *Comptes Rendue Acad. Sci. Paris III,* 293, 649–654.

Cowan, W. M. (1978). Aspects of neural development. In R. Porter (Ed.), *International review of physiology and neurophysiology III* (Vol. 17, pp. 149–191). Baltimore: University Park Press.

DeBoogd, T., & Nottebohm, F. (1981). Gonadal hormones induce dendritic growth in the adult avian brain. *Science, 214,* 202–204.

Dieter Dellman, H., and Jacobson, C. D. (1983). The sexually dimorphic nucleus of the preoptic area (SDN-POA): Ultrastructure. *Soc. Neurosci. Abstr., 9,* 516.

Döhler, K. D., Coquelin, A., Davis, F., Hines, M., Shryne, J. E., & Gorski, R. A. (1982). Differentiation of the sexually dimorphic nucleus in the preoptic area of the rat brain is determined by the perinatal hormone environment. *Neurosci. Lett., 33,* 295–298.

Döhler, K. D., Coquelin, A., Davis, F., Hines, M., Shryne, J. E. & Gorski, R. A. (1984). Pre- and postnatal influence of testosterone propionate and diethylstilbestrol on differentiation of the sexually dimorphic nucleus of the preoptic area in male and female rats. *Brain Res., 302,* 291–295.

Döhler, K. D., & Hancki, J. L. (1978). Thoughts on the mechanism of sexual brain differentiation. In G. Dorner & M. Kawakami (Eds.), *Hormones and brain development* (pp. 153–157). Amsterdam: Elsevier/North Holland Biomedical Press.

Döhler, K. D., Hines, M., Coquelin, A., Davis, F., Shryne, J. E., & Gorski, R. A. (1982). Pre- and postnatal influence of diethylstilbestrol on differentiation of the sexually dimorphic nucleus in the preoptic area of the female rat brain. *Neuroendocrinol. Lett., 4,* 361–365.

Döhler, K. D., Srivastava, S. S., Shryne, J. E., Jarzab, B., Sipos, A., & Gorski, R. A. (1984). Differentiation of the sexually dimorphic nucleus in the preoptic area of the rat brain is inhibited by postnatal treatment with an estrogen antagonist. *Neuroendocrinology, 38,* 297–301.

Doughty, C., & McDonald, P. G. (1974). Hormonal control of sexual differentiation of the hypothalamus in the neonatal female cat. *Differentiation, 2,* 275–285.

Gash, D., Sladek, J., & Sladek, C. (1980). Functional development of grafted vasopressin neurons. *Science, 210,* 1367–1369.

Gilman, D., and Hitt, J. (1978). Effects of gonadal hormones on pacing of sexual contacts by female rats. *Behav. Biol., 24,* 77–87.

Gorski, R. A. (1966). Localization and sexual differentiation of the nervous structures which regulate ovulation. *J. Reprod. Fertil., Suppl., 1,* 67–88.

Gorski, R. A. (1974). The neuroendocrine regulation of sexual behavior. In G. Newton & A. H. Riesen (Eds.), *Advances in Psychobiology* (Vol. II, pp. 1–58). New York: John Wiley & Sons.

Gorski, R. A. (1976). The possible neural sites of hormonal facilitation of sexual behavior in the female rat. *Psychoneuroendocrinology, 1,* 371–387.

Gorski, R. A. (1979). Long-term hormonal modulation of neuronal structure and function. In F. O. Schmitt & F. G. Worden (Eds.), *The neurosciences: Fourth study program* (pp. 969–982). Cambridge: MIT Press.

Gorski, R. A. (1983). Steroid-induced sexual characteristics in the brain. In: E. E. Muller & R. M. MacLeod (Eds.), *Neuroendocrine perspectives* (Vol. 2, pp. 1–35). Amsterdam: Elsevier.

Gorski, R. A. (1984a). Sexual differentiation of brain structure in rodents. In M. Serio, M. Sanisi, M. Motta, & L. Martini (Eds.), *Sexual differentiation: Basic and clinical aspects* (pp. 65–77). New York: Raven Press.

Gorski, R. A. (1984b). Critical role for the medial preoptic area in the sexual differentiation of the brain. *Progress in brain research, 61,* 129–146.

Gorski, R. A. (1985a). Sexual dimorphisms of the brain. *J. Animal Sci., 61,* Suppl. 3, 38–61.

Gorski, R. A. (1985b). Gonadal hormones as putative neuronotropic substances. In C. W. Cotman (Ed.), *Synaptic plasticity and remodeling* (pp. 287–310). New York: Guilford Press.

Gorski, R. A., Csernus, V. J., & Jacobson, C. D. (1981). Sexual dimorphism in the preoptic area. In: B. Flerko, G. Setalo, & L. Tima (Eds.), *Advances in physiological sciences: Reproduction and development* (Vol. 15, pp. 121–130). Budapest: Pergamon Press and Akademiai Kiado.

Gorski, R. A., Gordon, J. H., Shryne, J. E., & Southam, A. M. (1978). Evidence for a morphological sex difference within the medial preoptic area of the rat brain. *Brain Res., 148,* 333–346.

Gorski, R. A., Harlan, R. E., Jacobson, C. D., Shryne, J. E., & Southam, A. M. (1980). Evidence for the existence of a sexually dimorphic nucleus in the preoptic area of the rat. *J. Comp. Neurol., 193,* 529–539.

Gorski, R. A., & Wagner, J. W. (1965). Gonadal activity and sexual differentiation of the hypothalamus. *Endrocrinology, 76,* 226–239.

Goy, R. W., & McEwen, B. S. (1980). *Sexual differentiation of the brain* (223 pp). Cambridge: MIT Press.

Greenough, W. T., Carter, C. S., Steerman, C., & DeVoogd, T. J. (1977). Sex differences in dendritic patterns in hamster preoptic area. *Brain Res., 126,* 63–72.

Hamburger, V., & Oppenheim, R. W. (1982). Naturally-occurring neuronal death in vertebrates. *Neuroscience Commentaries, 1,* 39–55.

Hammer, R. P. (1984). The sexually dimorphic region of the preoptic area in rats contains denser opiate receptor sites in females. *Brain Res., 308,* 172–176.

Hancke, J. L., & Döhler, K. D. (1980). Postnatal estradiol treatment prevents tamoxifen-induced defeminization of the female rat brain. *Acta Endocrinol. Suppl., 234,* 102–103.

Handa, R. J., Corbier, P., Shyrne, J. E., Schoonmaker, J. N., & Gorski, R. A. (1985). Differential effects of the perinatal steroid environment on three sexually dimorphic parameters of the rat brain. *Biol. Reprod., 32,* 855–864.

Harlan, R. E., Gordon, J. H., & Gorski, R. A. (1979). Sexual differentiation of the brain: Implications for neuroscience. In *Reviews of neuroscience* (Vol. 4, pp. 31–71). New York: Raven Press.

Hodges, J. K., & Hearn, J. P. (1978). A positive feedback effect of oestradiol on LH release in the male marmoset monkey, *Callithrix jacchus. J. Reprod. Fert., 52,* 83–86.

Ifft, J. D. (1972). An autoradiographic study of the time of final division of neurons in rat hypothalamic nuclei. *J. Comp. Neurol., 144,* 193–204.

Jacobson, C. D., Arnold, A. P., & Gorski, R. A. (1982). Steroid accumulation in the sexually dimorphic nucleus of the preoptic area (SDN-POA). *Anat. Rec., 202,* 88A.

Jacobson, C. D., Csernus, V. J., Shryne, J. E., & Gorski, R. A. (1981). The influence of gonadectomy, androgen exposure, or a gonadal graft in the neonatal rat on the volume of the sexually dimorphic nucleus of the preoptic area. *J. Neurosci., 1,*(10), 1142–1147.

Jacobson, C. D., Davis, F. C., & Gorski, R. A. (1985). Formation of the sexually dimorphic nucleus of the preoptic area: neuronal growth, migration and changes in cell number. *Dev. Brain Res., 21,* 7–18.

Jacobson, C. D., & Gorski, R. A. (1981). Neurogenesis of the sexually dimorphic nucleus of the preoptic area of the rat. *J. Comp. Neurol., 196,* 519–529.

Jacobson, C. D., Shryne, J. E., Shapiro, F., & Gorski, R. A. (1980). Ontogeny of the sexually dimorphic nucleus of the preoptic area. *J. Comp. Neurol., 193,* 541–548.

Karsch, F. J., Dierschke, D. J., & Knobil, E. (1973). Sexual differentiation of pituitary function: apparent difference between primates and rodents. *Science, 179,* 484–486.

Korenbrot, C. C., Paup, D., & Gorski, R. A. (1975). Effects of testosterone propionate or dihydrotestosterone propionate on plasma FSH and LH levels in neonatal rats and on sexual differentiation of the brain. *Endocrinology, 97,* 709–717.

Krieger, D. T., Perlow, M. J., Gibson, M. J., Davies, T. F., Zimmerman, E. A., Ferin, M., & Charlton, H. M. (1982). Brain grafts reverse hypogonadism of gonadotropin releasing hormone deficiency. *Nature, 298,* 468–471.

Larriva-Sahd, J., & Gorski, R. A. (1984). Fine structure of the sexually dimorphic nucleus of the preoptic area (SDN-POA). *Anat. Rec., 208,* 100A–101A.

Larriva-Sahd, J., Gorski, R. A., & Micevych, P. (1985). Cholecystokinin synapses in the medial preoptic nucleus (MPN): an ultra-structural study. *Anat. Rec., 111,* 105A.

Lieberburg, I., Wallach, G., & McEwen, B. S. (1977). The effects of an inhibitor of aromatization (1,4,6-androstatriene-3,17-dione) and an antiestrogen (CI-628) on in vivo formed testosterone metabolites recovered from neonatal rat brain tissues and purified cell nuclei. Implications for sexual differentiation of the rat brain. *Brain Res., 128,* 176–181.

Loy, R., & Milner, T. A. (1980). Sexual dimorphism in extent of axonal sprouting in rat hippocampus. *Science, 208,* 1282–1284.

Matsumoto, A., & Arai, Y. (1980). Sexual dimorphism in "wiring pattern" in the hypothalamic arcuate nucleus and its modification by neonatal hormonal environment. *Brain Res., 190,* 138–242.

McEwen, B. S. (2983). Gonadal steroid influences on brain development and sexual differentiation. *Reprod. Physiol., 27,* 99–145.

McEwen, B. S., Biegon, A., Davis, P. G., Krey, L. C., Luine, V. N., McGinnis, M. Y., Paden, C. M., Parsons, B., & Rainbow, T. C. (1982). Steroid hormones: Humoral signals which alter brain cell properties and functions. In R. O. Greep (Ed.), *Recent progress in hormone research (Vol 38,* pp. 41–92). New York: Academic Press.

McEwen, B. S., Lieberburg, I., Chaptal, C., & Krey, L. C. (1977). Aromatization: Important for sexual differentiation of the neonatal rat brain. *Horm. Behav., 9,* 249–263.

Milner, T. A., & Loy, R. (1982). Hormonal regulation of axonal sprouting in the hippocampus. *Brain Res., 243,* 180–185.

Naess, O., Haug, E., Attramadal, A., Aakvaag, A., Hansson, V., & French, F. (1976). Androgen receptors in the anterior pituitary and central nervous system of the androgen "insensitive" (Tfm) rat: Correlation between receptor binding and effects of androgens on gonadotropic secretion. *Endocrinology, 99,* 1295–1303.

Naftolin, F., Ryan, K. J., Davies, I. J., Reddy, V. V., Flores, F., Petro, Z., & Kuhn, M. (1975). The formation of estrogens by central neuroendocrine tissues. *Recent Prog. Horm. Res., 31,* 295–319.

Nishizuka, M., & Arai, Y. (1981). Organizational action of estrogen on synaptic pattern in the amygdala: implications for sexual differentiation of the brain. *Brain Res., 213,* 422–426.

Nottebohm, F. (1981). A brain for all seasons: Cyclical anatomical changes in song control nuclei of the canary brain. *Science, 214,* 1368–1370.

Nottebohm, F., & Arnold, A. P. (1976). Sexual dimorphism in vocal control areas of the songbird brain. *Science, 194,* 211–213.

Nunez, E., Savu, L., Engelmann, F., Bennassayag, C., Crepy, O., & Jayle, M. F. (1971). Origine embryonnaire de la proteine serique fixant l'oestrone et l'oestradiol chez la Ratte impubere. *C. R. Acad. Sci. (Paris)*, 273, 242–245.

Ojeda, S. R., Kalra, P. S., & McCann, S. M. (1975). Further studies on the maturation of the estrogen negative feedback on gonadotropin release in the female rat. *Neuroendocrinology, 18*, 242–255.

Phoenix, C. H., Goy, R. W., Gerall, A. A., Young, W. C. (1959). Organizing action of prenatally administered testosterone propionate on the tissues mediating mating behavior in the female guinea pig. *Endocrinology, 65*, 369–382.

Plapinger, L., McEwen, B. S., & Clemens, L. E. (1973). Ontogeny of estradiol-binding sites in rat brain. II. Characteristics of a neonatal binding macromolecule. *Endocrinology, 93*, 1129–1139.

Raisman, G., & Field, P. M. (1973). Sexual dimorphism in the neuropil of the preoptic area of the rat and its dependence on neonatal androgen. *Brain Res., 54*, 1–29.

Raynaud, J. P., Mercier-Bodard, C., & Baulieu, E. E. (1971). Rat estradiol binding plasma protein (EBP). *Steroids, 18*, 767–788.

Rhoda, J., Corbier, P., Roffi, J., Castanier, M., & Scholler, R. (1983). Elévation du taux d'oestradiol dans l'hypothalamus du rat mâle, à la naissance. *Comptes Rendue Acad. Sci. Paris III, 296*, 405–408.

Schoonmaker, J. N., Breedlove, S. M., Arnold, A. P., & Gorski, R. A. (1983). Accumulation of steroid in the sexually dimorphic nucleus of the preoptic area in the neonatal rat hypothalamus. *Soc. Neurosci. Abstr., 9*, 1094.

Schoonmaker, J. N., & Gorski, R. A. (1984). Evidence for maturational changes in estrogen (E) accumulation in the sexually dimorphic nucleus of the preoptic area (SDN-POA) in the rat hypothalamus. *Biol. Reprod., 30*(Suppl. 1), 109.

Selmanoff, M. K., Brodkin, L. D., Weiner, R. I., & Siiteri, P. K. (1977). Aromatization and 5-alpha-reduction of androgens in discrete hypothalamic and limbic regions of the male and female rat. *Endocrinology, 101*, 841–848.

Silver, J. (1978). Cell death during development of the nervous system. In M. Jacobson (Ed.), *Handbook of sensory physiology: Vol. IX. Development of sensory systems* (pp. 419–436). Berlin: Springer-Verlag.

Simerly, R. B., Gorski, R. A., and Swanson, L. W. (1986). Neurotransmitters specificity of cells and fibers in the medial preoptic nucleus: an immunohistochemical study in the rat. *J. Comp. Neurol., 246*, 343–363.

Simerly, R. B., Swanson, L. W., & Gorski, R. A. (1984). Demonstration of a sexual dimorphism in the distribution of serotonin immunoreactive fibers in the medial preoptic nucleus of the rat. *J. Comp. Neurol.* 51–166.

Simerly, R. B., Swanson, L. W., & Gorski, R. A. (1985). Reversal of the sexually dimorphic distribution of serotonin immunoreactive fibers in the medial preoptic nucleus by treatment with perinatal androgen. *Brain Res., 340*, 91–98.

Swaab, D. F., & Fliers, E. (1985). A sexually dimorphic nucleus in the human brain. *Science, 228*, 1112–1115.

Taleisnik, S., Caligaris, L., & Astrada, J. J. (1971). Sex difference in hypothalamo-hypophysical function. In C. H. Sawyer and R. A. Gorski (Eds.), *Steroid hormones and brain function* (pp. 171–184). Los Angeles: University of California Press.

Tarttelin, M. F., Shryne, J. E., & Gorski, R. A. (1975). Patterns of body weight

change in rats following neonatal hormone manipulation: A "critical period" for androgen-induced growth increases. *Acta Endocrinol.*, *79*, 177–191.
Tobet, S. A., Gallagher, C. A., Zahniser, D. J., Cohen, M. H., & Baum, M. J. (1983). Sexual dimorphism in the preoptic/anterior hypothalamic area of adult ferrets. *Endocrinology, 112*(Suppl), 240.
Toran-Allerand, C. D. (1976). Sex steroids and the development of the newborn mouse hypothalamus and preoptic area in vitro: Implications for sexual differentiation. *Brain Res., 106,* 407–412.
Toran-Allerand, C. D. (1980a) Sex steroids and the development of the newborn mouse hypothalamus and preoptic area in vitro. II. Morphological correlates and hormonal specificity. *Brain Res., 189,* 413–427.
Toran-Allerand, C. D. (1980b). Coexistence of α-fetoprotein albumin and transferrin immunoreactivity in neurones of the developing mouse brain. *Nature, 286,* 733–735.
Toran-Allerand, C. D. (1982). Regional differences in intraneuronal localization of alpha-fetoprotein in developing mouse brain. *Dev. Brain Res.*, *5*, 213–217.
Toran-Allerand, C. D., Gerlach, J. L., & McEwen, B. S. (1980). Autoradiographic localization of ^{3}H-estradiol related to steroid responsiveness in cultures of the hypothalamus and preoptic area. *Brain Res., 184,* 517–522.
Weisz, J., & Gunsalus, P. (1973). Estrogen levels in immature female rats: True or spurious—ovarian or adrenal? *Endrocrinology, 93,* 1057–1065.
Weisz, J., & Ward, I. L. (1980). Plasma testosterone and progesterone titers of pregnant rat, their male and female fetuses, and neonatal offspring. *Endocrinology, 106,* 306–316.
Whalen, R. E., & Rezek, D. L. (1974). Inhibition of lordosis in female rats by subcutaneous implants of testosterone, androtenedione or dihydrotestosterone in infancy. *Horm. Behav., 5,* 125–128.
Yahr, P., & Commins, D. (1983). The neuroendocrinology of scent marking. In R. M. Silverstein and D. Muller-Schwarze (Eds.), *Chemical signals in vertebrates* (III, pp. 119–133). New York: Plenum Publishing Co.
Yu, W. A., and Yu, M. C. (1983). Acceleration of the regeneration of the crushed hypoglossal nerve by testosterone. *Exp. Neurol., 80,* 349–360.

5

Observations on Brain Sexual Differentiation: A Biochemist's View

Bruce S. McEwen

Gorski (Chapter 4) has described an elegant and extremely important series of studies that establish morphological sex differences in the mammalian brain and reveal their origin in the perinatal actions of testosterone (T) via its conversion to estradiol (E). These and related studies have provided the most dramatic evidence to date that hormones directly influence the brain in exerting their important effects on behavior. In this short essay I would like to expand on four aspects of Gorski's presentation that I believe deserve special emphasis.

Pathways Mediating T Actions on Brain Sexual Differentiation

The actions of T to bring about brain sexual differentiation are mediated by two pathways: one involving 5-alpha reduction of T and interaction of the product, 5-alpha dihydrotestosterone (DHT), with androgen receptors; the other involving aromatization of T and interaction of the product, estradiol, with estrogen receptors. Because such conversions of T undoubtedly operate in many vertebrate species and because they are largely independent of each other, it is important to consider what aspects of brain sexual differentation are involved and how each of the two pathways can be separately manipulated.

Sexual differentiation consists of at least two processes, defeminization and masculinization (Beach, 1975), which are separable both temporally and in terms of the two pathways noted. In the rat defemini-

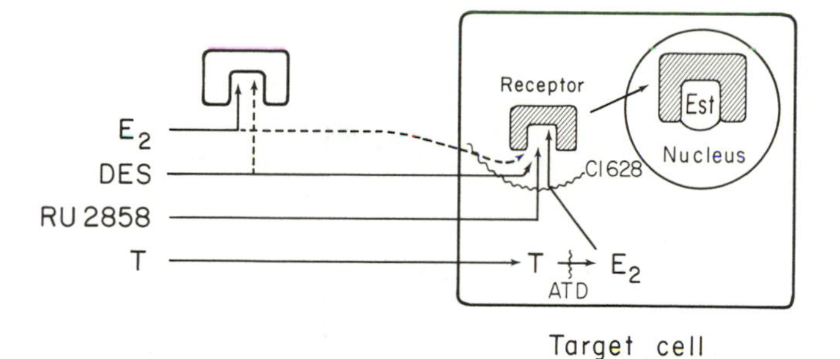

Figure 5–1. Schematic diagram of the protective role of alpha-fetoprotein and the ability of synthetic estrogens and testosterone to bypass this mechanism. E_2, estradiol; DES, diethylstilbestrol; RU2858, 11 β-methoxyethynylestradiol; T, testosterone; Est, various estrogens in nucleus. (Reprinted with permission from McEwen, Plapinger, Chaptal, Gerlach, & Wallach, 1975.)

zation is primarily a postnatal event that occurs through the process of aromatization, whereas masculinization depends on the action of both pathways and takes place more prenatally than postnatally (McEwen, 1983). How we know this has depended on the development of various procedures for manipulating the two pathways of T action. These include using a competitive inhibitor of aromatization, 1,4,6-androstatriene-3,17-dione (ATD) (see Fig. 5-1), and taking advantage of the absence of androgen receptors found in the androgen-insensitive (testicular feminizing mutation—Tfm) mutation. Tfm male brains have the capability for aromatization and contain estrogen receptors, and they undergo defeminization (see McEwen, 1983). Yet they lack normal masculine sexual behavior, as well as the spinal cord nucleus that innervates the penis, and they are deficient in rough-and-tumble play behavior (Fig. 5-2). This type of play behavior is induced in normal males by the perinatal actions of T or DHT, and it can be prevented from developing

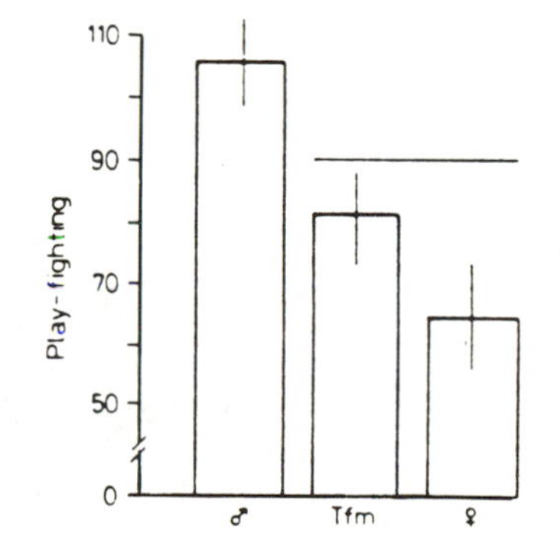

Figure 5–2. Mean (±standard error of the mean) frequency of play-fighting during the prepubertal period by normal male and female pups and testicular feminizing mutation (Tfm) male pups. Groups falling under the line do not differ significantly from one another. (Reprinted with permission from Meaney, Stewart, Poulin, & McEwen, 1983.)

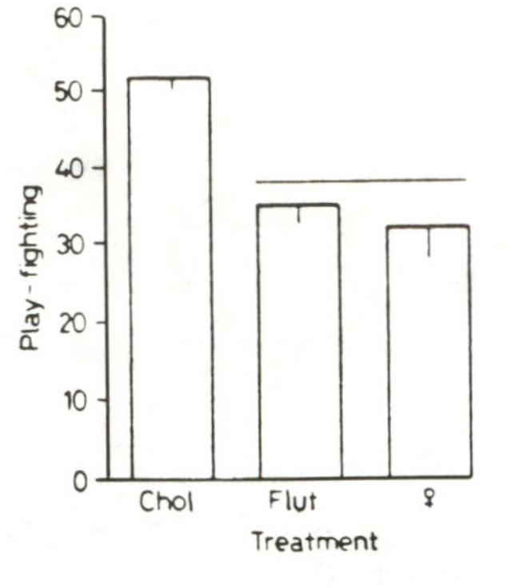

Figure 5–3. Mean (±SEM) frequency of play-fighting during the prepubertal period by flutamide- and cholesterol-treated males and untreated females. Groups falling under the line do not differ significantly from one another. (Reprinted with permission from Meaney et al., 1983.)

by the androgen receptor-blocking drug flutamide (Fig. 5-3). In contrast, male rats treated at birth with ATD are blocked from being defeminized and show high levels of feminine sexual behavior in response to estradiol and progesterone treatment. In fact, strictly postnatal ATD treatment produces male rats that are bisexual and show a high incidence of feminine sexual behavior (Table 5-1) as well as normal masculine sexual behavior.

Gorski has reviewed evidence that points to the aromatization pathway as the mediator of the enlargement of the sexually dimorphic nu-

Table 5-1
Feminine Sexual Behavior of ATD ($n=14$) and Control ($n=12$) Males Observed under Endogenous and Exogenous Hormonal Conditions

	X Mounts Received	*No. of Responders*	*X LQ Responders*	*No. Soliciting*
Precastration				
Test 1–3				
ATD	24.7	7	21	0
Control	14.5	0	—	—
Test 4				
ATD	10	9	77	7[a]
Control	10	2	55	0
Postcastration				
Test 1–2				
ATD	20	14	85	11
Control	20	5	59	1

Note: Every postnatal ATD treatment of male rats results in elevated feminine sexual responding precastration (no exogenous hormone) and also postcastration (after E + P priming). From "Independence of the Differentiation of Masculine and Feminine Sexual Behavior in Rats" by P. G. Davis, C. V. Chaptal, and B. S. McEwen, 1979, *Horm. Behav., 12,* 12–19. Reprinted by permission.
[a]One ATD male exhibited solicitation behaviors but no lordosis.

cleus (SDN) during perinatal development of the rat. This evidence includes the occurrence of a male-size SDN in Tfm male rats and the ability of estrogens such as diethylstilbestrol (DES) to mimic the ability of T to promote enlargement of the SDN in neonatal female rats. An important corollary of these studies on the SDN is that other anatomical loci and other features of the brain may undergo differentiation under the influence of T acting via estrogen or androgen receptors. The size of the spinal nucleus of the bulbocavernosus is an example of a morphological trait under androgen receptor mediation (Breedlove & Arnold, 1980, 1981). With respect to aromatization, we find that a biochemical trait, the ability of estradiol to induce cytosol progestin receptors, is larger in the ventromedial (VMN) region of female rats than it is in male rats (Rainbow, Parsons, & McEwen, 1982). Perinatal treatment of newborn females with E results in a malelike progestin receptor level in adulthood, whereas perinatal ATD treatment of male rat pups produces a femalelike progestin receptor induction in the VMN (Parsons, Rainbow, & McEwen, 1984).

Developmental Events in Sexual Differentiation

Because the events of sexual differentiation are mediated by specific receptors located in groups of neurons within various parts of the brain, it is important to consider the development of these receptor systems. Like many other specific features of cellular differentiation, the "steroid receptor phenotype" is expressed during early development under the influence of as yet unknown inducers. Presumably, the determination that certain cells will express such a trait is made at an earlier stage of development, during the phase of cell division. This is strongly suggested by the existence of clusters or nuclear groupings of cells bearing steroid receptors—as if they originated as a clone from some earlier single cell (see Fig. 5-4). Expression of traits such as the estrogen receptor phenotype is, however, delayed until after the final cell divisions are completed (Gerlach, McEwen, Toran-Allerand, & Friedman, 1984). In the rat and mouse brain, estrogen receptors rise dramatically during the last few days prior to birth, preceding the onset of the critical period during which defeminization occurs (Fig. 5-5 and 5-6). This perinatal increase is independent of the gonadal hormones, as indicated by the fact that it occurs equally in male and female fetuses (Fig. 5-5), in spite of differences in gonadal secretion perinatally. Moreover, hypothalamic tissue transplanted to the midbrain choroidal pia of adult hosts not only survives but also expresses estrogen and estrogen-inducible progestin receptors just as well in gonadectomized as in intact hosts (Stenevi et al., 1980; Paden, Silverman, McEwen, Stenevi, Bjorkland, & Thorngren, 1980).

Once the estrogen receptor phenotype is expressed, estradiol de-

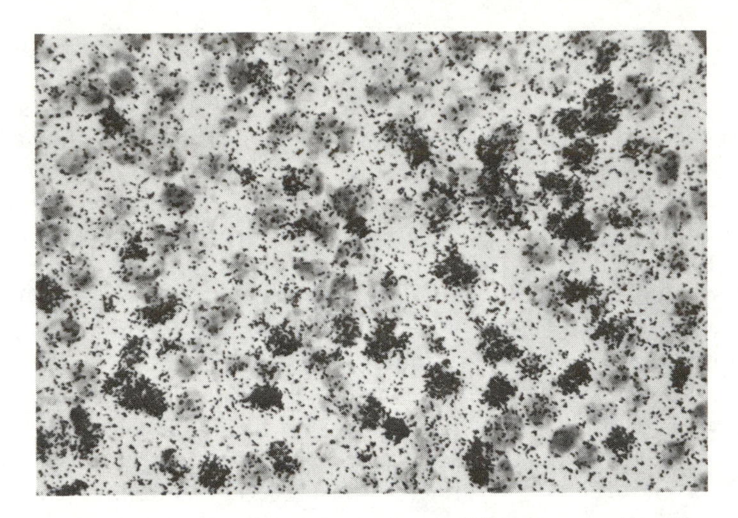

Figure 5–4. Radioautogram depicts binding of [^{3}H]moxestrol and/or metabolites in neurons of the ventromedial nucleus of the hypothalamus of a 15-day-old mouse. Although 2 days earlier labeling was scant, by E-15 the number of cells with estrogen receptors, as well as the population of receptors per cell, is relatively high, and both increase dramatically by E-18. (Radioautography as described in Gerlach, McEwen, Toran-Allerand, & Friedman, 1983.)

Figure 5–5. Perinatal ontogeny of soluble cytoplasmic estrogen receptor sites in the brains of male and female rats. Cytosols were prepared from either midbrain and brain stem (●), cerebral cortex (▲) or pooled hypothalamus, preoptic area, and amygdala (■) and were labeled in vitro with 2 nM [^{3}H]moxestrol. Receptor-bound radioactivity was measured by Sephadex LH-20 gel filtration. Each data point represents the mean (±SEM) of four determinations at each age. (Reprinted with permission from MacLusky, Lieberburg, & McEwen, 1979.)

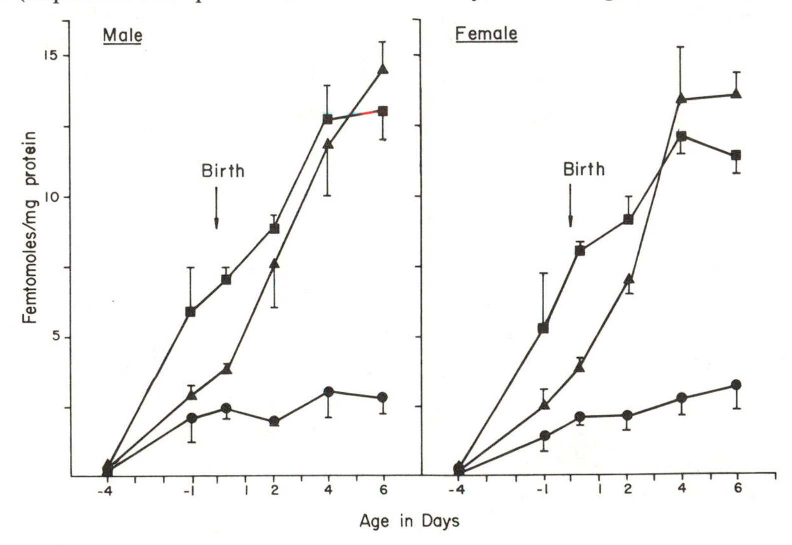

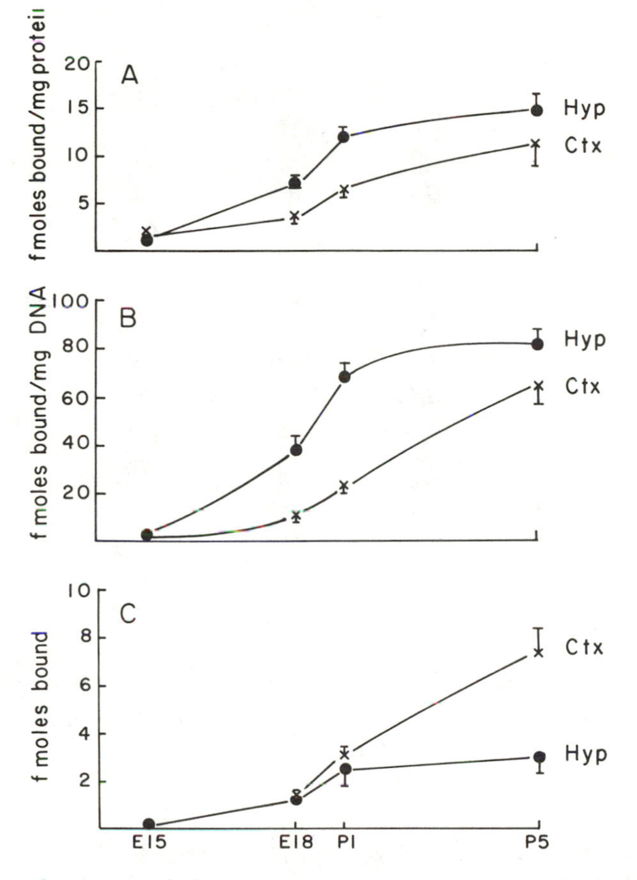

Figure 5–6. The perinatal changes in estrogen receptors in the hypothalamus and whole cortex shown by [^{3}H]moxestrol binding. C: relative magnitudes of the total amount of receptors in these regions. B: the receptor content corrected for cell number. A: the receptor content relative to protein mass. The increases in protein mass in the cortex attenuate the increase in receptor content that is evident in B and C. Four to five experiments were run at each age. In each experiment six to ten brains were pooled. (Reprinted with permission from Friedman, McEwen, Toran-Allerand, & Gerlach, 1983.)

rived from testosterone via aromatization occupies these receptors in males (Fig. 5-7), and this mediates the primary events that underlie sexual differentation. Gorski has noted that there may be more than one primary event, including hormonal influences on cell division, on cell survival, on growth, and on differentiation. One of these primary events, which has been clearly documented, as Gorski notes, is the process of neurite outgrowth stimulated by estrogens acting via estrogen receptors (Toran-Allerand, 1981). To account for the morphological

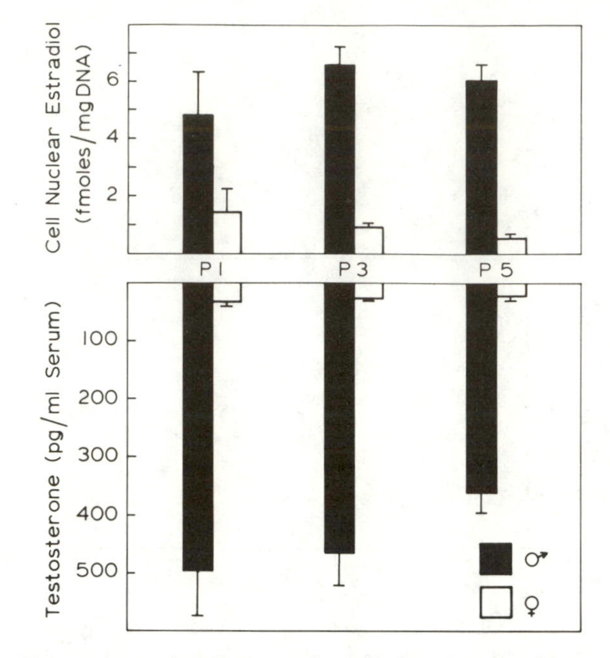

Figure 5–7. Testosterone levels in male and female rats (bottom panel) are compared with occupation of cell nuclear estrogen-receptor sites in brain resulting from aromatization of testosterone (top panel) at indicated postnatal ages. (Reprinted with permission from McEwen, 1983.)

sex differences Gorski has described in a nucleus such as the SDN, it is not necessary to suppose more than such outgrowth as the primary steroid-sensitive event. Greater neurite outgrowth might confer a competitive advantage on those neurons, allowing more of them to make successful synapses and survive, as well as leading to the development of larger cell bodies related to the larger field of terminals and processes. Thus, differences in synaptic contacts, cell soma size, and number all could emanate from one steroid-sensitive event. Nothing in this notion, however, precludes steroids from having more than one primary action on developing neurons.

Protection Mechanisms in the Fetus and Neonate

The fetus and the neonate must be protected to some degree from the potentially deleterious effects of gonadal steroids reaching the fetus through the placenta or the neonate via the mother's milk. Mammals have evolved several types of protection mechanisms (see McEwen, 1983). These include the placental barrier, which, in the case of E, markedly reduces the concentration of free E in the fetal circulation. Also important is progesterone, which acts to attenuate the actions of both androgens and estrogens on developing tissue. Finally, the rat and mouse use an estrogen-binding site on alpha-fetoprotein (AFP), a plentiful serum protein of the fetus and neonate, as a means of removing by mass action E that may enter the fetal or neonatal circulation (see Fig. 5-1). Considering the effectiveness of the placenta, AFP may actually play a more significant role in the neonate with respect to estrogens reaching the pup via the mother's milk. AFP, however, does not protect against all estrogens, because those such as DES do not bind to AFP and are, as a result, much more effective than E in defeminizing the rat brain. Curiously, DES also bypasses the placental barrier in primates and is thus more effective as an estrogen than E, even though the primate AFP does not bind estrogens.

Gorski notes the divergent views of AFP as a protector and as a possible deliverer of estradiol to developing cells. The delivery role is suggested by the observation that AFP occurs inside some neurons in the brain, as well as in the extracellular fluids. Cells with intracellular AFP, however, have virtually no intracellular estrogen receptors (Toran-Allerand, 1982), and thus the delivery of E would be to cells lacking the means to respond to the hormone. One further consideration is that in species such as the human, in which AFP lacks the estrogen-binding portion, AFP is also found inside some brain cells (Ali, Balapure, Singh, Shukla, & Sahib, 1981). Therefore, an alternative view of AFP's function in neural development is that it may have several different roles, and the role associated with its presence inside cells may not be directly related to E action.

Morphological versus Biochemical Sex Differences

Morphological sex differences imply that there will also be resultant functional differences at the level of synaptic transmission and the operation of specific neural circuits. Because in many other parts of the brain morphological sex differences have not yet been found and may not be found, we may appropriately ask whether there may be functional differences without a measurable structural difference, and if so what their basis might be.

In a recent review, a listing is given of biochemical and functional sex differences (McEwen, 1983). Some of these differences have been shown to persist in the absence of gonadal secretions in the adult, while others disappear. Those that persist may reflect intrinsic differences in brain cell structure and/or function, while those that disappear may simply reflect the activating effects of the different hormones normally produced by the testes and ovaries. In this connection, we should note

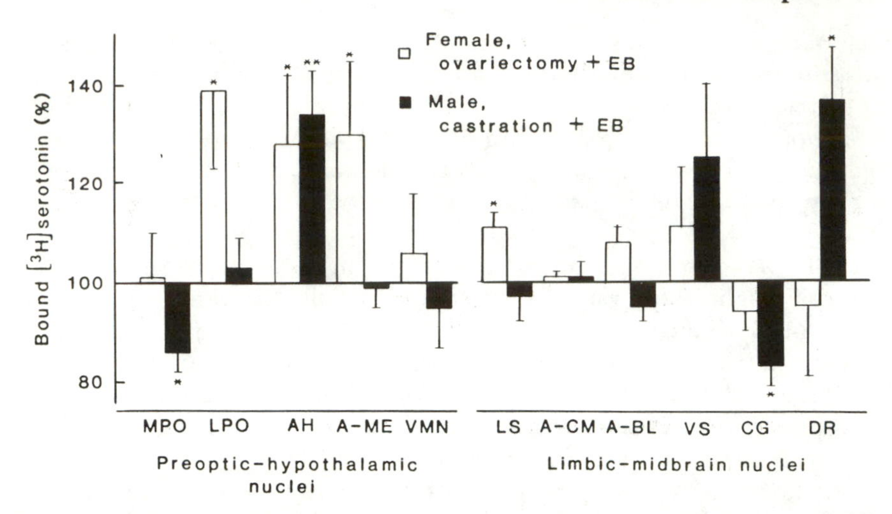

Figure 5–8. Mean (±SEM) tritiated serotonin specifically bound (as a percentage of control values) obtained from rats subjected to gonadectomy in microdissected preoptic, hypothalamic, and limbic-midbrain nuclei. (Abbreviations: MPO, medial preoptic area; LPO, lateral preoptic area; AH, anterior hypothalamic nucleus; A-ME, arcuate-median eminence; VMN, ventromedial nucleus of the hypothalamus; LS, lateral septum; ACM, corticomedial amygdala; ABL, basolateral amygdala; VS, ventral subiculum of the hippocampal formation; CG, central gray; DR, dorsal raphe nucleus.) In comparison with control rats subjected to gonadectomy, $*p<.05$, $**p<.01$ (two-tailed paired *t*-test). (Reprinted with permission from Fischette, Biegon, & McEwen, 1983.)

that the brains of males and females are both endowed with the same receptors for androgens, estrogens, and progestins (some of which are inducible by estrogens), as well as with similar complements of enzymes for aromatizing testosterone or producing 5 alpha reduced T metabolites. In spite of these similarities, the administration of estrogens to castrated (GDX) males and of androgens to ovariectomized (OVX) females does not always produce the same biochemical, pharmacological, or neuroendocrine response. For example, E induction of progestin receptors is markedly reduced in the ventromedial nucleus of male rats compared to the normal response in females (Rainbow et al., 1982; Parsons et al., 1984); E effects that elevate muscarinic cholinergic receptors in hypothalamus and preoptic area nuclei of female rats are absent in males (Dohanich, Witcher, Weaver, & Clemens, 1982; Rainbow, Snyder, Berck, & McEwen, 1984). Furthermore, E treatment of GDX male rats produces effects on serotonin-1 receptor levels that are different, in terms of brain regions involved and direction of effect, from those produced in OVX females (Fig. 5-8). For androgens, the administration of DHT is ineffective in OVX female rats in altering the "serotonin behavioral syndrome," even though this hormone effectively suppresses the syndrome in GDX males (Fischette, Biegon, & McEwen, 1984).

How might one account for these differences? We know of at least three possibilities. First, differences in level and distribution of estrogen or androgen receptors may exist. Some evidence for sex differences in estrogen receptor levels in rat preoptic area (Rainbow et al., 1982) exists, but known differences do not occur in brain areas that would account for the differences in estrogen effects described. For androgen receptors, there is no evidence of sex differences, but the data are limited. A second possibility is that differences in steroid metabolism and clearance might account for differences in response. This explanation might account for the results with DHT on the "serotonin behavioral syndrome" and might also account for a general lowering of progestin receptor induction by E in male guinea pig brains compared to females, but it does not account for the fact that E treatment can produce the same response in some male brain areas but not in others (e.g., see Fig. 5-8). A third possible explanation for highly regionalized sex differences that are not explained by the first two possibilities is that they may be the results of a differentiation in the "program" by which neurons respond to the same hormone in males and females (McEwen, Biegon, Fischette, Luine, Parsons, & Rainbow, 1984). The differentiated trait may be called the "regulatory phenotype." This concept represents a logical extension of the fact that in anatomically different groups of cells of the brain (even in the same sex), hormones such as estradiol activate different biochemical responses even though they both contain the same type of hormone receptors.

A regulatory phenotype implies differentiation of mechanisms and chemical structures, perhaps at the level of chromosomal proteins in the cell nucleus, that determine the genetic loci on which the steroid-receptor complex will act to alter gene expression. We are optimistic that we may be able to provide more direct evidence bearing on this regulatory phenotype. In the meantime, we must stress that just as changed structure implies altered function, so also does altered regulatory biochemistry imply that structural differences of some type—however subtle—may also exist. Thus, the importance of discussing neurochemical and morphological sex differences is not to raise a pointless semantic argument about structure versus biochemistry, but to point out the possibilities for future discoveries in the biochemical realm and to suggest how each category of sex difference, biochemical and morphological, may complement the other.

References

Ali, M., Balapure, K., Singh, D. R., Shukla, R. N., & Sahib, M. K. (1981). Ontogeny of alpha-fetoprotein in human foetal brain. *Brain Res., 207,* 459–464.

Beach, F. A. (1975). Hormonal modification of sexually dimorphic behavior. *Psychoneuroendocrinology, 1,* 3–23.

Breedlove, S. M., & Arnold, A. P. (1980). Hormone accumulation in a sexually dimorphic motor nucleus of the rat spinal cord. *Science, 210,* 564–566.

Breedlove, S. M., & Arnold, A. P. (1981). Sexually dimorphic motor nucleus in the rat lumbar spinal cord: Response to adult hormone manipulation, absence in androgen-insensitive rats. *Brain Res., 225,* 297–307.

Davis, P. G., Chaptal, C. V., & McEwen, B. S. (1979). Independence of the differentiation of masculine and feminine sexual behavior in rats. *Horm. Behav., 12,* 12–19.

Dohanich, G. P., Witcher, J. A., Weaver, D. R., & Clemens, L. G. (1982). Alteration of muscarinic binding in specific brain areas following estrogen treatment. *Brain Res., 241,* 347–350.

Fischette, C. T., Biegon, A., & McEwen, B. S. (1983). Sex differences in serotonin-1 receptor binding in rat brain. *Science, 222,* 333–335.

Fischette, C. T., Biegon, A., & McEwen, B. S. (1984). Sex steroid modulation of the serotonin behavioral syndrome. *Life Sci., 35,* (11), 1197–1206.

Friedman, W. J., McEwen, B. S., Toran-Allerand, C. D., & Gerlach, J. L. (1983). Perinatal development of hypothalamic and cortical estrogen receptors in mouse brain: Methodological aspects. *Devel. Brain Res., 11,* 19–27.

Gerlach, J. L., McEwen, B. S., Toran-Allerand, C. D., Friedman, W. J. (1983). Perinatal development of estrogen receptors in mouse brain assessed by radioautography, nuclear isolation and receptor assay. *Devel. Brain Res., 11,* 7–18.

MacLusky, N. J., Lieberburg, I., & McEwen, B. S. (1979). The development of estrogen receptor systems in the rat brain: Perinatal development. *Brain Res., 178,* 129–142.

McEwen, B. S. (1983). Gonadal steroid influences on brain development and sexual differentiation. In R. O. Greep (Ed.), *Reproductive physiology IV* (pp. 99–145). Baltimore: University Park Press.

McEwen, B. S., Biegon, A., Fischette, C. T., Luine, V. N., Parsons, B., & Rainbow, T. C. (1984). Sex differences in programming of responses to estradiol in the brain. In M. Serio, M. Motta & L. Martini (Eds.), *Sexual differentiation: Basic and clinical aspects (pp. 93–98). New York: Raven Press.*

McEwen, B. S., Plapinger, L., Chaptal, C., Gerlach, J. L., & Wallach, G. (1975). Role of fetoneonatal estrogen binding proteins in the association of estrogen with neonatal brain cell nuclear receptors. Brain Res., 96, 400–406.

Meaney, M. J., Stewart, J., Poulin, P., & McEwen, B. S. (1983). Sexual differentiation of social play in rat pups is mediated by the neonatal androgen receptor system. *Neuroendocrinology, 37,* 85–90.

Paden, C. M., Silverman, A.-J., McEwen, B. S., Stenevi, V., Bjorkland, A., & Thorngren, K. G. (1980). Hormonal effects on development of transplanted embryonic hypothalamus. *Peptides, 1,* 117–124.

Parsons, B., Rainbow, T. C., MacLusky, N. J., & McEwen, B. S. (1982). Progestin receptor levels in rat hypothalamic and limbic nuclei. *J. Neurosci, 2,* 1446–1452.

Parsons, B., Rainbow, T. C., & McEwen, B. S. (1984). Organizational effects of testosterone via aromatization on feminine reproductive behavior and neural progestin receptors in rat brain. *Endocrinology, 115,*(4), 412–417.

Rainbow, T. C., Parsons, B., & McEwen, B. S. (1982). Sex differences in rat brain oestrogen and progestin receptors. *Nature, 300,* 648–649.

Rainbow, T. C., Snyder, L., Berck, D. J., & McEwen, B. S. (1984). Correlation of muscarinic receptor induction in the ventromedial hypothalamic nucleus with the activation of feminine sexual behavior by estradiol. *Neuroendocrinology, 39,*(5), 476–480.

Stenevi, U., Björklund, A., Kromer, L. F., Paden, C. M., Gerlach, J. L., McEwen, B. S., & Silverman, A. J. (1980). Differentiation of embryonic hypothalamic transplants cultured on the choroidal pia in brains of adult rats. *Cell Tiss., 209,* 217–228.

Toran-Allerand, C. D. (1981). Cellular aspects of sexual differentiation of the brain. In G. Jagiello, H. J., Vogel (Eds.), *Bioregulators of reproduction* (pp. 43–57). New York: Academic Press.

Toran-Allerand, C. D. (1982). Regional differences in intraneuronal localization of alpha-fetoprotein in developing mouse brain. *Devel. Brain Res., 5,* 213–217.

III

EVOLUTIONARY PERSPECTIVES

6

Functional Associations in Behavioral Endocrinology

David Crews

> How many different fundamental . . . solutions of the same problem have been "invented" by different organisms. The broader the basis on which such inquiries are conducted, the more valuable will be the insight . . . . (F. Scharrer, 1946)

Sexuality has been the most and the least understood of biological phenomena. Because reproduction is the single most important aspect of an individual's life, here is where most biological disciplines converge. The many levels of analysis that must be utilized to unravel the causal mechanisms and functional outcomes of sexuality often result in confusion and even benign neglect by students of sexuality; in many instances the different perspectives have a way of hindering, instead of aiding, our understanding.

This paper considers some widely held assumptions in behavioral endocrinology. These assumptions have in common the belief that there are certain intrinsic functional associations in the reproductive process. I will first address three of these functional associations: (1) the functional association among gamete production, sex hormone secretion, and mating behavior; (2) the functional association between gonadal sex (i.e., male and female individuals) and behavioral sex; (3) the functional association among the components of sexuality. I will then consider the manner in which functional associations in behavioral endocrinology may develop and evolve. Finally, I will show how biological diversity can be used as a source for a broadened perspective guiding the selection and uses of animal models in the study of sexuality.

Functional Associations Among Gamete Production, Sex Hormone Secretion, and Mating Behavior

In seasonally breeding vertebrates, gamete maturation and maximum sex steroid hormone secretion are commonly found immediately preceding or coinciding with mating (Young, 1961; van Tienhoven, 1983) (Fig. 6-1). This pattern of gonadal activity in relation to mating may be termed an *associated* reproductive tactic (Crews, 1984). A large body of evidence demonstrates how in vertebrates exhibiting the *associated* reproductive tactic, the increase in circulating levels of sex steroid hormones accompanying gonadal growth stimulates mating behavior in both the male and the female. This hormone dependence of mating behavior is a central concept in vertebrate behavioral endocrinology.

Much evidence supports this causal relationship between gonadal hormones and mating behavior. Castration in many laboratory and domestic species results in a decline in mating behavior in both adult males and females, whereas replacement therapy with appropriate exogenous sex steroid hormones maintains or induces mating behavior in castrated individuals. Gonadal hormones have been shown to act on specific regions of the hypothalamus to alter the probability that an individual will behave in a particular manner. These same sex hormones also modulate the production and release of hypothalamic hormones and neurotransmitters, thereby affecting secretion of pituitary hormones. Secretion of these pituitary hormones regulates further production and secretion of sex steroid hormones as well as influences reproductive behavior. Similar results have been obtained from studies

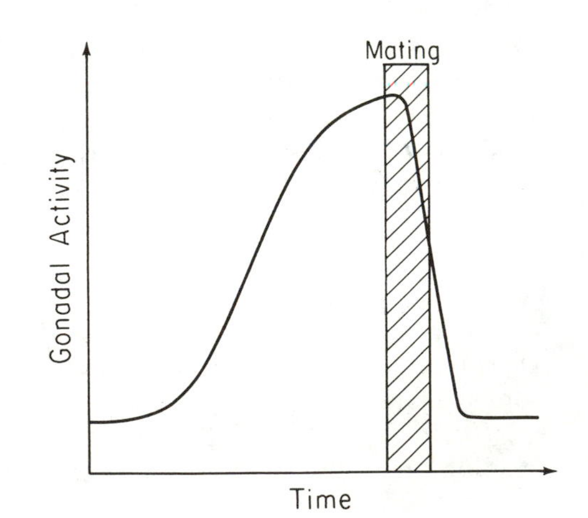

Figure 6–1. A common pattern of reproduction found in vertebrates. Gonadal activity is defined as the maturation of eggs and sperm and/or sex steroid hormone secretion. Note that maximum gonadal activity coincides with the occurrence of mating behavior.

of species representing each vertebrate class (Adkins-Regan, 1981; Crews, 1979; Kelley & Pfaff, 1978; McEwen, 1981; Stacey, 1983). They have been taken as evidence for a functional association among gamete production, sex hormone secretion, and mating behavior, thereby reflecting an evolutionary conservatism of hormone-behavior regulating mechanisms. Any similarities in the causal bases of mating behavior observed in different vertebrates, however, are more likely the result of repeated convergent evolutionary development (homoplasy and analogy) than of descent from a common ancestor (homology) (Campbell & Hodos, 1970; Hodos, 1976; Northcutt, 1984).

The paradigm that mating behavior is activated by a seasonal increase in circulating levels of gonadal hormones has persisted despite an increasing number of studies revealing variations to this rule. In the eider duck (*Somateria mollissima*) (Gorman, 1977), courtship behavior in the male is only indirectly correlated with circulating levels of testosterone. Neither maximum nor minimum concentration levels of sex steroid hormones appear to influence male courtship in at least five avian species. In the Australian zebra finch (*Taeniopygia guttata*), the male may maintain functional testes continuously for 3 years yet never breed; the stimulus for initiating mating behavior in many desert birds is rainfall (Marshall, 1970; Immelmann, 1973). In the ring dove (*Streptopelia risoria*), the male readily copulates during a phase of the reproductive cycle in which androgen levels are low (Silver & Barbiere, 1977). In the western gull (*Larus occidentalis*), males sustain constant circulating levels of androgens throughout the year yet exhibit a distinct annual spermatogenic and breeding cycle (Wingfield, Newman, Hunt, & Farner, 1982). In the mallard (*Anas platyrhynchos*), males copulate both when their testes are functionally active and inactive: in the first instance, mating coincides with the spring breeding season; in the latter instance, mating occurs during the winter (McKinney, 1975). In the white-crowned sparrow (*Zonotrichia leucophrys*), males with nonfunctional testes and basal circulating levels of androgens mount and copulate with estrogen-treated females (Moore, 1983; Moore & Krantz, 1983); even a castrated, sexually inexperienced male will copulate if he is solicited by an estrogen-treated female (Moore & Kranz, 1983). In both the horse (*Equus caballus*) (van Tienhoven, 1983) and the green sea turtle (*Chelonia mydas*) (Licht, 1984), the female has multiple ovulatory cycles but mates only once in the first follicular cycle. In the purple-throated Carib hummingbird (*Eulampis jugularis*), the female solicits and copulates with the male outside of the normal breeding season; these copulations appear to function to secure female access to male-defended food resources (Wolf, 1975). Mating outside of the species-typical breeding season has long been known to occur in many vertebrate species (Murton & Westwood, 1977; Morali & Beyer, 1979); the most outstanding example of independence of gonadal activity and mating behavior is the human species. Finally, a large body of literature shows in a variety of vertebrate species, including humans, that (1) gonadectomy has no effect on the expression of mating behavior (Tavolga, 1955; Aronson, 1959; Rosenblatt, 1965; Hart, 1974; McGill, 1978a, 1978b; Bloch & Davidson, 1968), and (2) circulating levels of sex hormones do not correlate with individual differences in sexual behavior (Young, 1961; Harding & Feder, 1976; Grunt & Young, 1953). Thus, the paradigm that gonadal hormones activate mating behavior in all vertebrates may be due more to bias in the species most studied than to a fundamental rule.

All of the data supporting the conclusion that there is a functional association among gamete production, sex hormone secretion, and mating behavior have been obtained from species in which mating in both the male and the female is associated with maximal gonadal activity. The hormone dependence of mating behavior among representative species of each vertebrate class has been taken as evidence of an association central to vertebrate reproduction. Is there an alternate reproductive tactic with which to test the hypothesis that the relationship among gamete production, sex hormone secretion, and mating behavior is fundamental?

Another, markedly different, pattern of gondal activity is exhibited in vertebrates. In some species, the gametes are stored, and mating occurs when the gonads are small and blood levels of sex steroid hormones are basal. This pattern of annual gonadal activity in relation to mating may be referred to as a *dissociated* reproductive tactic (Crews, 1984) (Fig. 6-2).

How common is it for gamete production, sex hormone secretion, and mating behavior to be dissociated? A review of the available literature reveals that a number of species exhibit a dissociated reproductive tactic (Crews, 1984). The actual number of such species cannot be estimated at this time because there are surprisingly few species for which the stage of gonadal activity at the time of mating is known for both the male and the female. Further, even in those species in which gonadal activity *is* associated with mating, we often do not know whether it is a cause, a consequence, or both. This point cannot be underestimated because, in a variety of species, mating initiates gonadal activity in the female and/or in the male (female: Licht & Bona-Gallo, 1982; Halpert, Garstka, & Crews, 1982; Hebard, 1951; Blanchard & Blanchard, 1941; Rahn, 1940; Clark, 1970; Richmond & Stehn, 1976; Dryden, 1969; male: Hebard, 1951; Blanchard & Blanchard, 1941; Rahn, 1940; Clark, 1970; Garstka & Crews, 1982; Vandenbergh, 1983a). In an equal number of species, copulatory stimuli are responsible for the final stages of gamete development and release (Crews, 1980; van Tienhoven, 1983). Significantly, sexual stimuli may have similar stimulatory and inhibitory effects on sexual maturation as well (Vandenbergh, 1983b).

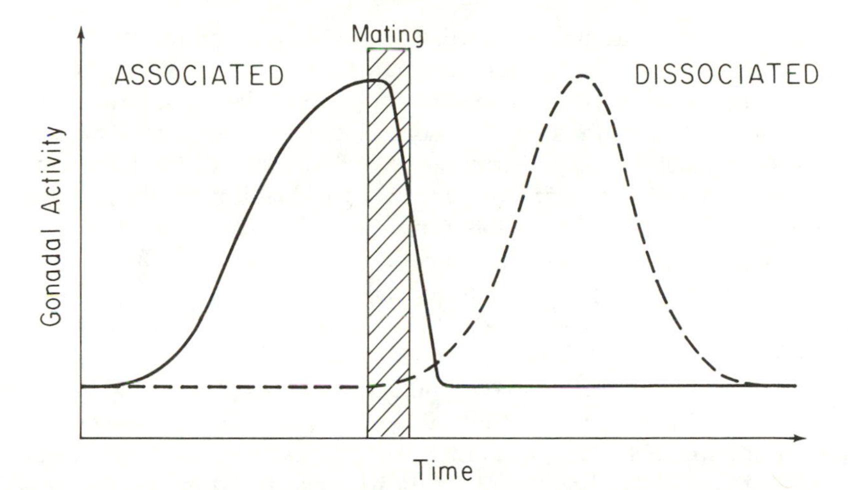

Figure 6–2. Two reproductive tactics, *associated* and *dissociated*, exhibited by vertebrates. Gonadal activity is defined here as the maturation of eggs and sperm and/or sex steroid hormone secretion and may or may not include the shedding of gametes or fertilization. Mating behavior is defined here as courtship and copulatory behavior. Note that in individuals exhibiting the associated reproductive tactic, the gonads are maximally active at the time of mating, whereas in individuals exhibiting a dissociated reproductive tactic, gonadal activity is minimal at the time of mating. (From Crews, 1984.)

If sexual species are categorized according to the time at which the gametes mature in relation to the time of mating in the male and in the female, we can predict four alternative reproductive tactics (Table 6-1). Notwithstanding the preceding examples, species (Type A/A) are known in which mating occurs when the gonads of both the male and the female are fully developed and circulating sex hormone levels are at their highest (Young, 1961; van Tienhoven, 1983; Reiter & Follett, 1980). In direct contrast, some species exhibit a Type D/D reproductive tactic

Table 6-1
Alternative Reproductive Tactics in Vertebrates

		MALE	
		Associated	*Dissociated*
FEMALE	*Associated*	Type A/A	Type A/D
	Dissociated	Type D/A	Type D/D

Note: It should be emphasized that the four reproductive tactics depicted represent extremes and other intermediate forms exist. (A = associated reproductive tactic; D = dissociated reproductive tactic.)

in which activity and mating are dissociated in both males and females. Species exhibiting this reproductive pattern are as follows: mammals—many hibernating vespertilionid and rhinolophid bats (Gustafson, 1979; Gustafson & Shemesh, 1976; Pearson, Koford, & Pearson, 1952; Saint-Girons, Brosset, & Saint-Girons, 1969; Oxberry, 1979); birds—none known; reptiles—the viviparous form of the Mexican lizard (*Sceloporus aeneus bicanthalis*) (Guillette, 1982), many snakes of various families including crotalids, elapids, and several colubrids (Saint-Girons, 1982), and the worm snake (*Carphophis vermis*) (Clark, 1970; Aldridge & Metter, 1973) amphibians—plethidontid salamanders (*Desmognathus* spp.) (Highton, 1956, 1962; Sayler, 1966; Houck, 1977a, 1977b); fishes—the plaice (*Pleuronectes platessa*) (Wingfield & Grimm, 1977), and the common carp (*Cyprinus carpio*) (Billard, 1981–1982). There are species in which gonadal activity and mating are dissociated in males but not in females (Type A/D). Some representative species exhibiting this reproductive tactic are as follows (I am not aware of any references to a mammalian or avian species exhibiting this reproductive tactic): reptiles—painted turtle (*Chrysemys picta*) (Callard, Lance, Salhanick, & Barad, 1978), the gopher tortoise (*Gopherus polyphemus*) (Taylor, 1982), the Indian lizard (*Agama tuberculatus*) (V. D. Joshi, personal communication), the timber rattlesnake (*Crotalus h. horridus*) (Saint-Girons, 1982), and the rough earth snake (*Virginia striatula*) (Clark, 1964); amphibians—the neotenic tiger salamander (*Ambystoma tigrinum*) (Norris, Norman, & Duvall, 1981); fish—the pike (*Esox lucius*) (Billard, 1981–1982). The complementary reproductive tactic is found in species in which gonadal activity and mating are associated in the male but not in the female (Type D/A). This has been documented in the following species: mammals—the Asian musk shrew (*Suncus murinus*) (Dryden, 1969, Dryden & Anderson, 1977); birds—the northern Fulmar (*Fulmarus glacialis*) (Marshall, 1949; Hatch, 1983); reptiles—an Australian skink (*Hemiergis peronii*) (Smyth & Smith, 1968), the American skink (*Eumeces egregius*) (Mount, 1963), a gecko (*Phyllodactylus marmoratus*) (King, 1977), a fence lizard (*Sceloporus grammicus microlepidotus*) (Guillette & Casas-Andren, 1981), the European viper (*Vipera aspis*) (Saint-Girons, 1982), and the eastern coral snake (*Micrurus fulvius*) (Quinn, 1979); fish—the shiner perch (*Cymatogaster aggregata*) (Wiebe, 1968; Shaw, 1971) and the catfish *Trachycorystes striatulus*) (von Ihering, 1937). There also is evidence that organisms other than vertebrates may show similar diversity in reproductive tactics (Barth & Lester, 1973; Englemann, 1970; Loher & Edson, 1973).

Gamete storage also varies in a predictable manner in each of the four reproductive tactics (Crews, 1984). For example, in species in which both the male and female show an association between gonadal activity and mating, gamete maturation coincides with mating. The interval between gametogenesis and mating may be many months in some species, however. An adaptation exhibited in species in which only one sex shows a dissociation between gonadal activity and mating is that the sex that shows the dissociation typically stores its gametes. Finally, in species in which both sexes exhibit a dissociation between mating and gonadal activity, both the male and the female store sperm.

Species that exhibit a dissociated reproductive tactic call into question the universality of a functional association among gamete production, sex hormone secretion, and mating behavior. In these species mating occurs when the individual's gonads are regressed and circulating levels of sex hormone secretion are low. Therefore, in these species, behavioral controlling mechanisms different from those currently known probably exist. Evidence to support this conclusion is in recent studies of two species exhibiting the *dissociated* reproductive tactic, the Asian musk shrew (*Suncus murinus*) and the red-sided garter snake (*Thamnophis sirtalis parietalis*). In the Asian musk shrew, development of reproductive structures and sexual receptivity in the female is independent of ovarian hormones; further, mating stimulates ovarian growth (Dryden & Anderson, 1977). A similar situation is found in the female red-sided garter snake (Licht & Bona-Gallo, 1982; Gartska, Tokarz, Halpert, Diamond, & Crews, 1984).

The most thorough investigation of the control of mating behavior of a species exhibiting a *dissociated* reproductive tactic has been conducted on the male red-sided garter snake. Courtship behavior in the adult male red-sided garter snake is independent of testicular and pituitary hormone control (Camazine, Garstka, Tokarz, & Crews, 1980; Crews, Camazine, Diamond, Mason, & Tokarz, & Garstka, 1984; Crews & Garstka, 1982; Garstka, Camazine, & Crews, 1982). Neither short-term nor long-term castration has an effect on courtship behavior in adult male garter snakes (Camazine et al., 1980; Crews et al., 1984; Garstka et al., 1982). As long as male garter snakes have passed through a hibernation period, castration either shortly after emergence in the spring or before entering winter hibernation in the fall has no effect on courtship. Similarly, castration before or during the testicular growth period of the previous summer fails to affect male courtship of females when they emerge from winter hibernation. Adrenalectomized and castrated males also court females on emergence. Further, hypophysectomy during or before hibernation does not prevent the male garter snake from exhibiting intense sexual activity on emergence from hibernation. Systemic administration of sex hormones, hypothalamic, and pituitary hormones, and a variety of neural and metabolic affectors does not elicit courtship behavior in noncourting males. Finally, recent studies indicate that implantation of testosterone directly into the hypothalamus has no effect on the behavior of noncourting males.

Given the fact that gonadal activity and mating behavior are not always associated, several conclusions can be drawn regarding the mechanisms controlling mating behavior. First, the lack of dependence of mating behavior on a seasonal increase in circulating levels of gonadal steroid hormones may be more common in vertebrates than previously thought. Second, the causal mechanisms regulating mating behavior, at least at the physiological level, may differ in vertebrates exhibiting these contrasting reproductive tactics. Third, male and female individuals of some species may have fundamentally different reproductive tactics (cf. Types A/D and D/A in Table 6-1).

Functional Association Between Gonadal Sex and Behavioral Sex

The majority of vertebrates are gonochoristic and rely on fertilization of female ova by male sperm for reproduction (Blackwelder & Shepherd, 1981). In gonochoristic species, gonadal sex differences are essential for reproduction, whereas behavioral sex differences facilitate reproduction (Lehrman, 1965; Hinde, 1965; Adler, 1978; Cheng, 1979; Crews, 1980, 1982). Thus, mating behavior both accomplishes the transfer of gametes and stimulates and coordinates the reproductive activity of conspecifics. This association of gamete type and specific behaviors is assumed to reflect an intrinsic functional association between gonadal sex and behavioral sex.

Gonadal sex and behavioral sex are distinct entities, however, a fact exemplified by hermaphroditic and parthenogenetic species. In sequentially hermaphroditic fish, sex reversals in behavior occur within hours (Shapiro, 1979), whereas sex reversals in morphology take weeks (Reinboth, 1975). Within the amniote vertebrates, at least 27 species of lizards representing seven taxonomic families reproduce by parthenogenesis (Cole, 1975; Cuellar, 1971). These species consist mostly or entirely of female individuals that produce clones of all-female offspring. Although these species lack male individuals, in at least two families, behavior patterns are exhibited regularly that are remarkably similar to the mating behavior of closely related sexual species (Crews & Fitzgerald, 1980; Moore, Whittier, Billy, & Crews, 1985; Werner, 1980). In the all-female whiptail lizard (*Cnemidophorus uniparens*), mounting and copulatory behavior as well as attractivity and receptivity are exhibited by the same female at different times of the ovarian cycle (Crews & Fitzgerald, 1980; Gustafson & Crews, 1981; Moore et al., 1984). Further, ovarian hormones influence the probability that both malelike and femalelike behaviors are exhibited (Gustafson & Crews, 1981; Crews, Gustafson, & Tokarz, 1983; Moore et al., 1985). Finally, these pseudomale behaviors facilitate reproduction (Gustafson & Crews, 1981) much as male courtship behavior does in gonochoristic species (Adler, 1978; Crews, 1980).

Study of the behavior of parthenogenetic species is particularly useful because it allows us to ask how fundamental the relationship is between gonadal sex and behavioral sex. In unisexual whiptail lizards, male individuals (and sperm) have been lost, yet malelike courtship and copulatory behaviors have been retained (Crews & Fitzgerald, 1980). This indicates that gonadal sex differences and behavioral sex differences can be uncoupled and are under different selection pressures. Significantly, in artificially selected, laboratory maintained strains of parthenogenetic *Drosophila,* female sexual behavior has been lost (Carson, Chang, & Lyttle, 1982). This result has been interpreted as evidence for the necessity of constant selection for a trait to persist. Our research with the whiptail lizards shows that in parthenogenetic lizards, courtship behavior increases fecundity, and thus may be the selective force Muller (1949) stipulated as necessary to maintain the genetic basis of a character.

Functional Associations Among the Components of Sexuality

Sexuality is not a unitary phenomenon. As Money pointed out 30 years ago (Money, 1955), the sexuality of the individual is a cumulative composite of separate "sexes" (Table 6-2). The first "sex" is the *chromosomal sex* of the individual—referring to the presence of heteromorphic chro-

Table 6-2
The Components of Sexuality

Component	*Identifying Characteristic*
Chromosomal sex	Presence of heteromorphic chromosomes
Gonadal sex	Differentiation of ovaries or testes
Physiological sex	Physiological, but primarily hormonal, differences between individuals of differing gonadal sexes
Morphological sex	Differentiation of internal and/or external morphologies between individuals of differing gonadal sexes
Behavioral sex	Differentiation of mating behaviors
Psychological sex (gender-role/identity)	Differentiation in gender role and orientation (identity)

Note: In today's parlance, the term *sex* can apply to any combination of the components shown. In the instances of chromosomal, gonadal, and (internal) morphological sex, these constitute true dimorphisms, whereas the other "sexes" reflect statistical differences in predisposition, preference, threshold, development, and frequencies.

Chromosomal SEX	Gonadal SEX	Physiological SEX	Morphological SEX		Behavioral SEX
Genotypic sex determination	Gonochorism (separate gonads in separate individuals)	Testes: T/e Ovaries: E/t	Accessory sex structures dependent on gonadal sex	Secondary sex characters dependent on physiological sex	Complementary mating behavior: Intromission and receptivity

Figure 6–3. Identifying characteristics of the components of sexuality in many laboratory and domesticated vertebrates.

mosomes. In the majority of species, the gonads of individuals with different heteromorphic chromosomes produce different gametes. The *gonadal sex* of the individual refers to the type of gonad present. The *physiological sex* of the individual refers principally to differences in the nature and pattern of gonadal hormone secretion between individuals of differing gonadal sex. The accessory sex structures and secondary sex characters together constitute the *morphological sex* of the individual. Differing behaviors between individuals of different gonadal sex are referred to as the *behavioral sex* of the individual. In humans there is firm evidence for a sixth component of an individual's sexuality, gender-identity/role or *psychological sex* (Money, 1974).

Clearly in some species the multiple components of sexuality are related causally and in a similar manner (Fig. 6-3) (Goy & McEwen, 1980; Money, 1974). Although only a few vertebrates have been examined in sufficient detail, it is now believed that in the "typical vertebrate," the moment of fertilization sets into motion a cascade of events culminating in the adult organism's mating and contributing its genes to subsequent generations. That is, the genetic constitution, and hence the chromosomal sex of the individual, is determined at conception. The sex chromosomes acting directly, or via H-Y antigen, in turn determine the gonadal sex of the individual. This is followed by differentiation and development of both gamete-delivery systems as well as secondary sex characters, a process mediated by hormonal differences present early in the two sexes. The gonadal hormones also influence the behaviors of each gonadal sex differently throughout development but particularly before and during mating as adults.

These deterministic relations of the different components of sexuality in the species studied to date are assumed to reflect a phylogenetically stable core system. Just as recent studies are questioning the generality of certain hormone-behavior relationships, the diversity in the expression of sexuality found in the vertebrates is clear proof that there is no intrinsic functional association among the components of sexuality.

Some vertebrates simply do not fit the "typical vertebrate" model (Fig. 6-4). One or more of the components of sexuality are "missing" in these species. That is, in some species sex is determined by the environment in which the embryo develops (Bull, 1980, 1983). Alterna-

Chromosomal SEX	Gonadal SEX	Physiological SEX	Morphological SEX		Behavioral SEX
Genotypic sex determination	Gonochorism (separate gonads in separate individuals)	Testes: T/e Ovaries: E/t	Accessory sex structures dependent on gonadal sex	Secondary sex characters dependent on physiological sex	Complementary mating behavior: Intromission and receptivity
Environmental sex determination	Hermaphroditism; Parthenogenesis	No sex specific gonadal steroids	Coelemic transport of gametes	Isomorphisms; "Reversed" dimorphism	Heterotypic sexual behavior in gonochoristic species; Reversed sex roles; Pseudosexual behavior in unisexual species

Figure 6–4. Some examples of biological diversity in the forms of sexuality. In many species a specific "sex" may be deleted or is different from that observed in vertebrates.

tively, in some species the gonads may reside in a single individual (=hermaphrodites) either sequentially or simultaneously throughout life. In still other species, reproduction is by parthenogenesis, instead of by sexual means. In some species, the sexes do not appear to differ in the nature or in the pattern of gonadal hormones secreted. In primitive vertebrates such as the lamprey and most teleost fish, there are no duct systems specialized for transporting the gametes to the external environment. Many species are monomorphic in appearance and in behavior, whereas in other species some males may mimic females. Finally, it is now well established that each sex has the capacity to exhibit as an adult heterotypical mating behavior as well as homotypical mating behavior (Beach, 1968). In some species this exhibition of heterotypical mating behavior is an effective alternative reproductive strategy.

Development of Functional Associations in Behavioral Endocrinology

Although sex differences in the neural substrates of behavior may be influenced by gonadal hormones during fetal development, the process by which this occurs probably differs among species (Baum, 1979). Still other patterns of neural development probably exist in those species in which the sexes differ in their reproductive tactics.

In type A/A species (see Table 6-1), gonadal hormones in the embryo clearly influence the organization of neural areas regulating mating behavior and reproductive physiology in the adult (Goy & McEwen, 1980). This action may be expressed at the structural level in the stimulation of both the growth (Gorski, 1979; Toran-Allerand, 1976) and destruction (Brawer, Schipper, & Robaire, 1983) of neurons. Hormonal mediation of central nervous system (CNS) organization may also be expressed at the functional level. For example, the stereotyped behaviors associated with mating in the rat can be elicited in the absence of hormones both in the neonate (Williams, 1986) and in the adult (Komisaruk, 1971a, b; Komisaruk, Terasawa, & Rodriguez-Sierra, 1981).

Experience can also have a permanent organizing effect on behavioral controlling mechanisms. In many domesticated birds and rodents, the longer the isolated castrated adult is deprived of hormones, the higher the threshold for eliciting mating behavior with exogenous hormone treatment (Hutchison, 1978). Whether this same change in sensitivity still occurs if castrated individuals are housed with sexually competent partners is not known. It is, however, well known that in adult vertebrates, environmental (Ewert, 1980) and social (Hinde, 1970) stimuli will selectively activate specific neural systems. Further, gonadal hormones may potentiate neuronal activity, particularly in those areas of the brain that concentrate sex steroid hormones (Komisaruk, 1971a, b, 1978; Komisaruk et al., 1981; Meyer & Zakon, 1982; Pfaff & McEwen, 1983). Other internal (Goy & McEwen, 1980; Greenough, 1976; Jacobson, 1978) and external (Coss & Globus, 1978) stimuli also affect the growth of those CNS structures, suggesting that such interaction of the individual during development with its internal and external environment is a major directive force in the maturation, and ultimately the evolution (Mayr, 1958, 1974; Domjan & Burkhard, 1982; Wilczynski, 1984), of neural systems. Thus, one can view hormones as allowing for the growth of neural connections, whereas the participation in particular behaviors establishes and maintains the functional integrity of these neural systems (Crews, Gustafson, & Tokarz, 1983; Nottebohm, 1981; Breedlove & Arnold, 1981; Arnold, 1981; Gottlieb, 1982; C. L. Moore, 1985; Dessi-Fulgheri, Lupo, Ciampi, Canonaco, & Larsson, 1983).

This observation, together with the discovery that gamete production, sex hormone secretion, and mating behavior may be uncoupled, could account for why in some species the effect of gonadectomy on mating behavior is less pronounced if the individual is sexually experienced (Young, 1961; Rosenblatt, 1965; Lehrman, 1962). Lehrman noted that "different species and the two sexes within the same species might differ with respect to the relative degree of dependence of their sexual behavior upon the presence of various hormones and upon various situational and experiential factors" (Lehrman, 1962). Mating behavior, and hence reproductive success, improves with experience (Rosenblatt, 1965; Lehrman, 1962; Lehrman & Wortis, 1960, 1967; Michel, 1977; Fisher & Hale, 1957; McGill, 1962; Michael, 1961; Whalen, 1963). Also, in species in which sexual experience plays a major role in the control of mating behavior, individuals frequently mate when the gonads are not producing gametes (Young, 1961; Valenstein & Goy, 1957; Parkes, 1976;

Beach, 1967; Kleiman & Mack, 1978). Mating when the gonad is regressed may indicate that the causal mechanisms underlying mating behavior are different at different times of the year or, alternatively, that in these species the mating behavior of one or both of the sexes is independent of hormones and more dependent on social context (Moore, 1983; Moore & Kranz, 1983). The potential interdependence of experiential and physiological factors is illustrated by the finding that castrated, sexually experienced rodents no longer exhibit a preference for the opposite sex, although they retain the ability to discriminate the opposite sex (Caroum & Bronson, 1971; Carr & Caul, 1962; Carr, Loeb, & Dissinger, 1965; Lydell & Doty, 1972; Stern, 1970). Finally, it is likely that certain functional associations allow for learning of the behavioral (Hailman, 1961; Ferris, 1967; Domjan, 1983), as well as the physiological (Anonymous, 1975; Bronson & Eleftheriou, 1965; Graham & Desjardins, 1980; Macrides, Bartke, & Dalterio, 1975; Coquelin & Bronson, 1980), responses attendant to mating. Therefore, one would expect development of neural differences to depend on both the physiological and the social environments.

Evolution of Functional Associations in Behavioral Endocrinology

The functional interrelationships among gamete production, gonadal hormone secretion, and mating behavior probably evolved to maximize reproductive synchrony (Fig. 6-5). How might this have occurred? A first stage might have been the functional association between gamete production and sex steroid hormone production at the level of the gonad. Sex steroid hormones serving as local regulatory mechanisms in germ cell maturation appear to be highly conserved in vertebrates and invertebrates (Kanatani & Nagahami, 1980). The second evolutionary stage was the development of hypothalamic feedback control of pituitary gonadotropin secretion. A third, or perhaps simultaneous, stage would be the functional association of gonadal hormone secretion and the development and later activation of gamete delivery systems (both acces-

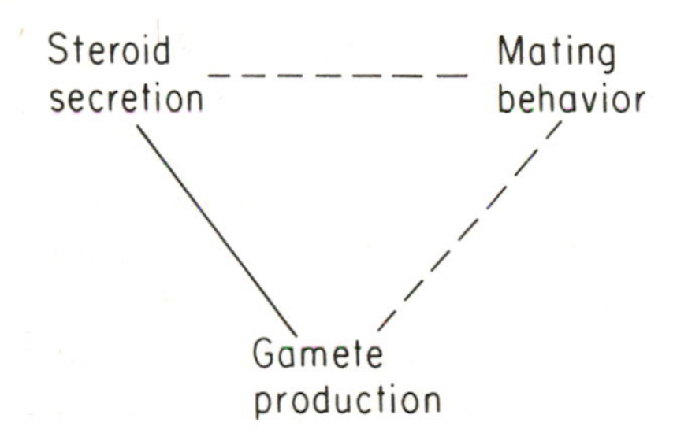

Figure 6–5. Primitive and derived functional associations among the three major components of the reproductive process. The solid line represents the only functional association shared by all living vertebrates. The dashed lines represent functional associations that have evolved independently in different lineages.

sory sex structures and secondary sexual characters). In many vertebrates this was followed by recruitment of behavioral integrative (limbic) areas in the CNS via expansion of the steroid-sensitive hypothalamic areas involved in the feedback control of pituitary function. [Note, however, that the presence of sex steroid concentrating neurones in behavioral integrative areas is not evidence a priori of a functional association between hormones and behavior. Neither garter snakes (Crews & Garstka, 1982; Crews et al., 1983) nor female Asian musk shrews (Dryden & Anderson, 1977) exhibit hormone-dependent sexual receptivity, yet sex steroid hormone-concentrating neurones have been identified in the anterior hypothalamus-preoptic area (AH-POA) of each species (Halpern, Morell, & Pfaff, 1982; Keefer & Dryden, 1982).] In some species there has been the further evolution of a functional association of specific sexual signals and secretion of reproductive hormones (Anonymous, 1975; Bronson & Eleftheriou, 1965; Graham & Desjardins, 1980; Macrides et al., 1975; Coquelin & Bronson, 1980). Speculations on the evolution of the mechanisms underlying other reproductive tactics must await study of each tactic.

The observation that all-female species engage in sexual behavior questions whether the phylogenies of gonadal sex and behavioral sex are synonymous. That the functional outcomes of these behaviors in all-female species are similar to those found in sexual vertebrate species leads one to ask whether behavioral sex is not more fundamental than gonadal sex (Crews, 1982). Many lineages have lost males, yet behaviors facilitating reproduction have persisted in unisexual and asexual organisms. This suggests that behavioral facilitation of reproduction may be ancestral to the existence of separate sexes.

Adaptation of Functional Associations in Behavioral Endocrinology

The behavioral differences among species are in accord with the environment in which the species live. A classic example of this principle is the observation that species living in extreme environments often begin to breed within minutes or hours after a change in a specific environmental cue (Marshall, 1970; Immelmann, 1973; Bragg, 1955–1956; Pianka, 1970; Wells, 1977). Just as behavior has been shaped by the conditions in which organisms live, so too have the mechanisms that control mating behavior. Of particular interest are the environmental constraints that gave rise to each of the reproductive tactics depicted in Table 6-1 (Crews, 1984). In this context I anticipate that the following variables will prove important: (1) the differential cost of reproduction for males and females, (2) the availability of mating opportunities, (3) the predictability of ultimate factors, (4) the survivorship of young, (5) the constraints imposed by the time required for maturation of the gametes, zygote, and juvenile, and (6) temperature regulation. Prudent behavioral ecological studies will provide new directions for investigations of behavioral controlling mechanisms just as laboratory studies of the causes and functions of mating behavior will contribute to a better understanding of behavioral ecology.

Conclusion

Whereas gamete production, sex hormone secretion, and mating behavior coincide in some vertebrates, this is not true for all. We find species in which mating behavior does not depend on increased levels of gonadal steroid hormones. Not only can mating behavior be uncoupled from gametogenesis and sex hormone secretion, but gonadal sex and behavioral sex can vary independent of one another. Finally, many alternate combinations of the components of sexuality are found in modern vertebrates.

By studying such diverse species, each of which must be considered successful biologically, we gain information regarding natural variation. This knowledge reveals which characters are shared, and hence ancestral, and which characters are not shared, and hence derived. Shared characters are likely to be more fundamental than are derived characters (Mayr, 1982).

Although species that do not fit the well-established concepts have only recently received the attention they deserve, some patterns are beginning to emerge. These studies indicate that the functional associations among gamete production, sex steroid secretion, and mating behavior—and the functional associations between gonadal sex and behavioral sex—are probably not fundamental reproductive processes. Indeed, the only evolutionarily conserved functional association is one in which steroid hormones serve as local regulatory mechanisms in germ cell development (Kanatani & Nagahami, 1980) (Fig. 6-5). Thus, by examining diverse reproductive systems, we will be better able to evaluate whether these are coequal evolutionary alternatives or merely interesting exceptions that prove the rule.

Acknowledgments

The ideas in this paper have their roots with two teachers encountered early in my career, William Hodos and Jay S. Rosenblatt, both of whom taught me the value of the comparative approach. I dedicate this paper to them.

Andrew Halpert and John Money provided helpful comments on an earlier draft of this manuscript. The research reported in this paper was supported by the following grants: NICHHD 16687; NSF BNS8002531, and NIMH Research Scientist Development Award 00135.

References

Adkins-Regan, E. T. (1981). Early organizational effects of hormones: An evolutionary perspective. In N. T. Adler (Ed.), *Neuroendocrinology of reproduction* (pp. 159–228). New York: Plenum Press.

Adler, N. T. (1978). On the mechanisms of sexual behaviour and their evolutionary constraints. In J. B. Hutchison (Ed.), *Biological determinants of sexual behaviour* (pp. 657–695). Chichester: John Wiley & Sons.

Aldridge, R. D., & Metter, E. E. (1973). The reproductive cycle of the western worm snake *Carphophis vermis* in Missouri. *Copeia, 1973,* 472–477.

Anonymous. (1975). Effects of sexual activity on beard grown in man. *Nature, 226,* 869–871.

Arnold, A. P. (1981). Logical levels of steroid hormone action in the control of vertebrate behavior. *American Zoologist, 21,* 233–242.

Aronson, L. R. (1959). Hormones and reproductive behavior: Some phylogenetic considerations. In A. Gorbman (Ed.), *Comparative Endocrinology* (pp. 98–120). New York: John Wiley & Sons.

Barth, R. H., & Lester, L. J. (1973). Neuro-hormonal control of sexual behavior in insects. *Annual Review of Entomology, 18,* 445–472.

Baum, M. J. (1979). Differentiation of coital behavior in mammals: A comparative analysis. *Neurosciences Biobehavioral Review, 3,* 265–284.

Beach, F. A. (1967). Cerebral and hormonal control of reflexive mechanisms involved in copulatory behavior. *Physiological Reviews, 47,* 289–316.

Beach, F. A. (1968). Factors involved in the control of mounting behavior by female mammals. In M. Diamond (Ed.), *Perspectives in reproduction and sexual behavior* (pp. 83–131) Bloomington: Indiana University Press.

Beach, F. A. (1971). Hormonal factors controlling the differentiation, development, and display of copulatory behavior in the ramstergig and related species. In E. Tobach, L. R. Aronson, & E. Shaw (Eds.), *The biopsychology of development* (pp. 249–296). New York: Academic Press.

Billard, R. (1981–1982). The reproductive cycle in teleost fish. *Estratto dalla Rivista Italian di Piscicoltura e Ittiopatologia,* (p. 32). Torino, Italy.

Blackweler, R. E., & Shepherd, B. A. (1981). *The diversity of animal reproduction.* Boca Raton: CRC Press.

Blanchard, F. N., & Blanchard, F. C. (1941). Factors determining time of birth in the garter snake Thamnophis sirtalis sirtalis (Linnaeus). *Papers of the Michigan Academy of Science, Arts and Letters, 26,* 161–176.

Bloch, G. J., & Davidson, J. M. (1968). Effects of adrenalectomy and experience on postcastration sex behavior in the male rat. *Physiology and Behavior, 3,* 461–465.

Bragg, A. N. (1955–1956). In quest of the spadefoots. *New Mexico Quarterly, 25,* 345–358.

Brawer, J., Schipper, H., & Robaire, B. (1983). Effects of long term androgen and estradiol exposure on the hypothalamus. *Endocrinology, 112,* 194–199.

Breedlove, S. M., & Arnold, A. P. (1981). Hormone accumulation in a sexually dimorphic motor nucleus of the rat spinal cord. *Science, 210,* 564–566.

Bronson, F. H., & Eleftheriou, B. E. (1965). Adrenal responses to fighting in mice: Separation of physical and psychological causes. *Science, 147,* 627–628.

Bull, J. J. (1983). *Evolution of sex determining mechanisms.* Menlo Park, Ca: Benjamin/ Cummings Publishing Company.

Bull, J. J. (1980). Sex determination in reptiles. *Quarterly Review of Biology, 55,* 3–21.

Callard, I. P., Lance, V., Salhanick, A. R., & Barad, D. (1978). The annual ovarian cycle of *Chrysemys picta:* Correlated changes in plasma steroids and parameters of vitellgenesis. *General and Comparative Endocrinology, 35,* 245–257.

Camazine, B., Garstka, W. R., Tokarz, R. R., & Crews, D. (1980). Effects of castration and androgen replacement on male courtship behavior in the red-sided garter snake *(Thamnophis sirtalis parietalis). Hormones and Behavior, 14,* 358–372.

Campbell, C. B. B., & Hodos, W. (1970). The concept of homology and the evolution of the nervous system. *Brain, Behavior, and Evolution, 3,* 353–367.

Caroum, D., & Bronson, F. H. (1971). Responsiveness of female mice to preputial attractant: Effects of sexual experience and ovarian hormones. *Physiology and Behavior, 7,* 659–662.

Carr, W. J., & Caul, W. F. (1962). The effect of castration in rat upon the discrimination of sex odours. *Animal Behaviour, 10,* 20–27.

Carr, W. J., Loeb, L. S., & Dissinger, M. L. (1965). Responses of rats to sex odours. *Journal of Comparative and Physiological Psychology, 59,* 370–377.

Carson, H. L. Chang, L. S., & Lyttle, T. W. (1982). Decay of female sexual receptivity under parthenogenesis. *Science, 218,* 68–70.

Cheng, M.-F. (1979). Progress and prospects in ring dove research: A personal view. In J. S. Rosenblatt, R. A. Hinde, C. G. Beer, & M. C. Busnel (Eds.), *Advances in the study of behavior.* (Vol. 9, pp. 97–129). New York: Academic Press.

Clark, D. R. (1964). Reproduction and sexual dimorphism in a population of the rough earth snake, *Virginia striatula. Texas Journal of Science, 16,* 265–295.

Clark, D. R. (1970). Ecological study of the worm snake, *Carphophis vermis (Kennicott). University of Kansas Publications of the Museum of Natural History, 19,* 87–194.

Cole, C. J. (1975). Evolution of parthenogenetic species of reptiles. In R. Reinboth (Ed.), *Intersexuality in the animal kingdom* (pp. 340–355. Berlin: Springer-Verlag.

Coquelin, A., & Bronson, F. H. (1980). Secretion of luteinizing hormone in male mice: Factors that influence release during sexual encounters. *Endocrinology, 106,* 1224–1229.

Coss, R. G., & Globus, A. (1978). Spine stems on tectal interneurons in jewel fish are shortened by social stimulation. *Science, 200,* 787–790.

Crews, D. (1979). The neuroendocrinology of reproduction in reptiles. *Biology of Reproduction, 20,* 51–73.

Crews, D. (1980) Interrelationships among ecological, behavioral and neuroendocrine processes in the reproductive cycle of *Anolis carolinensis* and other reptiles. In J. S. Rosenblatt, R. A. Hinde, C. G. Beer, & M. C. Busnel (Eds.), *Advances in the study of behavior* (Vol. 11, pp. 1–74). New York: Academic Press.

Crews, D. (1982). On the origin of sexual behavior. *Psychoneuroendocrinology, 7,* 259–270.

Crews, D. (1984). Gamete production, sex hormone secretion, and mating behavior uncoupled. *Hormones and Behavior, 18,* 22–28.

Crews, D., Camazine, B., Diamond, M., Mason R. T., Tokarz, R. R., & Gar-

stka, W. R. (1984). Hormonal independence of courtship behavior in the male garter snake. *Hormones and Behavior, 18,* 29–41.
Crews, D., & Fitzgerald, K. T. (1980). "Sexual" behavior in parthenogenetic lizards (*Cnemidophorus*). *Proceedings of the National Academy of Sciences, 77,* 499–502.
Crews, D., & Garstka, W. R. (1982). The ecological physiology of the garter snake. *Scientific American, 247,* 158–168.
Crews, D., Gustafson, J. E., & Tokarz, R. R. (1983). Psychobiology of parthenogenesis in reptiles. In R. Huey, T. Schoener, & E. Pianka (Eds.), *Lizard ecology* (pp. 205–231). Cambridge: Harvard University Press.
Cuellar, O. (1971). Reproduction and the mechanism of meiotic restitution in the parthenogenetic lizard *Cnemidophorus uniparens. Journal of Morphology, 133,* 139–166.
Demski, L. S. (1984). Evolution of neural systems in the vertebrates: Functional anatomical approaches. *American Zoologist, 24*(3), 691.
Dessi-Fulgheri, F., Lupo, C., Ciampi, G. M., Canonaco, M., & Larsson, K. (1983). Exposure of odour during development and hypothalamic metabolism of testosterone. In J. Balthazart, E. Prove, & R. Gilles (Eds.), *Hormones and behaviour in higher vertebrates* (pp. 305–313). Berlin: Springer-Verlag.
Dewsbury, D. A. (1978). The comparative method in studies of reproductive behavior. In T. E. McGill, D. A. Dewsbury, & B. D. Sachs (Eds.), *Sex and behavior: Status and prospectus* (pp. 83–114). New York: Plenum Press.
Domjan, M. (1983). Biological constraints on instrumental and classical conditioning: Implications for general process theory. In G. H. Bower (Ed.), *The psychology of learning and motivation.* New York: Academic Press.
Domjan, M., & Burkhard, B. (1982). *Principles of learning and behavior.* Monterey: Brooks/Cole.
Doty, R. L. (1974). A cry for the liberation of the female rodent: Courtship and copulation in Rodentia. *Psychological Bulletin, 81,* 159–172.
Dryden, G. L. (1969). Reproduction in *Suncus murinus. Journal of Reproduction and Fertility (Supplement), 6,* 377–396.
Dryden, G. L., & Anderson, J. N. (1977). Ovarian hormone: Lack of effect on reproductive structures of female Asian musk shrews. *Science, 197,* 782–784.
Englemann, F. (1970). *The physiology of insect reproduction.* New York: Pergamon Press.
Ewert, J.-P. (1980). *Neuroethology.* Berlin: Springer-Verlag.
Ferris, H. E. (1967). Classical conditioning of courting behavior in the Japanese quail. *Journal of the Experimental Analysis of Behavior, 10,* 213–218.
Fisher, A. E., & Hale, E. B. (1957). Stimulus determinants of sexual and aggressive behavior in male domestic fowl. *Behaviour, 10,* 309–323.
Fox, W. (1952). Seasonal variation in the male reproductive system of Pacific coast garter snakes. *Journal of Morphology, 90,* 481–553.
Fox, W. (1954). Genetic and environmental variation in the timing of the reproductive cycles of male garter snakes. *Journal of Morphology, 95,* 415–450.
Garstka, W. R., & Crews, D. (1982). Female control of male reproductive function in a Mexican snake. *Science, 217,* 1159–1160.
Garstka, W. R., Camazine, B., & Crews, D. (1982). Interactions of behavior and physiology during the annual reproductive cycle of the red-sided garter snake, *Thamnophis sirtalis parietalis. Herpetologica, 38,* 104–123.
Garstka, W. R., Tokarz, R. R., Halpert, A., Diamond, M., & Crews, D. (1984). Behavioral and physiological control of yolk synthesis and deposition in

the female red-sided garter snake *(Thamnophis sirtalis parietalis). Hormones and Behavior, 19,* 137–153.
Gorman, M. L. (1977). Sexual behaviour and plasma androgen concentrations in the male eider duck *(Somateria mollissima). Journal of Reproduction and Fertility, 49,* 225–230.
Gorski, R. (1979). The neuroendocrinology of reproduction: An overview. *Biology of Reproduction, 20,* 111–127.
Gottlieb, G. (1982). Development of species identification in ducklings: IX. The necessity of experiencing normal variations in embryonic auditory stimulation. *Developmental Psychobiology, 15,* 507–517.
Goy, R. W., & McEwen, B. S. (1980). *Sexual differentiation of the brain.* Cambridge: MIT Press.
Graham, J. M., & Desjardins, C. (1980). Classical conditioning: Induction of luteinizing hormone and testosterone secretion in anticipation of sexual activity. *Science, 210,* 1039–1041.
Greenough, W. T. (1976). Enduring brain effects of differential experience and training. In M. R. Rosenzweig & E. L. Bennett (Eds.), *Neural mechanisms of learning and memory* (pp. 255–278). Cambridge: MIT Press.
Gregory, P. T. (1976). Life history parameters of the red-sided garter snake *(Thamnophis sirtalis parietalis)* in an extreme environment, the Interlake region of Manitoba. *National Museum of Natural Sciences, Ottawa, Publications in Zoology, Number 13.*
Grunt, J. A., & Young, W. C. (1953). Consistency of sexual behavior patterns in individual male guinea pigs following castration and androgen therapy. *Journal of Comparative and Physiological Psychology, 46,* 138–144.
Guillette, L. J. (1982). The evolution of viviparity and placentation in the high elevation Mexican lizard *Sceloporus aeneus. Herpetologica, 38,* 94–103.
Guillette, L. J., & Casas-Andren, G. (1981). Fall reproductive activity in the high altitude Mexican lizard, *Sceloporus grammicus microlepidotus. Herpetologica, 14,* 143–147.
Gustafson, A. W. (1979). Male reproductive patterns in hibernating bats. *Journal of Reproduction and Fertility, 56,* 317–331.
Gustafson, A. W., & Shemesh, M. (1976). Changes in plasma testosterone levels during the annual reproductive cycle of the hibernating bat, *Myotus lucifugus lucifugus* with a survey of plasma testosterone levels in adult male vertebrates. *Biology of Reproduction, 15,* 9–24.
Gustafson, J. E., & Crews, D. (1981). Effect of group size and physiological state of a cagemate on reproduction in the parthenogenetic lizard, *Cnemidophorus uniparens* (Teiidae). *Behavioral Ecology and Sociobiology, 8,* 267–272.
Hailman, J. P. (1961). How an instinct is learned. *Scientific American, 221,* 98–106.
Halpern, M., Morrell, J., & Pfaff, D. W. (1982). Cellular ^{3}H-estradiol and ^{3}H-testosterone localization in the brains of garter snakes: An autoradiographic study. *General and Comparative Endocrinology, 46,* 211–224.
Halpert A., Garstka, W. R., & Crews, D. (1982). Sperm transport and storage and its relation to the annual sexual cycle of the female red-sided garter snake, *Thamnophis sirtalis parietalis. Journal of Morphology, 174,* 149–159.
Harding, C. F., & Feder, H. H. (1976). Relation between individual differences in sexual behavior and plasma testosterone levels in the guinea pig. *Endocrinology, 98,* 1198–1205.
Hart, B. L. (1974). Gonadal androgen and sociosexual behavior of male mammals: A comparative analysis. *Psychological Bulletin, 81,* 383–400.

Hatch S. A. (1983). Mechanism and ecological significance of sperm storage in the northern fulmar with reference to its occurrence in other birds. *Auk, 100,* 593–600.

Hawley, A. W. L., & Aleksiuk, M. (1975). Thermal regulation of spring mating behavior in the red-sided garter snake *(Thamnophis sirtalis parietalis). Canadian Journal of Zoology, 53,* 768–776.

Hawley A. W. L., & Aleksiuk, M. (1976). The influence of photoperiod and temperature on seasonal testicular recrudescence in the red-sided garter snake *(Thamnophis sirtalis parietalis). Comparative Biochemistry and Physiology, 53A,* 215–221.

Hebard, W. B. (1951). Notes on the life history of the Puget Sound garter snake, *Thamnophis ordinoides. Herpetologica, 7,* 177–179.

Highton, R. (1956). The life history of the slimy salamander, *Plethodon glutinosus,* in Florida, *Copeia, 1956,* 75–93.

Highton, R. (1962). Geographic variation in the life history of the slimy salamander. *Copeia, 1962,* 597–612.

Hinde, R. A. (1970). *Animal behaviour* (2nd Ed). New York: McGraw Hill.

Hinde, R. A. (1982). *Ethology.* New York: Oxford University Press.

Hinde, R. A. (1965). Interaction of internal and external factors in integration of canary reproduction. In F. A. Beach (Ed.), *Sex and behavior* (pp. 381–415). New York: John Wiley & Sons.

Hodos, W. (1976). The concept of homology and the evolution of behavior. In P. B. Masterton, W. Hodos, & J. Jerison (Eds.), *Evolution, brain, and behavior* (pp. 153–168). Hillsdale: Lawrence Erlbaum Associates.

Houck, L. D. (1977a). Reproductive biology of a neotropical salamander, *Bolitoglossa rostrata* (Caudata: Plethodontidae). *Copeia, 1977,* 70–83.

Houck, L. D. (1977b) Life history patterns and reproductive biology of neotropical salamanders. In D. H. Taylor & S. I. Guttman (Eds.), *The reproductive biology of amphibians* (pp. 43–72). New York: Plenum.

Hutchison, J. B. (1978). *The biological determinants of sexual behaviour.* Chichester: John Wiley & Sons.

Immelmann, K. (1973). Role of the environment in reproduction as source of "predictive" information. In D. Farner (Ed.), *Breeding biology of birds* (pp. 121–147). Washington: National Academy of Sciences.

Jacobson, M. (1978). *Developmental neurobiology.* New York: Plenum Press.

Kanatani H., & Nagahami, Y. (1980). Mediators of oocyte maturation. *Biomedical Research, 1,* 273–291.

Keefer, D. A., & Dryden, G. L. (1982). Nuclear uptake of radioactivity by cells of pituitary, brain, uterus, and vagina of the Asian musk shrew *(Suncus murinus)* following [^{3}H] estradiol administration. *General and Comparative Endocrinology, 47,* 125–130.

Kelley, D. B., & Pfaff, D. W. (1978). Generalizations from comparative studies on neuroanatomical and endocrine mechanisms of sexual behavior. In J. B. Hutchison (Ed.), *The Biological determinants of sexual behaviour* (pp. 225–254). Chichester: John Wiley & Sons.

King, M. (1977). Reproduction in the Australian gekko *Phyllodactylus marmoratus* (Gray). *Herpetologica, 33,* 7–13.

Kleiman, D., & Mack, D. S. (1978). A peak in sexual activity during midpregnancy in the golden lion tamarin, *Leontopithecus rasalia* (Primates: Callitrichidae). *Journal of Mammalogy, 58,* 657–660.

Komisaruk, B. K. (1971a). Induction of lordosis in overiectomized rats by stimulation of the vaginal cervix: hormonal and neural interrelationships. In C. H. Sawyer & R. A. Gorski (Eds.), *Steroid hormones and brain function* (pp. 127–141). Los Angeles: UCLA Press.

Komisaruk. B. K. (1971b). Strategies of neurendocrine neurophysiology. *American Zoologist, 11*, 741–754.

Komisaruk, B. K. (1978). The nature of the neural substrate of female sexual behaviour in mammals and its hormonal sensitivity: Review and speculations. In J. B. Hutchison (Ed.), *Biological determinants of sexual behaviour* (pp. 349–393). Chichester: John Wiley & Sons.

Komisaruk, B. R., Terasawa, E., & Rodriguez-Sierra, J. F. (1981). How the brain mediates ovarian responses to environmental stimuli: Neuroanatomy and neurophysiology. In N. T. Adler (Ed.), *Neuroendocrinology of reproduction* (pp. 349–376). New York: Plenum Press.

Lehrman, D. S. (1962). Interaction of hormonal and experiential influences on development of behavior. In K. L. Bliss (Ed.), *Roots of behavior* (pp. 142–156). New York: Hoeber Medical Division of Harper & Bros.

Lehrman, D. S. (1965). Interaction between internal and external environments in the regulation of the reproductive cycle of the ring dove. In F. A. Beach (Ed.), *Sex and behavior* (pp. 355–380). New York: John Wiley & Sons.

Lehrman, D. S. & Wortis, R. P. (1960). Previous breeding experience and hormone-induced incubation behavior in the ring dove. *Science, 132*, 1667–1668.

Lehrman, D. S., & Wortis, R. P. (1967). Breeding experience and breeding efficiency in the ring dove. *Animal Behaviour, 15*, 223–228.

Licht, P. (1984). Seasonal cycles in reptilian reproductive physiology. In E. Lamming (Ed.), *Marshall's physiology of reproduction* (pp. 206–282). Edinburgh: Churchill Livingstone.

Licht, P., & Bona-Gallo, A. (1982). Mating-induced follicular growth in garter snakes. *American Zoologist, 22*, 857.

Lincoln, G. A., Guinniss, F., & Short, R. V. (1972). The way in which testosterone controls the social and sexual behavior of the red deer stag *(Cervus elaphus) Hormones and Behavior, 3*, 375–396.

Lofts, B. (1977). Patterns of spermatogenesis and steroidogenesis in male reptiles. In J. H. Calaby & C. H. Tyndale-Biscoe (Eds.), *Reproduction and evolution* (pp. 127–136). Canberra: Australian Academy of Sciences.

Loher, W., & Edson, E. K. (1973). The effect of mating on egg production and release in the cricket. *Entomologica Experimentalis Applicata, 16*, 483–490.

Lydell, K., & Doty, R. L. (1972). Male rat odor preferences for female urine as a function of sexual experience, urine age, and urine source. *Hormones and Behavior, 3*, 205–212.

Macrides, F., Bartke, A., & Dalterio, S. (1975). Strange females increase plasma testosterone levels in male mice. *Science, 189*, 1104–1106.

Marshall, A. J. (1949). On the function of the interstitium of the testis: The sexual cycle of a wild bird, *Fulmarus glacialis* (L.). *Quarterly Journal of the Microscopical Sciences, 90*, 265–281.

Marshall, A. J. (1970). Environmental factors other than light involved in the control of sexual cycles in birds and mammals. In J. Benoit & I. Assenmacher (Eds.), *La photoregulation de la reproduction chez les oiseaux et les mammiferes* (pp. 53–69. Paris: Colloques International aux du Centre National de la Recherche Scientifique.

Mayr, E. (1958). Behavior and systematics. In A. Roe & C. G. Simpson (Eds.), *Behavior and evolution* (pp. 341–362). New Haven: Yale University Press.

Mayr, E. (1974). Behavior programs and evolutionary strategies. *American Scientist, 62*, 650–659.

Mayr, E. (1982). *The growth of biological thought.* Boston: Belknap Press of Harvard University Press.

McClintock, M. K. (1983). The behavioral endocrinology of rodents: A functional analysis. *BioScience, 33,* 573–577.

McEwen, B. S. (1981). Neural gonadal steroid actions. *Science, 211,* 1303–1311.

McGill, T. E. (1962). Reduction in "head-mounts" in the sexual behavior of the mouse as a function of experience. *Psychological Reports, 10,* 284.

McGill, T. E. (1978a). Genetic factors influencing the action of hormones on sexual behavior. In J. B. Hutchison (Ed.), *Biological determinants of sexual behaviour* (pp. 7–28). New York: John Wiley & Sons.

McGill, T. E. (1978b). Genotype-hormone interactins. In T. E. McGill, D. A. Dewsbury, & B. D. Sachs (Eds.), *Sex and behavior: Status and prospectus* (pp. 161–187). New York: Plenum Press.

McKinney, F. (1975). The evolution of duck displays. In G. Baerends, C. G. Beer, & A. Manning (Eds.), *Function and evolution of behavior* (pp. 331–361). Oxford: Clarendon Press.

Meyer, J. H., & Zakon, J. (1982). Androgens alter the timing of electroreceptors. *Science, 217,* 635–637.

Michael, R. P. (1961). Observations upon the sexual behaviour of the domestic cat (*Felis catus* L.) under laboratory conditions. *Behaviour, 18,* 1–24.

Michel, G. F. (1977). Experience and progesterone in ring dove incubation. *Animal Behaviour, 25,* 281–285.

Money, J. (1955). Hermaphroditism, gender and precocity in hyperandrenocorticism: Psychological findings. *Bulletin of Johns Hopkins Hospital, 96,* 253–264.

Money, J. (1974). Intersexual and transexual behavior and syndromes. In S. Arieti & E. B. Brady (Eds.), *American handbook of psychiatry* (Vol. 3, pp. 253–264). New York: Basic Books.

Moore, C. L. (1985). Another psychobiological view of sexual differentiation. *Developmental Review, 5,* 18–25.

Moore, M. C. (1982). Hormonal response of free-living male whitecrowned sparrows to experimental manipulation of female sexual behavior. *Hormones and Behavior, 16,* 323–329.

Moore, M. C. (1983). Effect of female sexual displays on the endocrine physiology and behaviour of male white-crowned sparrows *Zonotrichia leucophrys. Journal of Zoology (London), 199,* 137–148.

Moore, M. C., & Kranz, R. (1983). Evidence of androgen independence of male mounting behavior in white-crowned sparrows *(Zonotrichia leucophrys gambelii). Hormones and Behavior, 17,* 414–423.

Moore, M. C., Whittier, J. M., Billy, A. J., & Crews, D. (1985). Male-like behaviour in an all-female lizard: relationship to ovarian cycle. *Animal Behaviour, 33,* 284–289.

Morali, G., & Beyer, C. (1979). Neuroendocrine control of mammalian estrous behavior. In C. Beyer (Ed.), *Endocrine control of sexual behavior* (pp. 33–75). New York: Raven Press.

Morrell, J. I., & Pfaff, D. W. (1978). A neuroendocrine approach to brain function: Localization of sex steroid concentrating cells in vertebrate brains. *American Zoologist, 18,* 447–460.

Mount, R. H. (1963). The natural history of the red-tailed skink *Eumeces egregius* Baird. *American Midland Naturalist, 70,* 356–385.

Muller, H. J. (1949). The Darwinian and modern conceptions of natural selection. *Proceedings of the American Philosophical Society, 93,* 459–470.

Murton, R. K., & Westwood, N. J. (1977). *Avian breeding cycles.* Oxford: Clarendon Press.

Norris, D. O., Norman, F., & Duvall, D. (1981). Seasonal variations in repro-

ductive parameters of male neotenic *Ambystoma tigrinum*. *American Zoologist, 21*, 949.
Northcutt, R. G. (1984). Evolution of the vertebrate central nervous system: Patterns and processes. *American Zoologist, 24*, 701–716.
Nottebohm, F. (1981). A brain for all seasons: Cyclical anatomical changes in song control nuclei of the canary brain. *Science, 214*, 1368–1370.
Oxberry, B. A. (1979). Female reproductive patterns in hibernating bats. *Journal of Reproduction and Fertility, 56*, 359–367.
Parkes, A. S. (1976). *Patterns of sexuality and reproduction*. London: Oxford University Press.
Pearson, O. P., Koford, M. R., & Pearson, A. R. (1952). Reproduction of the lump-nosed bats *(Corynorhinus rafinesquei)* in California. *Journal of Mammalogy, 33*, 273–320.
Pfaff, D. W., & McEwen, B. S. (1983). Actions of estrogens and progestins on nerve cells. *Science, 219*, 808–814.
Pianka, E. R. (1970). On R and K selection. *American Naturalist, 104*, 592–597.
Quinn, H. R. (1979). Reproduction and growth of the Texas coral snake. *Copeia, 1979*, 453–463.
Rahn, H. (1940). Sperm viability in the uterus of the garter snake, *Thamnophis*. *Copeia, 1940*, 109–115.
Reiter, R. J., & Follett, B. K. (Eds.). (1980). *Seasonal reproduction in higher vertebrates*. Basel: S. Karger.
Reinboth, R. (Ed.). (1975). *Intersexuality in the animal kingdom*. New York: Springer-Verlag.
Richmond, M. & Stehn, R. (1976). Olfaction and reproductive behavior in microtine rodents. In R. Doty (Ed.), *Mammalian olfaction, reproductive processes, and behavior* (pp. 198–218). New York: Academic Press.
Rosenblatt, J. S. (1965). Effects of experience on sexual behavior in male cats. In F. A. Beach (Ed.), *Sex and behavior* (pp. 416–439). New York: John Wiley & Sons.
Saint-Girons, H. (1982). Reproductive cycles of male snakes and their relationships with climate and female reproductive cycles. *Herpetologica, 38*, 5–16.
Saint-Girons, H., Brosset, A., & Saint-Girons, M. C. (1969). Contribution à la connaissance du cycle annuel de la chauve-souris *Rhinolophus ferrumequinum* (Schreber, 1774). *Mammalia, 33*, 357–470.
Sayler, S. A. (1966). The reproductive ecology of the red-backed salamander, *Plethodon cinereus* in Maryland. *Copeia, 1966*, 183–193.
Scharrer, E. (1946). Anatomy and the concept of analogy. *Science, 103*, 578–579.
Shapiro, D. Y. (1979). Social behavior, group structure, and the control of sex reversal in hermaphroditic fish. In J. S. Rosenblatt, R. A. Hinde, C. Beer, & M.-C. Busnel (Eds.), *Advances in the study of behavior* (Vol. 10, pp. 43–104). New York: Academic Press.
Shaw, E. (1971). Evidence of sexual maturation in young adult shiner perch, *Cymatogaster aggregata* Gibbons *(Perciformes, Embiotocidae)*. *American Museum Novitates, 2479*, 1–10.
Silver, R., & Barbiere, C. (1977). Display of courtship and incubation behavior during the reproductive cycle of the male ring dove *(Streptopelia risoria)*. *Hormones and Behavior, 8*, 8–21.
Smyth, M., & Smith, J. J. (1968). Obligatory sperm storage in the skink *Hemiegris peronii*. *Science, 161*, 575–576.
Stacey, N. E. (1983). Hormones and pheromones in fish sexual behavior. *BioScience, 33*, 552–555.

Stern, J. J. (1970). Response of male rats to sex odors. *Physiology and Behavior, 5,* 519–524.

Tavolga, W. N. (1955). Effects of gonadectomy and hypophysectomy on prespawning behavior in males of the gobiid fish *Bathygobius soporator. Physiological Zoology, 28,* 218–231.

Taylor, R. (1982). *Reproductive and behavioral ecology of the gopher tortoise.* Unpublished doctoral thesis, University of Florida, Gainesville.

Toran-Allerand, C. D. (1976). Gonadal hormones and brain development: Cellular aspects of sexual differentiation. *American Zoologist, 18,* 553–565.

Valenstein, E. S., & Goy, R. W. (1957). Further studies on the organization and display of sexual behavior in male guinea pigs. *Journal of Comparative and Physiological Psychology, 50,* 115–119.

van Tienhoven, A. (1983). *Reproductive physiology of vertebrates.* Ithaca: Cornell University Press.

Vandenbergh, J. G. (1983a). Social factors controlling puberty in the female mouse. In J. Balthazart, E. Prove, & R. Gilles (Eds.), *Hormones and behaviour in higher vertebrates* (pp. 342–349). Berlin: Springer-Verlag.

Vandenbergh, J. G. (1983b). Pheromonal regulation of puberty. In J. G. Vandenbergh (Ed.), *Pheromones and reproduction in mammals* (pp. 95–112). New York: Academic Press.

von Ihering R. (1937). Oviducal fertilization in the South American catfish, *Trachycorystes. Copeia, 1937,* 201–205.

Volsoe H. (1944). Structure and seasonal variation of the male reproductive organs of *Vipera berus* (L.). *Spolia Zoologica Museuma Hauniensis, 5,* 1–172.

Werner, Y. (1980). Apparent homosexual behavior in an all-female population of a lizard, Lepidodactylus lugubris and its probable interpretation. *Zeitschrifft fur Tierpsychologie, 54,* 144–150.

Wells, K. D. (1977). The social behaviour of anuran amphibians. *Animal Behaviour, 25,* 666–693.

Whalen, R. E. (1963). Sexual behavior of cats. *Behaviour, 20,* 321–342.

Wiebe, J. P. (1968). The reproductive cycle of the viviparous seapearch, *Cymatogaster aggregata* Gibbons. *Canadian Journal of Zoology, 46,* 1221–1234.

Wilczynski, W. (1984). Central neural systems subserving a homoplasous periphery. *American Zoologist, 24,* 755–764.

Williams, C. L. (1986). A reevaluation of the concept of separable periods of organizational and activational actions of estrogens in the development of brain and behavior. In B. Komisaruk (Ed.), *Reproduction: A behavioral and neuroendocrine perspective.* New York: Annals of the New York Academy of Sciences.

Wingfield, J. C., Newman, A. L., Hunt, G. L., & Farner, D. S. (1982). Endocrine aspects of female-female pairing in the western gull *(Larus occidentalis wymani). Animal Behaviour, 30,* 9–22.

Wingfield, J. C. & Grimm, A. S. (1977). Seasonal changes in plasma cortisol, testosterone and oestradiol-17beta in the plaice, *Pleuronectes platessa* L. *General and Comparative Endocrinology, 31,* 1–11.

Wolf, L. L. (1975). "Prostitution" behavior in a tropical hummingbird. *Condor, 77,* 140–144.

Young, W. C. (1961). The hormones and mating behavior. In W. C. Young (Ed.), *Sex and internal secretions* (pp. 1173–1239). Baltimore: Williams & Wilkins.

7

The Importance of Chemical Messengers in Mammalian Reproduction

Milos Novotny

Chemical communication, in its broadest sense, occurs within many living structures. At the molecular level, a variety of important processes are coordinated through hormones, hormone mediators, neurotransmitters, etc., to attain a high degree of sophistication for the total organism's effective survival in its environment. Those are the most subtle forms of "chemical communication." Communication of a similar kind also occurs among different organisms. Highly specific substances, excreted by individuals for the purpose of communicating to other members of the same species, have been called *pheromones.*

To a large degree, the role of pheromones in the lives of different organisms appears to depend on the sophistication of such organisms to perceive and integrate other forms of communication (visual, auditory, tactile, etc.). Theoretical aspects, various mechanisms of communication, and their relative importance for various animals have been reviewed in a book by Sebeok (1977). Although pheromone communication is widely documented throughout the literature on insects, considerably less is known about these forms of communication in mammals, for two principal reasons. First, projections of the role of olfactory signals in mammals to the possible role in humans tend to deemphasize this way of communication in comparison to the importance of vision and audition. Second, suitable techniques to pursue such investigations are lacking. To some investigators, the concept of pheromone communication, as applied so widely in the insect world, appears too rigid to extrapolate to mammals; yet even casual observers of mammals in nature are fully aware of the importance of olfactory signals to reproduction. The role of chemical communication in reproduction has been observed in certain primates, and a few reports even speculate on the possible existence of pheromones in humans.

Different chemicals are excreted by mammals for a variety of communication functions. We should distinguish between pheromonal communication and chemical communication in general, although the dividing line can be difficult at times. Production of a pheromone by organisms appears to be an inherited trait, with the same substance being used by successive generations. The response to such a substance should be genetically controlled and not learned (Karlson & Butenandt, 1959; Beauchamp, Doty, Moulton, & Mugford, 1976). This distinction should separate pheromones from a large group of chemical substances, the perception of which could be readily learned. As an example, Albone, Gosden, & Ware (1977) suggested that animals may recognize individuals based on the odors produced by their bacteria. Thus, various combinations of microbes and substrates may give each individual a unique odor, while the animal is not directly responsible for its odor. This certainly qualifies as chemical communication, but hardly fits the definition of a pheromone.

According to the widely accepted terminology (Karlson, 1959; Wilson & Bossert, 1963), pheromones can be divided into "primers" and "releasers." Primer pheromones are substances that cause long-term hormonally mediated changes in a recipient, while releaser (signaling) pheromones produce an immediate behavioral response. To this date, the primer effects have been best documented in small rodents, although their parallels appear to exist in many other mammals. Effects such as male-mediated puberty acceleration and estrus synchronization in females, and a counteracting (female-to-female) estrus-delaying phenomenon, are among the best known examples with a significant biological rationale. Examples of releaser effects include intermale aggression, territorial dominance, sexual attraction, and signaling of stress. With regard to behavioral complexity of mammals, the same chemical substance (or a mixture of chemicals) could be used to communicate a variety of messages, with the specific meaning being determined by the circumstances surrounding transmission and reception. Although the term *releaser* was originally coined for use with insects with rigid behavioral patterns, the same concept applying to mammals must be considered in more liberal terms.

A strong interrelationship exists between hormones and pheromones. This relationship forms a basis for the current investigations into the molecular nature of mammalian pheromones. Consequently, genetic and endocrinological manipulations can be followed biochemically in both qualitative and quantitative terms. Following the structural elucidation of specific messengers, such substances are synthetically prepared in the laboratory and tested on animals under laboratory conditions and, preferably, in the natural environment. The structural knowledge of pheromones, as well as their availability for testing, will undoubtedly answer numerous important questions on chemical communication in general.

The present article is primarily concerned with the mammalian pheromones related to gender and reproduction. The pheromones are undoubtedly among the most important expressions of sex differences in mammals. Because they are directly related to mating behavior throughout the animal kingdom, at least some pheromones (in both a qualitative and a quantitative sense) should be acceptable as criteria of masculinity and femininity. Examples for this discussion have been chosen from the author's research area and relate primarily to the chemical communication in the house mouse and certain canids. While potentially interesting, any direct projections to other animal species are currently unwarranted. As expressed appropriately by Bronson (1979), "while the brain-pituitary-gonadal axis forms the endocrinological core for reproduction in all mammals, considerable species-to-species variation is evidence of the fine functioning of this axis. Specific adaptations also occur in the interactions of this axis with other reproductive tissues and supportive behaviors, in its interactions with other physiological systems, and in its regulation by environmental factors." With these facts in mind, we assess the importance of chemical communication not only for different species, but even within one species under different environmental conditions.

Relations Between Hormones and Pheromones

A strong axis between hormones and pheromones is a general characteristic of a mammalian system. In numerous cases, hormones are known to exert a powerful influence over pheromone expression, while these same pheromones are known to cause hormonal changes in the recipients—or to modify their behavior. Endocrinologically altered animals often lose the capability to excrete a pheromone, and proper hormone supplementation will renew its production. In some instances, castration of animals is likely to result in degeneration of pheromone receptors (Carr, Solberg, & Pfaffman, 1962). Although no conclusive evidence has yet been obtained on the biochemical basis of behavioral differences between dominant and subordinate animals, variations in pheromone expression are once again strongly suspected.

Although little biochemical information is currently available, many indirect observations have been made on endocrine influences over pheromone production. Many male animals exhibit odors that are both qualitatively and quantitatively different from the odors of females in the same species. Such odors are frequently suppressed by castration—hence the rationale for castrating certain domestic animals to eliminate odors unpleasant to humans.

In many species, females produce odorous substances during the follicular phase of their cycles; these odors are readily recognized by males. Beach and co-workers (Beach & Leboeuf, 1967; Beach & Merari, 1968, 1970; Beach, Rogers, & Leboeuf, 1968) have implicated olfactory cues in coital behavior of dogs and associated them with hormonal levels. Male dogs are more interested in investigating vaginal secretions from estrogen- and progesterone-treated bitches than in investigating secretions from untreated females (Beach & Merari, 1968, 1970). Male dogs with adrenal and Sertoli cell tumors that are known to secrete estrogens also become attractive to other male dogs (Pierrepoint, Galley, Griffiths, & Grant, 1967).

Vaginal secretions of estrous rhesus monkeys (Michael & Keverne, 1970), cows (Hart, Mead, & Reagan, 1946), hamsters (Singer, Agosta, O'Connell, Pfaffmann, Bowen, & Field, 1976), etc., were found to produce responses in males toward copulatory behavior. In addition, various female animals are known to display preferences for the odors of intact males as compared to those of castrates. Whether previous sexual experience is imperative for the display of preference is still under discussion. Obviously, more careful experimentation is needed to draw the line between pheromones, the production or perception of which is a clear, inherited trait, and certain "trivial" odors that were previously associated with sexual rewards.

Animals display their odors widely and in a variety of ways. Although urine is frequently a source of a specific signal or a primer pheromone, other sources have been implicated: glandular secretions, contents of both sebaceous and apocrine glands, saliva, feces, etc. Although a serious search for the specific substances in these complex materials has barely been initiated, many of their biological effects have been widely observed. Hormonal control of the production of such compounds is currently beyond dispute, but little information is available at present as to the mechanism of these processes. At the biochemical level, appropriate hormones will induce specific enzymes and promote the formation of pheromones (in most part, relatively small molecules) from certain substrates. A variety of sexually dimorphic tissues, including the liver, can participate in such metabolic processes. Sex-related differences in the blood concentration of some common biochemical substrates (e.g., amino acids) are well known. As the specific compounds with pheromonal activities will gradually be discovered, elucidation of the corresponding biogenetic pathways will undoubtedly follow, providing further information on the relationship between hormones and pheromones.

After a pheromone has been perceived at its target tissue, a sequence of biochemical processes must occur prior to the brain's regulation of an appropriate physiological or behavioral response. Although a variety of schemes have been postulated, little concrete evidence exists to date. It is reasonable to assume that some of the olfactory epithelial structures are concerned with the perception of reproduction-oriented pheromone molecules. Biochemical and electrophysiological changes that follow the absorption of olfactorily active molecules on the mucous membrane are relayed to the olfactory bulbs. Numerous mammals, however, seem to have an additional route of perception: the so-called vomeronasal (Jacobson's) organ. Recent investigations strongly suggest that the vomeronasal organ may be responsible for perception of primer pheromones (Kaneko, Debski, Wilson, & Whitten, 1980). Electrophysiological manipulations of the olfactory bulb tissue and the nerves between these various structures and the hypothalamus result in differential responses to pheromones in mice (Kaneko et al., 1980) and voles (Dluzen, Ramirez, Carter, & Getz, 1981). Relatively large molecules can penetrate into this organ from the environment (Wysocki, Wellington, & Beauchamp, 1980). Once again, it appears that availability of the primer pheromones as defined chemical substances can significantly enhance our understanding of these receptor structures.

Although many isolated facts are now available from the literature on the brain-pituitary-gonadal system of numerous mammals, studies of these phenomena in the house mouse seem the most complete. An overview of the pheromonal cuing system of this mammal has been provided in reviews by Bronson (1979, 1984). As shown in Figure 7-1 (Bronson, 1979), the pheromones excreted in the urine of both males and females are implicated in a number of physiologically important processes. Mutual stimulation of the hormonal systems by the respective pheromones in both male and female mice is a general characteristic of the system. In addition, female-female interactions provide "competition" with the male influence, but only among prepubertal females.

After exposure to urine from the normal males, the female mice responded by producing their luteinizing hormone (LH) and estrogens [estradiol was observed to rise dramatically at a level about 20 times higher than basal levels, just 12 hours after exposure (Bronson & Desjardins, 1974)]. This rapid rise in estrogens, when possibly supplemented by other events, could explain several effects readily observed in female mice, depending on their endocrine status (i.e., immature, mature, or pregnant). Correspondingly, puberty acceleration (Vandenbergh, 1967) could be induced in young female mice, while promotion of a regular 4-day estrous cycle will occur in mature females (Whitten, 1956). In addition, unusually high levels of estrogen could prevent preparation of the uterus for implantation [Bruce & Parrott, 1960)] in a

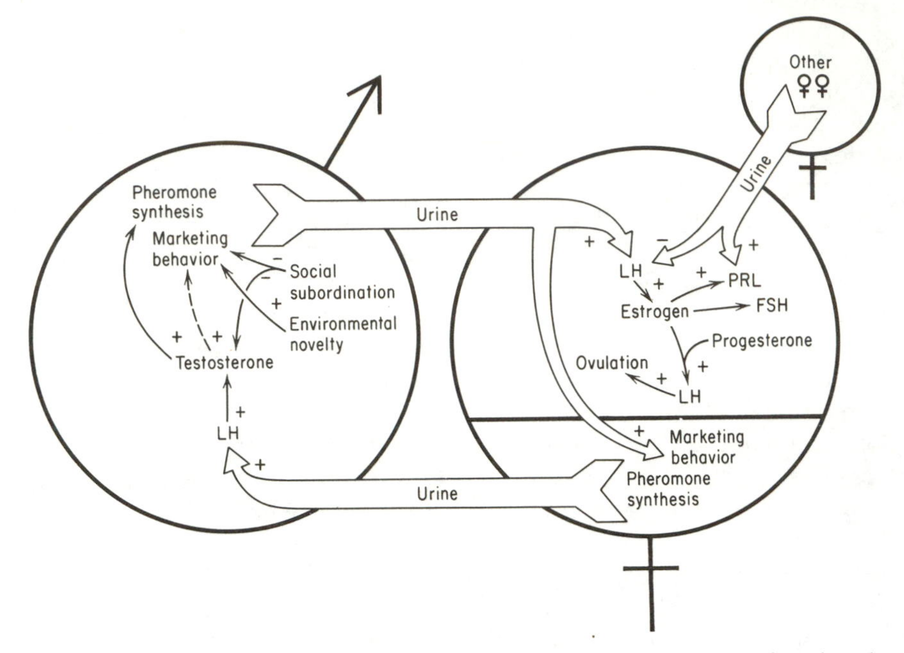

Figure 7–1. Male-female pheromone communication in mice. (Reproduced with permission of Stony Brook Foundation, Inc., from Bronson, 1979.)

pregnant mouse. Some of these effects could be counteracted by other mature females (McIntosh & Drickamer, 1977; Drickamer, McIntosh, & Rose, 1978).

An increased urine marking behavior by females is also induced after exposure to the male. Not surprisingly, the female urine carries an olfactory signal to the male, who responds by an increased formation of his own LH and testosterone (Maruniak & Bronson, 1976). Without a sufficient testosterone level, the necessary pheromone biosynthesis will not take place; indeed, a distinction between the dominant and subordinate males is likely to be related to their higher or lower testosterone levels, and the corresponding excretion of pheromones that are under the hormone's control.

As discussed by Bronson (1979, 1984), this pheromonal cuing system may have a wide range of ecological implications (dispersal of the young, territorial maintenance, habituation under a variety of conditions, colonization, etc.). To what degree such factors are operative in other mammals under field conditions is largely unknown at present.

Research Methodologies

Investigations in the field of mammalian pheromones are very dependent on interdisciplinary research; they may involve close collaboration of scientists from endocrinology, psychobiology, and at least two major areas of chemistry (bioanalytical and organic). The understanding of mammalian pheromones has been inhibited by the limited availability of adequate research techniques in the past, but recent advances in pheromone identification methodology will now accelerate the overall scientific effort in this area.

Once the existence of a specific response to a chemical signal becomes established, it is important to trace the signal source (urine, glandular secretions, special organs, etc.). Knowing the circumstances under which a pheromone is dispersed into the environment (the phase of reproductive cycle, breeding season, etc.) is often helpful. Furthermore, understanding the hormone-pheromone dependencies is essential, because various hormonal manipulations can significantly help to follow the putative pheromones under various circumstances by chemical methods. Ideally, there is a coincidence of their concentrations in the signal source with the observed biological phenomena. Certain male mammals lose their ability to produce a pheromone following castration, while subsequent supplementation with testosterone may renew the effect on females. Comparing the biological samples from normal versus castrated males by suitable chemical techniques is often the beginning of pheromone structural identification. Similarly, samples obtained from normal females versus ovariectomized or estrogen-treated animals can be compared.

Modern analytical measurement methodology has been extremely important to progress in this area. Because urine samples or glandular secretions of mammals may contain hundreds of organic components, it is essential to employ chemical separation techniques with extraordinary resolving power, so that the putative pheromones can be properly isolated from the rest of the metabolic products that may serve little purpose in chemical communication, but may interfere in the determination of these biologically active substances. Furthermore, because these pheromones may be present as only trace components (parts per million or less) in their source compound matrix, the chemical analytical techniques must be highly sensitive. Provided that the pheromones under study are *volatile substances* (as they are known to be in the insect world), capillary gas chromatography is the most powerful analytical technique for the separation and simultaneous sensitive measurement of such compounds.

In a typical sample of a physiological fluid or glandular secretion, capillary gas chromatography can separate around 100 volatile components. As shown with the example of the urinary volatile profiles (Fig. 7-2) of estrous and immature female mice (Schwende, Wiesler, & Novotny, 1984), these materials can be exceedingly complex.

Once the sample has been separated into its individual components, their structural identification must be accomplished. Although the sub-

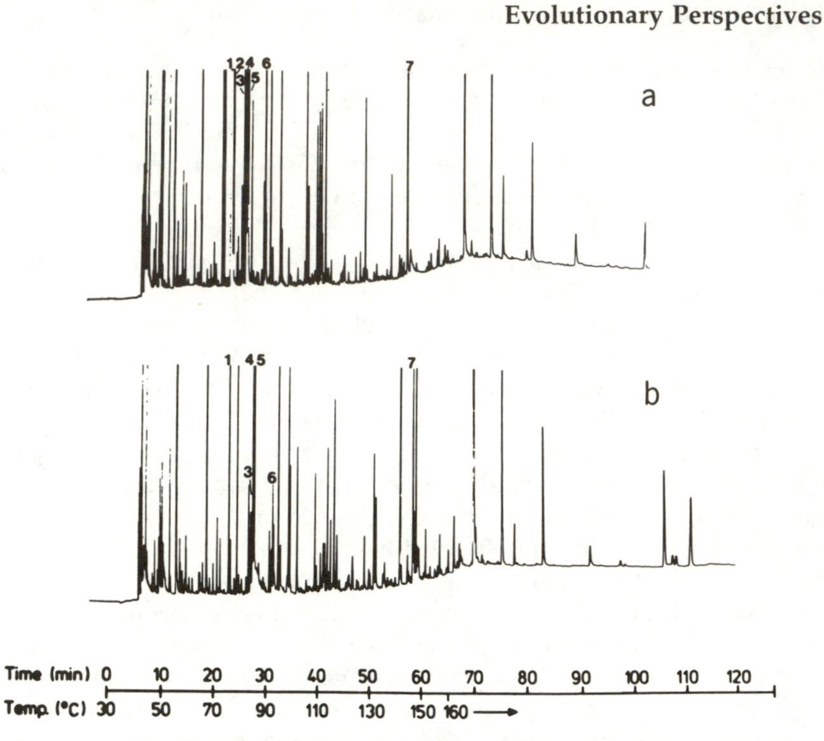

Figure 7–2. Differences in the excretion of urinary volatile substances in mice, as measured by capillary gas chromatography. a. Females with induced estrus. b. Immature females. (Reproduced with permission of Birkhauser Verlag from Schwende, Wiesler, & Novotny, 1984.)

stances that appear under endocrine control are the primary target of identification work, chemical knowledge of the other constituents of the entire chromatographic profile may also be important for at least two reasons: (1) in some cases, the pheromones may exhibit biological activity only if perceived in the overall context of the source odor, and (2) correct quantitative proportions of several mixture components may be essential in producing a response. Examples of both cases will be discussed here in relation to communication in the house mouse and canid species.

Prior to capillary gas chromatography, the volatile constituents of urinary or glandular secretion samples need to be preconcentrated and separated from the nonvolatile matrix. Although extraction into organic solvents may be feasible to accomplish this important step, headspace preconcentration with porous polymers (Novotny, Lee, & Bartle, 1974; Novotny, McConnell, Lee, & Farlow, 1974) appears preferable. The literature shows some indications that certain pheromones may not be volatile; examples are the puberty-accelerating primer pheromone of male mice (Vandenbergh, Whitsett, & Lombardi, 1975; Novotny et al., 1980), as well as some biologically active fractions of hamster vaginal secretions (Singer, Macrides, & Agosta, 1980). In such cases, entirely different analytical techniques must be employed. Modern techniques of liquid chromatography appear most suitable for separation purposes.

Component identification can be accomplished through a variety of modern structural techniques such as mass spectrometry, Fourier-transform infrared spectroscopy, or nuclear magnetic resonance spectrometry. A particular technique is seldom sufficient for positive compound identification; most typically, a combination of several approaches leads to identification. This is frequently the stage of research where the chemist's expertise is crucially important and, while some identifications may be straightforward, others could present relatively difficult investigations in terms of both time and resources.

Once a putative pheromone becomes chemically characterized, a synthetic effort is initiated to verify the structure as well as to test such an artificial compound for biological activity. The difficulty of this task may vary with the type of molecular structure. Just as with known insect pheromones, some biologically active compounds are relatively simple and readily available, while others may need involved syntheses (including various stereochemical molecular arrangements). For example, one of the pheromonal components of the male mouse, *exo*-3, 4-dehydrobrevicomin, identified recently in our laboratory (Novotny, Schwende, Wiesler, Jorgenson, & Carmack, 1984), required a nine-step synthesis (Wiesler, Schwende, Carmack, & Novotny, 1984).

As various endocrinologically and behaviorally active substances become identified and chemically synthesized, their availability for both laboratory experiments and field testing will undoubtedly add an important dimension to the rapidly developing field of chemical communication.

Mouse Pheromones

An improved understanding of reproductive physiology and behavior in rodents is highly important for several reasons: (1) their special significance to humans because of destructive influences on food consumption, (2) their importance as models of reproductive efficiency and dynamics of population growth, and (3) their extensive use as laboratory animals. The common house mouse has received considerable attention with respect to pheromones and chemical communication. Its nocturnal behavior and relatively poor vision create a strong rationale for using chemical communication extensively. The mouse has a variety of known and potential signal sources in different specialized glands;

the best documented priming and signaling phenomena have been attributed to urine. Although the vast majority of work has been carried out with inbred laboratory mice, similar phenomena have been observed in natural mouse populations.

Primer Activities

The primer phenomena best documented in laboratory mice include the estrus-synchronizing (Whitten) effect, the pregnancy-blocking (Bruce) effect, and the puberty-accelerating (Vandenbergh) effect. In all three effects, the urine of normal males is the source. The Whitten effect is related to the estrous cycle of females: in the absence of males or male urine, female mice show irregular cycles; however, exposure to male urine induces a regular 4–5 day estrous cycle. The Bruce effect concerns the action of male urine on pregnant females. After a female is mated with a male, she will continue her pregnancy regardless of whether or not the male is removed from her presence. If she is brought in contact with the urine of a "strange" male (a male other than the stud male) prior to implantation, however, she has a high probability of returning to the estrous cycle. In the Vandenbergh (puberty-acceleration) effect, urine of males is able to induce early maturation in prepubertal female mice. Castration of a male mouse will eliminate his ability to produce all three pheromones. For a review of these primer effects, see Bronson (1979).

Although the suggestion has been made that the three male primer effects may be mediated by the same substance or group of substances, the preliminary results of our laboratory indicate the contrary. A long-lasting controversy over whether the substance(s) is a large or a small (volatile) molecule may now be resolved. We have extensively fractionated male mouse urine and purified the fraction that is active in the puberty-accelerating bioassay. In agreement with the initial work by Vandenbergh et al. (1975), we attribute this activity to a relatively large polar molecule, although structural elucidation of this pheromone has not yet been completed. Conversely, substances causing the estrus-synchronization effect appear volatile in our recent investigations (Jemiolo, Alberts, Sochinski-Wiggins, Harvey, & Novotny, 1985; Novotny, Harvey, Jemiolo, & Alberts, 1985), supporting the validity of earlier work by Whitten, Bronson, and Greenstein (1968) using a wind-tunnel experimental arrangement. Controversies over the nature of the pheromone causing the Bruce effect still exist (Hoppe, 1975; Marchlewska-Koj, 1980).

The best known female primer activity involves estrus suppression (van der Lee & Boot, 1955) in grouped females, without the presence of males (the Lee-Boot effect). When confined to all-female groups, the females were observed to become anestrous or to develop spontaneous pseudopregnancies. Evidence for this pheromone has been indirect, in that the cycle suppression is reduced following removal of the olfactory bulbs and is independent of visual and physical contact (Whitten, 1959). The source of this pheromone remains uncertain.

More recently, increasing emphasis has been placed on a female pheromone that causes puberty delay in other females. In juvenile females, puberty is delayed by the presence of mature female mice (Vandenbergh, Drickamer, & Colby, 1972). A similar effect was observed when premature females were exposed to the urine collected from adult females (McIntosh & Drickamer, 1977; Drickamer, 1974). The presence of this "puberty-delaying pheromone" in urine was not affected by ovariectomy, but the effect disappeared after removal of the adrenal glands (Drickamer et al., 1978). During recent experiments in our laboratory (Novotny, Jemiolo, Harvey, Wiesler, & Marchlewska-Koj, 1986), biochemical differences were found between urines from the groups of normal and adrenalectomized females. Six major urinary components, including three long-chain, unsaturated ketones, two acetate esters, and a pyrazine derivative, are significantly reduced following the removal of the adrenal glands. Moreover, when the urine of adrenalectomized animals is supplemented with these six synthetic substances (at the levels mimicking their natural content in the normal urine), the biological activity appears fully recovered. Further studies of this primer pheromone system are likely to be interesting.

Although much remains to be learned about the nature of primer pheromones and the mechanism of their action, numerous investigations conducted in the house mouse to this date begin to shape important biological hypotheses of mammalian reproduction. For instance, the timing of puberty and the estrous cycle in normal female mice appear to be significantly influenced by the population density and presence of males. While the male-originated substances will accelerate puberty in immature females and synchronize estrus in mature female mice, the other females may, through their signals, contribute both to delaying such puberty and to lengthening of the cycle. Thus, for each male substance, the females appear capable of producing a pheromone with opposing properties. These pheromones could contribute to reproductive efficiency in wild populations under different circumstances.

Releaser Pheromones

Releaser pheromones may have both volatile and non-volatile components. While the actions of primer and releaser pheromones may be well defined biologically, the distinction does not require that separate substances be used, as suggested by Bronson (1971). A single urinary component could act as both a primer and a releaser pheromone, depending on the sex and endocrine status of the pheromone recipient. The notion is attractive in its simplicity and, as seen below, it appears to be operative in our recent experiments (Jemiolo et al., 1985; Novotny et al., 1985).

Numerous male- and female-originated signaling effects have been widely documented in the literature. Among various observed phenomena, the urine-marking behavior of female mice increases in the presence of males or male urine (Maruniak, Owen, Bronson, & Desjardins, 1975), suggesting the possibility of a signaling function. This function may include sexual attraction (Davies & Bellamy, 1972), an ultrasound-eliciting signal (Dizinno, Whitney & Nyby, 1978), or aggression-reducing signal (Mugford & Nowell, 1970a). W. K. Whitten (personal communication, 1977) also observed that male mice are more interested in sniffing urine from estrous females than diestrous mice. We have recently established (Schwende, Wiesler, & Novotny, 1984) that the female mice that had been made estrous through hormone implantation excreted in their urine significantly enhanced levels of n-pentyl acetate, *cis*-2-penten-1-yl acetate, p-toluidine, 2-heptanone, and three unsaturated ketones. Although some signaling function may be associated with these compounds in the female urine, no specific responses to them were found as yet.

The territorial marking behavior of male mice appears related to the application of an aversive signal. Markings by dominant males are avoided by other males under both the laboratory conditions (Harrington, 1976) and in wild populations (Berry, 1970). Normal male urine is also a source of aggression pheromone, which causes attack from other males; the pheromone is apparently under androgen control, as even androgenized females elicit attacks from intact males (Mugford & Nowell, 1970b). Male mice also produce a pheromone signaling fear when subjected to stress (Carr, Martorano & Krames, 1970); a suggestion has been made that the signal may be associated with the differences between dominant and subordinate male odors. Finally, it has been demonstrated (Scott & Pfaff, 1970) that female mice are more attracted to the odor of intact male urine than that of a castrate, suggesting the presence of a sex attractant.

The recent results of endocrinological manipulations and chromatographic analyses in our laboratory (Novotny et al., 1980; Novotny et al., 1984; Schwende, Wiesler, Jorgenson, Carmack, & Novotny, 1986) revealed that there are two conspicuous constituents of the normal male mouse urine, 2-*sec*-butyl dihydrothiazole (I), and *exo*-3,4-dehydrobrevicomin (II).

I

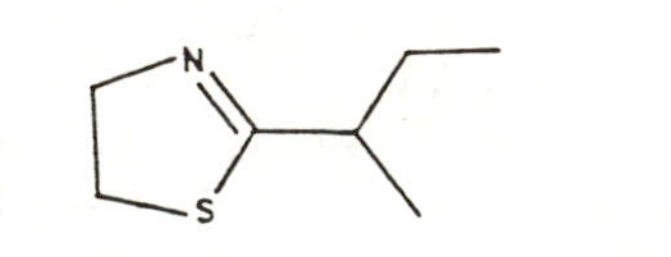

II

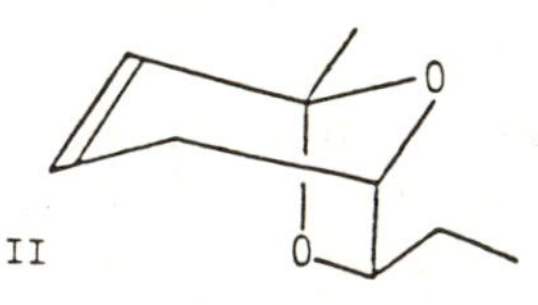

Their androgen dependency has now been established (Novotny et al., 1984; Schwende et al., 1986). The availability of these synthetic compounds has further enabled us to carry out various biological experiments. In the first set of experiments (Novotny et al., 1985), the two synthetic compounds were found to act synergistically (when added to castrate male urine but not in water) to provoke intermale aggression that is quantitatively and qualitatively comparable to that elicited by intact male urine. These two androgen-dependent substances are also active (Jemiolo, 1985) with respect to the female recipients, yet in two entirely different ways: attraction as well as an increased frequency of estrous cycles. Again, these two synthetic compounds are synergistic in the context of castrate urine, producing an olfactory message that mimics behaviorally and physiologically the activity of the normal biological signal. Tentative evidence for the hypothesis (Bronson, 1971) that some chemical substances could provide a variety of effects, depending on a test situation, seems thus provided.

Interesting observations were made by Yamazaki et al. (1976) that the major histocompatibility genetic complex (H-2) on the 17th chromosome pair of the mouse exerts an influence over mate selection. Males of a given H-2 genotype showed a preference either for or against females of their own type, depending on the genotypes investigated. Further experiments with a Y-maze arrangement (Yamazaki et al., 1979) showed that urine was the source of odor that the trained animals could discriminate, as based on H-2 differences. Although the genetic complex probably does not directly control the production of specific olfactants, even subtle quantitative differences in odors, caused by small immunogenetic factors, could be perceived by olfactorily sensitive mice (Bowers & Alexander, 1967). Through precise quantitative measurements of urinary volatile profiles by capillary gas chromatography, we have been able to provide tentative chemical support (Schwende, Jorgenson, & Novotny, 1984) for this histocompatibility-related mating preference. This sort of "genetic signaling" may constitute a new and unusual type of chemical signaling system with important implications to symbiosis, immunology, and recognition of self and nonself (Thomas, 1975).

Chemical Communication in Canids

Overall communication patterns in canids can be complex (Fox & Cohen, 1977). This complexity varies directly with the degree of sociability of the species such that a gregarious animal like the wolf (*Canis lupus*) appears to have a more complex communication system than the more solitary red fox (*Vulpes vulpes*). The wolf widely utilizes visual, vocal, tactile, and chemoolfactory modes of communication, but olfactory signals appear of utmost importance to the red fox. Two major sources of olfactory signals in canids are urine and secretions from the anal glands.

Most important olfactory communication in the red fox includes territorial marking, markers in food scavenging, and recognition of dominance and reproductive state (Henry, 1977; Fox, 1971). The three main sources of odor known in the red fox are the anal sac, the supracaudal gland, and urine. It is generally believed that the anal sac is of significance, because histophysiology and secretory activities of the anal gland appear to correlate with locomotor and sexual activity (Spannhof, 1969). Chemical studies thus far performed (Albone, Eglington, Walker, & Ware, 1974; Albone and Perry, 1976; Albone, Robins, & Patel, 1976) on secretions from the anal gland reveal only the presence of certain "trivial" volatile constituents (acids and amines) that are most likely produced by the action of microorganisms. The supracaudal (tail) gland of the red fox has also been investigated (Albone, 1975; Joffre, 1976) because of its "ambrosial" odor. Several terpenelike compounds were found.

The fox uses urine as its principal means of scent marking, while the urinary odor is intensified and changed in quality during the mating season. This marking behavior appears to coincide in the male fox with the period when spermatogenic activity, testicular and prostatic weights, testosterone levels, etc., are also increased (Joffre, 1976). Communication of sexual identity and endocrine status is thus likely. In order to elucidate the chemical nature of fox-specific urinary constituents occurring during the mating season, urine samples were collected and analyzed in our laboratory (Jorgenson et al., 1978). Differences between the excretion of volatiles in male versus female foxes were noted. Two unique sulfur-containing compounds, Δ^3-isopentenyl methyl sulfide and 2-phenylethyl methyl sulfide, were later successfully tested for their attractiveness to foxes in their natural environment (Whitten, Wilson, Jorgenson, Novotny, & Carmack, 1980). The obtained data indicate that these sulfur compounds are a necessary and perhaps sufficient factor in eliciting marking behavior in wild foxes.

A major mode of chemoolfactory communication (scent marking) in the wolf is urination; the way urine is deposited relates to behavior displayed during marking (Peters & Mech, 1975; Rothman & Mech, 1979).

Following the successful characterization and behavioral testing of the chemical scent constituents in red fox urine, a similar study was initiated in the wolf, as related to sex, endocrine status, and seasonality. Using capillary gas chromatography/mass spectrometry and Fourier-transform infrared spectroscopy to identify and quantitate typical wolf

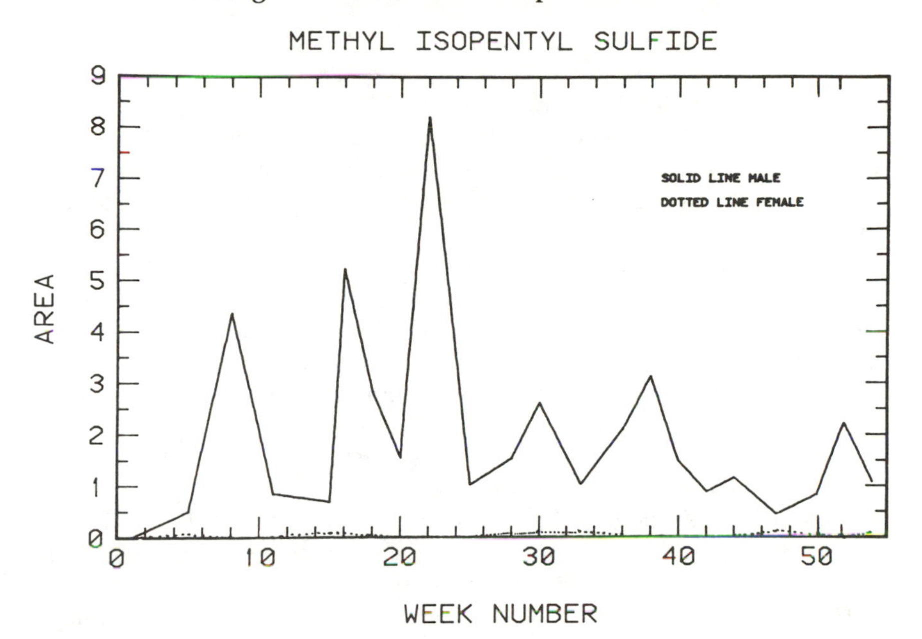

Figure 7–3. Comparative urinary excretion of isopentyl methyl sulfide by normal male (solid line) and normal female (dotted line) wolves throughout the year. (Reproduced with permission of Birkhauser Verlag from Wiesler, Novotny, Asa, Seal, & Mech, 1984.)

urinary components, several unique urinary constituents were discovered (Raymer, Wiesler, Novotny, Asa, Seal, & Mech, 1984). Just as in the red fox, sulfur-containing compounds appear important in the wolf. The appearance of some of these compounds is obviously sex dependent, as shown in the example of isopentyl methyl sulfide during the winter season (Fig. 7-3). A series of structurally unusual, highly branched ketones also appeared to be unique to male wolves. Further experiments, using hormonal treatments, verified that these compounds are endocrinologically dependent (Raymer, Wiesler, Novotny, Asa, Seal, & Mech, 1985); behavioral studies with synthetic compounds are currently in progress.

The functions that have been hypothesized for anal gland secretions in the canids include (1) sexual attraction (Albone et al., 1974; Preti, Muetterties, Furman, Kenelly, & Johns, 1976); (2) individual recognition and territorial demarcation (Albone et al., 1974; Peters & Mech, 1975; Preti et al., 1976); and (3) alarm function (Albone et al., 1974; Preti et al., 1976; Doty & Dunbar, 1974). A study by Doty and Dunbar (1974) utilizing beagles found no support for the sexual attraction hypothesis, but indicated some evidence for an alarm function. Until recently, only "trivial" odorants were identified as the constituents of various canids. In a more recent study (Raymer et al., 1985), further components (alcohols, aldehydes, and ketones) were characterized. Statistical comparisons of various groups (males, females, castrated animals, etc.) indicated that there may be a chemical basis for the use of these substances in wolf secretions.

Many of the aspects of urine marking, and communication in general, are commonly shared between wolves and domestic dogs. Wolves live in packs, usually consisting of a dominant "alpha" breeding pair and their offspring. When domestic dogs become strays, they often revert to pack living. Territorial demarcation is commonly observed in both wolves and dogs.

Attraction of male dogs to the vaginal secretions of estrous bitches was mentioned earlier. Goodwin, Gooding, and Regnier (1979) reported that methyl *para*-hydroxy benzoate is the chemical responsible for this activity. The validity of their conclusion was recently challenged (Kruse & Howard, 1983), however.

Much remains to be learned about the behavioral and endocrinological aspects of chemical communication in wild and domestic canids; additional studies are highly desirable because of their unique social interactions and the related ecological considerations.

Possible Existence of Human Messengers

The existence of a relationship between body odors and sexual processes in humans appears widely supported by numerous anthropological reports [for a review, see Doty (1977)]. In particular, various customs in relatively primitive cultures may suggest the existence of biological information on the basis of body odors. Few scientific data exist to support certain anectdotal and often exaggerated reports on the importance of olfactory communication in humans.

Research on the subject may prove exceedingly difficult for a variety of reasons. Because a learned response to certain sexual odors could be involved in these processes, it may be misleading to look for any specific substances. General sex-related odors come to mind; the volatile steroid metabolites are good examples. Kloek (1961) pointed out that perception of the odor of such compounds differs in men and women; this sensitivity also varies with the stages of the menstrual cycle. When 5α-androst-16-en-3-one, a steroid present in men's axillary sweat (Claus & Alsing, 1976), was sprayed onto a seat previously avoided by women in a waiting room, the incidence of women using the seat supposedly increased (Kirk-Smith & Booth, 1980). With a higher dose of this substance, the incidence of men using the seat decreased. From several studies on the detection of sex differences in sweat odors (Doty, 1977; Russell, 1976), it would be hard to dispute the fact that humans are olfactorily capable of discriminating sex and individuality. Interestingly, the male odors were usually described as "musky" and the female odors as "sweet" (Russell, 1976).

Following reports (Curtiss, Ballantine, Keverne, Bonsall, & Michael, 1971; Michael, Zumpe, Keverne, & Bonsall, 1972) that vaginal secretions of female rhesus monkeys contain olfactory signals that induce sexual activity in rhesus males, investigations were also conducted on vaginal odors in humans throughout the menstrual cycle (Doty, Ford, Preti, & Huggins, 1975) concerning intensity and "pleasantness" of such odors. Because of marked day-to-day and cycle-to-cycle variations, no discernible trends were observed.

While most investigators in the field of mammalian communication may consider the existence of primer effects in humans to be too far-fetched, at least two reports are particularly noteworthy: menstrual synchrony and suppression in normally cycling human females, with and without male contact (McClintock, 1971), and the effect of men's presence on the occurrence of ovulation in women (Veith, Buck, Getzlaf, van Dalfsen, & Slade, 1983).

Olfactory communication in humans remains a controversial subject at present, and only a few would advocate its value in modern society; at least the evolutionary aspects of this question are worth further exploration. Any research in this area will continue to be hindered by both ethical and technical difficulties. In addition, from chemical and physiological points of view, the biochemical variability of humans may present enormous difficulties of its own.

References

Albone, E. S. (1975). Dihydroactinidiolide in the supracaudal scent gland secretions of the red fox. *Nature, 256,* 575.

Albone, E.S., & Flood, P.F. (1976). The supracaudal scent gland of the red fox, *Vulpes vulpes. Journal of Chemical Ecology, 2,* 167–175.

Albone, E. S., & Perry, G. C. (1976). Anal sac secretion of the red fox, *Vulpes vulpes.* Volatile fatty acids and diamines. Implications for a fermentative hypothesis of chemical recognition. *Journal of Chemical Ecology, 2,* 101–111.

Albone, E. S., Gosden, P. E., & Ware, G. C. (1977). Bacteria as a source of chemical signals in mammals. In D. Müller-Schwarze & M. M. Mozell (Eds.), *Chemical signals in vertebrates* (pp. 35–43). New York: Plenum Press.

Albone, E. S., Robins, S. P., & Patel, D. (1976). 5-Aminovaleric acid, a major free amino acid component of the anal sac secretion of the red fox, *Vulpes vulpes. Comparative Biochemistry and Physiology. 55*B, 483-486.

Albone, E. S., Eglington, G., Walker, J.M. & Ware, G. C. (1974). The anal sac secretion of the red fox *Vulpes vulpes:* Its chemistry and microbiology. A comparison with the anal secretion of the lion, *Panthera leo. Life Science, 14,* 387–400.

Beach, F. A., & Leboeuf, B. L. (1967). Coital behavior in dogs. I. Preferential mating in the bitch. *Animal Behaviour, 15,* 546–558.

Beach, F. A., & Merari, A. (1968). Coital behavior in dogs. IV. Effects of progesterone in the bitch. *Proceedings of the National Academy of Sciences of the United States of America, 61,* 442–446.

Beach, F.A., & Merari, A. (1970). Coital behavior in dogs. V. Effects of estrogen and progesterone on mating and other forms of social behavior in the bitch. *Journal of Comparative and Physiological Psychology, 70,* 1–22.

Beach, F. A., Rogers, C. M., & Leboeuf, B. L. (1968). Coital behavior in dogs. II. Effects of estrogen on mounting by females. *Journal of Comparative and Physiological Psychology, 66,* 296–307.

Beauchamp, G. K., Doty, R. L., Moulton, D. G., & Mugford, R. A. (1976). The pheromone concept in mammalian chemical communication: A critique. In R. Doty (Ed.), *Mammalian olfaction, reproductive process and behavior* (pp. 143–160). New York: Academic Press.

Berry, R. J., (1970). The natural history of the house mouse. *Field Studies, 3* (2), 219–262.

Bowers, J. M., & Alexander, B. K. (1967). Mice: Individual recognition by olfactory cues. *Science, 158,* 1208–1210.

Bronson, F. H. (1971), Rodent pheromones. *Biology of Reproduction, 4,* 344–357.

Bronson, F. H. (1979). Reproductive ecology of the house mouse. *Quarterly Review of Biology, 54,* 265–299.

Bronson, F. H. (1984). The adaptability of the house mouse. *Scientific American, 250*(3), 116–125.

Bronson, F. H., & Desjardins, C. (1974). Relationship between scent marking by male and the pheromone-induced secretion of the gonadotropic and ovarian hormones that accompany puberty in female mice. In W. Montagna & W. A. Sadler (Eds.), *Reproductive behavior* (pp. 157–178). New York: Plenum Press.

Bruce, H. M. & Parrott, D. M. V. (1960). Role of olfactory sense in pregnancy block by strange male. *Science, 131,* 1526.

Carr, W. J., Martorano, R. D., & Krames, L. (1970). Responses of mice to odors associated with stress. *Journal of Comparative Physiological Psychology, 71,* 223–228.

Carr, W. J., Solberg, B., & Pfaffman, C. (1962). The olfactory threshold for estrous female urine in normal and castrated male rats. *Journal of Comparative Physiological Psychology, 55,* 415–417.

Claus, R., & Alsing, W. (1976). Occurrence of 5-androst-16-en-3-one, a boar pheromone, in man and its relationship to testosterone. *Journal of Endocrinology, 68,* 483–484.

Curtis, R. F., Ballantine, J. A., Keverne, E. B., Bonsall, R. W., & Michael, R. P. (1971). Identification of primate sexual pheromones and the properties of synthetic attractants. *Nature, 232,* 396–398.

Davies, V. J., & Bellamy, D. (1972). The olfactory response of mice to urine and effects of gonadectomy. *Journal of Endocrinology, 55,* 11–20.

Dizinno, G., Whitney, G., & Nyby, J. (1978). Ultrasonic vocalizations by male mice *(Mus musculus)* to female sex pheromone: Experiential determinants. *Behavioral Biology, 22,* 104–113.

Dluzen, D. E., Ramirez, V. D. Carter, C. S., & Getz, L. L. (1981). Male vole urine changes luteinizing hormone-releasing hormone and norepinephrine in female olfactory bulb. *Science, 212,* 573–575.

Doty, R. L. (1977). A review of recent psychophysical studies examining the possibility of chemical communication of sex and reproductive state in humans. In D. Müller-Schwarze & M. M. Mozell (Eds.), *Chemical signals in vertebrates* (pp. 273–286). New York: Plenum Press.

Doty, R. L., & Dunbar, I. (1974). Attraction of beagles to conspecific urine, vaginal and anal sac secretion odors. *Physiology and Behavior, 12,* 825–833.

Doty, R. L., Ford, M., Preti, G., & Huggins, G. R. (1975). Changes in the intensity and pleasantness of human vaginal odors during the menstrual cycle. *Science, 190,* 1316–1318.

Drickamer, L. C. (1974). Sexual maturation of female house mice: Social inhibition. *Development Psychobiology, 7,* 257–265.

Drickamer, L. C., McIntosh, T. K., & Rose, E. A. (1978). Effects of ovariectomy on the presence of a maturation-delaying pheromone in the urine of female mice. *Hormones and Behavior, 11,* 131–137.

Fox, M. W. (1971). *Behavior of wolves, dogs, and related canids.* New York: Harper & Row.

Fox, M. W., & Cohen, J. A. (1977). In T. A. Sebeok (Ed.), *How do animals communicate?* (pp. 728–736). Bloomington and London: Indiana University Press.

Goodwin, M., Gooding, K. M., & Regnier, F. (1979). Sex pheromone in the dog. *Science, 203,* 559–561.

Harrington, J. E. (1976). Recognition of the territorial boundaries by olfactory cues in mice. *Zeitschrift fur Tierpsychologie (Journal of Comparative Ethology), 41,* 295–306.

Hart, G. H., Mead, S. W., & Reagan, W. M. (1946). Stimulating the sex drive of bovine male in artificial insemination. *Endocrinology, 39,* 221–223.

Henry, J. D. (1977). The use of urine marking in the scavenging behavior of the red fox, *Vulpes vulpes. Behaviour, 61,* 82–106.

Hoppe, P. C. (1975). Genetic and endocrine studies of the pregnancy-blocking pheromone of mice. *Journal of Reproduction and Fertility, 45,* 109–115.

Jemiolo, B., Alberts, J., Sochinski-Wiggins, S., Harvey, S., & Novotny, M. (1985). Behavioral and endocrine responses of female mice to synthetic analogues of volatile compounds in male urine. *Animal Behavior, 33,* 1114–1118.

Joffre, M. (1976). Puberté et cycle génital saisonnier du renard mâle *(Vulpes vulpes)* [Puberty and seasonal sexual cycle of the wild male red fox *(Vulpes vulpes).*] *Annales de Biologie animale Biochimie Biophysique, 16,* 503–520.

Jorgenson, J. W., Novotny, M., Carmack, M., Copland, G. B., Wilson, S. R., Whitten, W. K., & Katona, S. (1978). Chemical scent constituents in the urine of the red fox *(Vulpes vulpes)* during the winter season. *Science, 199,* 796–798.

Kaneko, N., Debski, E. A., Wilson, M. C., & Whitten, W. K. (1980). Puberty acceleration in mice. II. Evidence that the vomeronasal organ is a receptor for the primer pheromone in male mouse urine. *Biology of Reproduction, 22,* 873–878.

Karlson, P., & Butenandt, A. (1959), Pheromones (ecto hormones) in insects. *Annual Review of Entomology, 4,* 39–58.

Kirk-Smith, M. D., & Booth, D. A.(1980). *7th International Symposium on Olfaction and Taste,* p. 397.

Kloek, J. (1961). The smell of some steroid sex-hormones and their metabolites. *Psychiatria Neurologia Neurochircergia, 64,* 309–344.

Marchlewska-Koj, A. (1980). Partial isolation of pregnancy block pheromone in mice. In D. Müller-Schwarze & R. M. Silverstein (Eds.), *Chemical signals: vertebrates and aquatic invertebrates* (pp. 413–414). New York: Plenum Press.

Maruniak, J. A., & Bronson, F. H. (1976). Gonadotropic responses of male mice to female urine. *Endocrinology, 99,* 963–969.

Maruniak, J. A., Owen, K., Bronson, F. H., & Desjardins, C. (1975). Urinary marking in female house mice: Effects of ovarian steroids, sex experience, and type of stimulus. *Behavioral Biology, 13,* 211–217.

McClintock, M. K. (1971). Menstrual synchrony and suppression. *Nature, 229,* 244–245.

McIntosh, T. K., & Drickamer, L. C. (1977). Excreted urine, bladder urine and the delay of sexual maturation in female house mice. *Animal Behaviour, 25,* 999–1004.

McKenna Kruse, S., & Howard, W. E. (1983). Canid sex attractant studies. *Journal of Chemical Ecology, 9,* 1503–1510.

Michael, R. P., & Keverne, E. B. (1970). Primate sex pheromones of vaginal origin. *Nature, 775,* 84–85.

Michael, R. P. Zumpe, D., Keverne, E. B., & Bonsall, R. W. (1972). Neuroendocrine factors in the control of primate behavior. *Recent Progress in Hormone Research, 28,* 665–704.

Mugford, R. A., & Nowell, N. W. (1970a). Pheromones and their effect on aggression in mice. *Nature, 226,* 967–968.

Mugford, R. A., & Nowell, N. W. (1970b). The aggression of male mice against androgenized females. *Psychonomic Science, 20,* 191–192.

Novotny, M., Harvey, S., Jemiolo, B., & Alberts, J. (1985). Synthetic pheromones that promote inter-male aggression in mice. *Proceedings of the National Academy of Sciences of the United States of America, 82,* 2059–2061.

Novotny, M., Jemiolo, B., Harvey, S., Wiesler, D., & Marchlewska-Koj, A. (1986) Adrenal-mediated endogenous metabolites inhibit puberty in female mice. *Science, 231,* 722–725.

Novotny, M., Jorgenson, J. W., Carmack, M., Wilson, S. R., Boyse, E. A., Yamazaki, K., Wilson, M., Beamer, W., & Whitten, W. K. (1980). Chemical studies of the primer mouse pheromones. In D. Müller-Schwarze & R. M. Silverstein (Eds.), *Chemical signals: Vertebrates and aquatic invertebrates* (pp. 377–390). New York: Plenum Press.

Novotny, M., Lee, M. L., & Bartle, K. D. (1974). Some analytical aspects of the chromatographic headspace concentration method using a porous polymer. *Chromatographia, 7,* 333–338.

Novotny, M., McConnell, M. L., Lee, M. L., & Farlow, R. (1974). High resolution gas chromatographic analyses of the volatile constituents of body fluids, with use of glass capillary columns. *Clinical Chemistry, 20,* 1105–1110.

Novotny, M., Schwende, F. J., Wiesler, D., Jorgenson, J. W., & Carmack, M. (1984). Identification of a testosterone-dependent unique volatile constituent of male mouse urine: 7-*exo*-ethyl-5-methyl-6,8-dioxabicyclo[3.2.1]-3-octane. *Experientia, 40,* 217–219.

Peters, R. P., & Mech, L. D. (1975). Scent-marking in wolves. *American Scientist, 63,* 628–637.

Pierrepoint, C. G., Galley, J. McI., Griffiths, K., & Grant, J. K. (1967). Steroid metabolism of a Sertoli cell tumour of testis of a dog with feminization and alopecia of normal canine testis. *Journal of Endocrinology, 38,* 61–69.

Preti, G., Muetterties, E., J., Furman, J. M., Kenelly, J. J., & Johns B. E. (1976). Volatile constituents of dog *(Canis familiaris)* and coyote *(Canis latrans)* anal sacs. *Journal of Chemical Ecology, 2,* 177–186.

Raymer, J., Wiesler, D., Novotny, M., Asa, C., Seal, U. S., & Mech, L. D. (1984). Volatile constituents of wolf *(Canis lupus)* urine as related to gender and season. *Experientia, 40,* 707–709.

Raymer, J., Wiesler, D., Novotny, M., Asa, C., Seal, U. S., & Mech, L. D. (1985). Chemical investigations of wolf *(Canis lupus)* anal-sac secretion in relation to breeding season. *Journal of Chemical Ecology, 11,* 593–608.

Raymer, J., Wiesler, D., Novotny, M., Asa, C., Seal, U. S., & Mech, L. D. (1986). Chemical scent constituents in urine of wolf *(Canis lupus)* and their dependence on reproductive hormones. *Journal of Chemical Ecology, 12,* 297–314.

Rothman, R., J., & Mech, L. D. (1979). Scent marking in lone wolves and newly formed pairs. *Animal Behaviour, 27,* 750–760.

Russell, M. J. (1976). Human olfactory communication. *Nature, 260,* 520–522.

Schwende, F. J., Jorgenson, J. W., & Novotny, M. (1984). Possible chemical basis for histocompatibility-related mating preference in mice. *Journal of Chemical Ecology, 10,* 1603–1615.

Schwende, F. J., Wiesler, D., Jorgenson, J. W., Carmack, M., & Novotny, M. (1986). Urinary volatile constituents of the house mouse, *Mus musculus,* and their endocrine dependency. *Journal of Chemical Ecology, 12,* 277–296.

Schwende, F. J., Wiesler, D., & Novotny, M. (1984). Volatile compounds associated with estrus in mouse urine: potential pheromones. *Experientia, 40,* 213–215.

Scott, J. W., & Pfaff, D. W. (1970). Behavioral and electrophysiological responses of female mice to male urine odors. *Physiology and Behavior, 5,* 407–411.

Sebeok, T. A., (Ed.). (1977). *How do animals communicate?* Bloomington and London: Indiana University Press.

Singer, A. G., Agosta, W. C., O'Connell, R. J., Pfaffmann, G., Bowen, D. V., & Field, F. H. (1976). Dimethyl disulphide; an attractant pheromone in hamster vaginal secretion. *Science, 191,* 948–950.

Singer, A. G., Macrides, F., & Agosta, W. C. (1980). Chemical studies of hamster reproductive pheromones. In D. Müller-Schwarze & R. M. Silverstein (Eds.), *Chemical signals: Vertebrates and aquatic invertebrates* (pp. 365–375). New York: Plenum Press.

Spannhof, I. (1969). The histophysiology and function of the anal sac of the red fox, *Vulpes vulpes. Forma et functio, 1,* 26–45.

Thomas, L. (1975). Symbiosis as an immunologic problem. In E. Neter & F. Milgrom (Eds.), *The immune system and infectious disease* (pp. 2–11). 4th International Convocation on Immunology, Karger.

Vandenbergh, J. G. (1967). Effect of the presence of a male on the social maturation of female mice. *Endocrinology, 81,* 345–349.

Vandenbergh, J. G., Drickamer, L. C., & Colby, D. R. (1972). Social and dietary factors in the sexual maturation of female mice. *Journal of Reproduction and Fertility, 28,* 397–405.

Vandenbergh, J. G., Whitsett, J. M., & Lombardi, J. R. (1975). Partial isolation of pheromone accelerating puberty in female mice. *Journal of Reproduction and Fertility, 43,* 515–523.

van der Lee, S., & Boot, L. M. (1955). Spontaneous pseudopregnancy in mice. I. *Acta Physiologica et Pharmacologica Neerlandica, 4,* 442–443.

Veith, J. L., Buck, M., Getzlaf, S., van Dalfsen, P., & Slade, S. (1983). Exposure to men influences the occurrence of ovulation in women. *Physiology and Behavior, 31,* 313–315.

Whitten, W. K. (1956). Modification of the oestrus cycle of the mouse by exter-

nal stimuli associated with the male. *Journal of Endocrinology, 13,* 399–404.
Whitten, W. K. (1959). Occurrence of anoestrus in mice caged in group. *Journal of Endocrinology, 18,* 102–107.
Whitten, W. K. (1977). *personal communication.* Bar Harbor, Maine: Jackson Laboratory.
Whitten, W. K., Bronson, F. H., & Greenstein, J. A. (1968). Estrus-inducing pheromone of male mice: Transport by movement of air. *Science, 161,* 584–585.
Whitten, W. K., Wilson, M. C, Jorgenson, J. W., Novotny, M., & Carmack, M. (1980). Induction of marking behavior in wild red foxes *(Vulpes vulpes)* by synthetic urinary constituents. *Journal of Chemical Ecology, 6,* 49–51.
Wiesler, D, Schwende, F. J., Carmack, M., & Novotny, M. (1984). Structural determination and synthesis of a chemical signal of the male state and a potential multipurpose pheromone of the mouse *Mus musculus. Journal of Organic Chemistry, 49,* 882–884.
Wilson E. O., & Bossert, W. A. (1963). Chemical communication among animals. *Recent Progress in Hormone Research, 19,* 673–716.
Wysocki, C. J., Wellington, J. L., Beauchamp, G. K. (1980). Access of urinary nonvolatiles to the mammalian vomeronasal organ. *Science, 207,* 781–783.
Yamazaki, K., Boyse, E. A., Mike, V., Thaler, H. T., Mathieson, B. J., Abbot, J., Boyse, J., Zayas, Z. A., & Thomas, L. (1976). Control of mating preferences in mice by genes in the major histocompatibility. *Journal of Experimental Medicine, 144,* 1324–1335.
Yamazaki, K., Yamaguchi, M., Baranoski, L., Bard, J., Boyse, E. A., & Thomas, L. (1979). Recognition among mice. Evidence from the use of a Y-maze differentially scented by congenic mice of different major histocompatibility types. *Journal of Experimental Medicine, 150,* 755–760.

8

Behavioral Dimorphisms in the Sexual Initiative of Great Apes

Ronald D. Nadler

This paper describes the sexually dimorphic behavior that adult male and female great apes exhibit during their sexual interactions, discusses species differences in terms of ultimate causation, and suggests how proximate factors, especially environmental ones, may influence the frequency, timing, and quality of such behavior. The approach thus includes analyses typical of comparative psychology within a biological framework.

Investigators of animal sexual behavior now recognize that behavioral dimorphisms are not mutually exclusive (Beach, 1978). Sexually dimorphic behavior for present purposes is defined, therefore, as "any observable response that it displayed more frequently, more readily, or more intensely by one sex than by the other" (Goy & Goldfoot, 1973, p. 169). The aspects of sexually dimorphic behavior considered are the masculine and feminine roles of great apes in the initiation of sexual interactions as derived from studies *in the natural habitat* (i.e., species-typical behavior). The environmental factors that are proposed to influence these behavioral dimorphisms are those inherent in the laboratory studies of such behavior, especially spatial, temporal, and social factors.

Identification of behavioral dimorphisms by observation and measurement in the natural environment does not meet the criteria previously proposed by Goy and Goldfoot (1973). These investigators maintained that such identification requires that the environmental conditions in which the behavior of males and females is assessed must be identical. The environmental conditions to which they referred related to characteristics of the stimulus animal with which males and females are tested and the physical aspects of the test. In order to test for sex differences, in mounting behavior, for example, the stimulus animals for both females and males should be the same and the tests should be conducted in the same physical surroundings. These criteria, however, go beyond "identification" of sexually dimorphic behavior. They are, in fact, the appropriate criteria for "analysis of the determinants of dimorphic behavior," as subsequently stated by Goy and Goldfoot (1973 p. 169), but they need not be included in the initial identification of such behavior and they are not required for all such analyses. For the set of variables these investigators considered (i.e., hormonal), their approach is entirely appropriate. For the present paper, however, the differences in environments are the focus of interest.

The environmental conditions that are analyzed herein are those found in the natural habitat and in two different laboratory test paradigms: (1) the free-access test or traditional pair test, and (2) the restricted-access test permitting female choice. Under natural conditions, masculine and feminine roles of the great apes in sexual initiation conform to predictions based on the concept of sexual selection (Darwin, 1871), applied originally to the sexual behavior of great apes by Short (1977) and formalized by Harcourt (1981). In general, the greater the number of males in competition for an estrous female, the more pronounced and elaborate is the masculine role in initiating sexual activity. In the free-access test, both the male and female are placed in a single test cage for the duration of the test. Under these conditions, the male is generally dominant over the female. In the restricted-access test, the male and female are placed in separate, adjoining cages, and the female is given the option of determining whether and when contact and copulation can occur. Admittedly, comparisons between the natural habitat and these laboratory tests lack the rigor that can be achieved in controlled studies conducted entirely within the laboratory. Use of relatively small numbers of apes in the laboratory, moreover, limits the confidence with which generalizations can be made regarding species characteristics. Furthermore, the analyses are post hoc because the elucidation of sexually dimorphic behavior per se was not the focus of the orginal studies from which the data were obtained.

Despite these caveats, we have reason to propose that identification of masculine and feminine behavior from observations in the natural environment is an appropriate and useful starting point for analysis. Such descriptions of behavior have validity in that they characterize the behavior animals actually exhibit under natural conditions. The same approach can be used for all species, thereby permitting cross-species comparisons. Analysis of naturalistic observations permits interpretation of biological contributions to behavior, thereby expanding our perspective with respect to ultimate causation. For behavioral scientists whose research is conducted in the laboratory, this approach encourages interdisciplinary exchange and the potential for new insights from investigators in a related area of research. The use of naturalistic observations for the initial identification of behavioral dimorphisms, moreover, in no way compromises the subsequent application of principles for studying the determinants of such behavior as described by Goy and Goldfoot (1973) and others (e.g., Beach, 1978). Over the past 5–10 years, field studies of the great apes have contributed sufficient data for adequate descriptions of species-typical patterns of sexual interaction, and laboratory data are available for assessing the influence of environmental factors on these patterns.

Sexual Initiative in the Natural Habitat

Chimpanzees

In their natural habitat, chimpanzees live in relatively large communities of individuals of mixed age and sex (van Lawick-Goodall, 1968; Nishida, 1968; Sugiyama, 1969). Such communities are generally divided into smaller groupings or bands, the composition of which changes routinely over time. In this multimale mating system, three distinct mating patterns have been described: opportunistic, possessive, and consort (Tutin, 1979; Tutin & McGinnis, 1981). Opportunistic mating comprised 73% of all copulations observed at the Gombe Stream Reserve. In this pattern females may copulate with any or all males in the community with little or no aggressive competition among the males. Possessive mating comprised 25% of all copulations observed and was described as those interactions in which the male showed persistent attention toward a particular female for periods longer than 1 hour. It is characterized by the maintenance of proximity to a female, by following, leading, gazing, or waiting for the female, and by attempts to interrupt copulation between the female and another male. Consort mating comprised only 2% of the observed copulations and is characterized by a male leading a female away from the main group and copulating with her exclusively. The pair maintains close proximity and avoids contact with other chimpanzees, the female clearly cooperating in this activity.

Despite the differences among the mating patterns, the form of sexual initiative is essentially the same in all three and reflects a predominance of male activity. The male typically approaches the female with an erect penis and performs a courtship display consisting of varied components, including branch shaking, hair erection, directing the erect penis toward the female, and gazing intently at the female. Because the male's courtship includes aggressive components, such as hair erection, Tutin and McGinnis (1981) proposed that the maintenance of some distance between the animals is important. This permits the female to regulate her approach toward the male in a situation with approach-avoidance characteristics. The female generally responds to the male by glancing in his direction, approaching, and presenting for copulation. Most copulations occur during the phase of maximal female gential swelling, which comprises approximately 10 days of the 36-day female cycle.

Among chimpanzees, therefore, numerous males are typically in competition for an estrous female, and the male plays the major role in sexual initiation by the performance of a conspicuous courtship display. The female responds to this display by approaching and presenting. Female involvment and cooperation are apparent, however, during consort matings, suggesting that partner perferences are determined primarily by females.

Gorillas

The wild mountain gorilla studied originally by Schaller (1963) and subsequently by scientists of the Karisoke Research Center (Harcourt, 1979; Harcourt, Stewart, & Fossey, 1981; Fossey, 1982) lives in a relatively stable harem characterized as a one-male system. An adult silverbacked male is the leader of the group, but other adult silverbacks or blackbacked males may also be members. Despite these other males, only the dominant leader copulates with adult parous females (see Figure 8-1). The leader's dominance is attributed, in part, to the fact that he is generally the sire of other males in the group. There is thus little or no intermale competition at estrus among gorillas. Adult cycling females typically avoid proximity with the silverback except during a few to several days of estrus. During such times the females establish and maintain proximity to the male and periodically throughout the day approach closely and stare intently at the male. Although the female may sometimes present to the male independently, she typically appears to wait for some signal from the male. A signal described by Harcourt et al. (1981) consisted of the male spreading or raising his arms in preparation for grasping the female's flanks prior to copulation. Other possible signals observed were components of the male's chest-beat display, which frequently preceded the female's final approach and presentation (personal observations).

In this species, therefore, intermale competition at estrous is minimal, and females play the major role in sexual initiation. Females determine the time (in the cycle) when copulation may occur by (1) establishing proximity to the male only during a relatively few number of

Figure 8–1. Dorso-ventral copulation by wild mountain gorillas. Note the presence of a 4-year-old male offspring in proximity to the copulating pair.

days and (2) approaching and presenting for copulation. The male's role is a relatively passive one of waiting for the female to approach, but it frequently includes some signal to the female immediately before the female's final approach to present.

Orangutans

The orangutan possesses an unusual social structure for a primate with adults living a semisolitary existence (Mackinnon, 1971; Rodman, 1973; Horr, 1975). Fully adult males generally live alone most of their lives in home ranges that overlap to some degree and that subsume the smaller home ranges of a few to several adult females. Adult females eschew the presence of other adults, much as the males do, but are accompanied by one or two dependent offspring. Much has been written about the forcible copulations carried out by male orangutans. Although this form of mating was initially described as an optimal strategy for a spatially dispersed species, that no longer seems to be the case. It is now apparent that forcible copulations are carried out primarily by subadult males without home ranges of their own, and evidence suggests that these matings are not fruitful (Rijksen, 1978; Galdikas, 1981; Schürmann, 1981). The more recent data indicate that copulations among adults come about as a result primarily of female initiative. When in estrus, the female seeks out and consorts with the dominant male (or males) in the vicinity and solicits copulation in varied ways. The female in some cases actually mounts the male, which may remain inactive and present his erect penis for such mounting. The male can play a more active role as well by positioning the female for copulation, with her cooperation. Which of these forms of sexual initiative predominates is not clear. It is also not certain whether the orangutan's mating system is one-male or multimale, but if the latter, fewer males are in competition or available at the time of estrus than in the case of chimpanzees. Female sexual initiative is prominent in this species, but male orangutans probably express more initiative than male gorillas under natural conditions.

Biological Contributions to Sexual Initiative

As stated earlier, the different patterns of sexual initiative among the apes have been attributed to the number of males in competition for a female at the time of estrus (Harcourt, 1981).

Chimpanzees

A multimale mating system, such as that characteristic of the chimpanzee, requires that males compete for estrus females to maximize their reproductive success. The lack of aggressive competition among male chimpanzees *within a community* is attributed to their kinship. Because male chimpanzees do not transfer between communities, those within a given community are related to some degree. Competition for females among adult male chimpanzees at the time of estrus is expressed nonaggressively, therefore, in terms of the courtship display, which, among the apes, is most elaborate in this species. Consistent with this lack of aggressive competition between individual males is the limited sexual dimorphism in this species, in comparison to the gorilla and orangutan. Aggressive intermale competition is reported, but it occurs in relation to territorial boundaries *between communities* and involves a *group of males* from one community attacking a smaller group or an individual from another community.

Although a male display may be necessary for reproductive success in chimpanzees, it may not be sufficient, as suggested by the female's role in choosing a particular male as a consort. The female's choice of a consort is related to several aspects of the male's social competence, instead of the quality of his display per se. The importance of the choice is reflected in the disproportionately high number of the conceptions that occur as a result of consortships, in comparison to opportunistic and possessive matings.

Gorillas

The lack of conspicuous sexual initiative in male gorillas is attributed to their one-male mating system. Because male gorillas, like male chimpanzees, do not transfer between groups, males other than the silver-backed leader of the group are thought to be the leader's offspring. As a result presumably of these kinship ties, no aggressive interactions are found among male gorillas *within a group* (harem), and the leader is the only male observed mating with parous females. No competition with the leader is seen from other males at the time a female is in estrus, so no selective pressure has been placed on male gorillas to evolve conspicuous sexual initiative. Female gorillas apparently choose the silverback with which they will mate at the time they join a harem and when in estrus overtly solicit copulation only from that silverback. The pronounced sexual dimorphism of gorillas is attributed to aggressive intermale competition *between groups* for permanent possession of females, or more commonly between leaders of groups and lone males without harems.

Orangutans

The roles of males and females in sexual initiation are least certain for orangutans, among the great apes. Clearly, however, females play a more important role than previously recognized, because they seek out fully adult, dominant males when they are in estrus. The home ranges of these males overlap, so possibly (not documented), female orangutans may have more than a single male available to them at such times. If this is the case, their mating system would be intermediate to those of the chimpanzee and gorilla. Consistent with this interpretation is the finding that male orangutans exhibit courtship behavior consisting of a "penile display" that is less elaborate than that of the male chimpanzee, but more so than that of the male gorilla (Nadler, 1982). Adult male orangutans, moreover, are more active in initiating sexual activity than are male gorillas.

Among orangutans, therefore, the number of the males competing for estrous females is one, or at most two or three. The balance between male and female sexual initiative is closer to that of the gorilla than the chimpanzee and reflects a predominance of the feminine role. The interpretation is therefore similar to that given for the gorilla. In the absence of significant male competition *at the time of estrus,* there was little pressure on males to develop elaborate courtship behavior and sexual initiative. The pronounced sexual dimorphism of orangutans is attributed to aggressive intermale competition proposed to take place in the establishment and maintenance of male home ranges.

Sexual Initiative in the Traditional Free-Access Pair Test

Conceptually, the behavioral dimorphisms under consideration can be described in terms of three dimensions: qualitative, temporal, and quantitative. Qualitatively, they consist of specific motor patterns characteristic of each sex during sexual interactions, especially those patterns that precede and serve to initiate copulation. The temporal dimension relates to the phase of the female cycle during which the motor patterns are performed—in general terms, the periovulatory phase. Quantitatively, the behavioral dimorphisms are defined by the frequency with which the motor patterns are exhibited on the days of sexual activity.

Chimpanzees

Three studies of chimpanzee sexual behavior have used the free-access paradigm (Yerkes & Elder, 1936; Yerkes, 1939; Young & Orbison, 1944), but the two conducted by Yerkes contain the most complete description and interpretation of sexual initiative. The description of masculine and feminine roles was, in fact, quite similar to those reported for wild chimpanzees many years later. Although these authors reported that females instead of males generally initiated sexual interactions, their conclusion seems to represent a semantic rather than an actual difference from the field data. In the tests, the male and female were in different cages at the beginning of the test and generally separated by a fence through which the animals could see each other. A guillotine door between the cages was raised to initiate the test, and the sexual initiator was defined as the animal that first entered the cage of the other. Females entered the cage of the males more frequently than the reverse, and therefore were described as the primary initiators of sexual activity. Because the male courtship display (see Fig. 8-2) was described as a routine occurence in these tests and because the females had visual access to the males prior to the start of the test, one can reasonably propose that the females were responding to this initiative on the part of the males, much as they do in the wild. This then would not represent a difference from the species-typical behavior.

The foregoing description of chimpanzee sexual initiative in the laboratory suggests, therefore, that there was no clear qualitative difference from that displayed in the wild, apart from the absence of those activities that require physical aspects of the environment such as branches or trees. Yerkes and Elder (1936) reported, however, that males initiated copulation at times in the cycle when females were not maximally swollen, as well as during gestation and lactation. The interpretation presented for such mating was that "if and when copulation occurs, it is by reason of the dominance or impulsiveness of the male

Figure 8–2. Courtship display by an adult male chimpanzee. The light-skinned erect penis is conspicuous against the male's dark hair.

and the desire of the female to avoid risk of injury by obeying his command'' (Yerkes & Elder, 1936, p. 34). Copulation at such times is uncommon in the wild and reflects a temporal difference in the male's role in the laboratory tests. In the laboratory tests, male sexual initiative occurred at phases of the female cycle and during other physiological states that are inappropriate for conception. Whether or not differences occur in the quantitative dimension cannot be assessed in the chimpanzees because the tests were too short in duration.

Gorillas

Two laboratory studies on sexual behavior of gorillas have provided somewhat different results regarding sexual initiative. In the initial study with young adults, female sexual initiative was prominent and was displayed for a relatively brief 1–4 day period, as described for wild gorillas (Nadler, 1976). In the second study, conducted when the animals were several years older, a different picture emerged (Nadler, Collins, Miller, & Graham, 1983). Copulation occurred at times temporally dissociated from the periovulatory period, and male sexual initiative was more conspicuous. In fact, male sexual initiative accounted for two thirds of all the copulations recorded. In approximately half of these copulations, the males solicited directly, whereas in the other half the male performed some variant of the chest-beat display (see Fig. 8-3). Al-

Figure 8–3. Chest-beat display by an adult male lowland gorilla.

though initially considered in the context of aggression (Nadler & Miller, 1982), such variants of the chest-beat display were subsequently observed to precede copulation in the wild. As such, they appear to qualify as a component of male sexual initiative. The noteworthy difference in the laboratory tests was that the displays were performed repeatedly in the tests, irrespective of cycle phase. Although these displays were effective in inducing copulation primarily by only one of the two males tested, both males performed the displays at relatively high frequencies. In the gorilla, therefore, pair testing resulted in both a temporal and quantitive change in male sexual initiative, but no obvious qualitative change.

Orangutans

The sexual behavior of orangutans during free-access pair testing exhibited the most dramatic difference from the wild among the great apes (Nadler, 1977). Essentially all copulations were initiated forcibly by the males irrespective of cycle phase. (see Fig. 8-4). These sexual interactions resembled the forcible copulations of wild subadult males which, as described earlier, differ from those of fully mature individuals. Female sexual initiative or proceptivity, significant in the wild, was infrequent in these pair tests. Sexual initiative of orangutans in these laboratory tests therefore was altered on at least two dimensions. It

Figure 8–4. Forcible copulation by an adult male orangutan during a free-access pair test.

was qualitatively different in that the male's initiative was forcible instead of coordinated with the female's, and female proceptivity was all but lacking. There was also a temporal difference, in that males initiated copulation throughout the cycle, instead of in the 5–6 days reported for wild orangutans. Insufficient data have been reported regarding the daily frequency of copulation in orangutans to evaluate the possibility of quantitative changes.

Sexual Initiative in the Restricted-Access Test with Female Choice

Only preliminary data have been reported for the restricted-access tests on great apes, but these help to clarify some of the laboratory/field differences described (Nadler, 1981). The quantity of data available for the different species differs and is least for the chimpanzee. The overall picture is quite similar for the three, however, and reflects the increased prerogatives of the female. The females entered the male's cage during a more restricted portion of the cycle, essentially the periovulatory phase of the cycle, and sexual initiative was more similar to that observed in the wild. The clearest difference from the free-access tests was found in the orangutans, in which forcible copulation by the males was largely eliminated and female proceptivity readily apparent (see Figs. 8-5 and 8-6).

Figure 8–5. Adult male orangutan directing a "penile display" toward an adult female (not visible) during a restricted-access pair test.

The data on sexual behavior of the great apes provide support for the hypothesis proposed originally by Yerkes (1939), on the basis of his laboratory research with chimpanzees. This hypothesis asserts, in effect, that for species in which females regulate mating, copulation is restricted to the periovulatory phase of the cycle, whereas for those in which males are in control, copulation can be dissociated from that phase. Under natural conditions, males initiate mating among chimpanzees, and copulation begins 10 or more days before ovulation. Among gorillas, in which females initiate mating, copulation is restricted to a few days of the periovulatory period. Orangutans appear to be intermediate to the other apes under these conditions. The increased distribution of mating in the cycle found during the free-access laboratory pair tests in these species was also attributed to the male's influence. These results suggest that the masculine characteristics that contribute

Figure 8–6. Adult female orangutan "mounted" on an adult male during a restricted-access pair test.

to nonperiovulatory mating are present in each of these species to some degree, and that their expression depends on other factors that compromise a female's options for regulating copulation. This conclusion was supported by the results from the restricted-access tests in the sense that restoration of female prerogatives led to less frequent copulation more closely associated with the presumptive time of ovulation.

Environmental Influences on Sexual Initiative

From the research described in the preceding section, we see that differences between sexual initiative in the field and in the free-access test reflect primarily an increased influence of the male and a reduced influence of the female in the regulation of sexual activity under those test conditions. In this section, environmental factors are examined for their possible relationship to the differences in behavior.

In an analysis of this research, we should distinguish between the factors that *permit* the male to exert this type of influence over the female from those that might *stimulate* such behavior. Two factors that permit such male behavior are (1) male dominance over females and (2) confinement in a test cage. Adult male apes are generally dominant over females in a one-on-one situation. Confinement in a spatially restricted area, moreover, prevents the female from avoiding and/or escaping from the male, and also isolates the female from conspecifics that in the natural habitat might exert a moderating influence on the male. These factors alone would suffice to account for the results if sexual aggressiveness, as described, was intrinsic to the male great apes but was normally suppressed by social controls within the group. Such an assumption appears to underlie Yerkes' (1939) hypothesis, which attributes nonperiovulatory mating to "the male's influence" and also Maslow's (1936) construct of a dominance drive. A similar view of masculine sexuality derives from the biological perspective that males are selected for relatively frequent copulation in comparison to females (Bateman, 1948), consistent with the male's generally limited parental investment (Trivers, 1972). Greater insights into causal mechanisms are likely achieved, however, when such behavior is viewed as an interaction between internal and external variables (Tinbergen, 1968).

In his review of sex differences in aggression, Moyer (1974) defined sexual aggression (sex-related aggression, in his terms) as "aggressive behavior that is elicited by the same stimuli that elicit sexual behavior" (p. 357). Whether it is useful to conceptualize the behavior of the male apes in the free-access tests as aggression is not clear, but it appears superficially to conform to Moyer's definition. From the examples given, however, Moyer's referent was apparently species-typical behavior, in particular, the aggressive components of "normal" mating behavior exhibited by many species under natural conditions. This would account for his proposal that sexual aggression "should be considered a separate class of aggressive behavior in that the stimulus situation eliciting it is relatively specific and because it most probably differs in physiological basis from other types of aggression" (Moyer, 1974, pp 356–357). This constraint need not apply to the present analysis, therefore, since the sexual aggression of the male apes in the free-access tests is an alteration in species-typical mating. As such, this issue with respect to the male apes can be examined from two perspectives: (1) sexual aggression as aggression per se and (2) sexual aggression as altered sexual behavior. In the first sense, environmental conditions can be analyzed for the same factors known to stimulate aggression generally (i.e., with some degree of cross-species generality), whereas in the second, analysis can proceed in terms of factors that have been shown to stimulate sexual interaction.

Sexual Aggression as Aggression

At least three aspects of the pair-test paradigm are related to conditions previously shown to stimulate aggression in certain other species: (1)

social isolation of the male, (2) introduction of a stranger, and (3) crowding. In laboratory tests of sexual behavior, it is common practice to isolate the male socially in order to limit sexual activity to the time of testing. Social isolation prior to testing, however, is also a standard procedure used to induce aggression in some species (Scott, 1966). The introduction of a strange conspecific is another potent stimulus for aggression (Southwick, Siddiqi, Farooqui, & Pal, 1974). In the tests of sexual behavior, the female spends only a short period of time with the male each day. The female may be considered a relative stranger to the male, therefore, and this mechanism may possibly stimulate aggressive behavior. Crowding can also stimulate aggression through increased interaction (Calhoun, 1962), especially in the presence of strangers. Thus, confinement of the apes in a single cage may constitute a condition of crowding for these species. Determining whether any or all of these factors actually stimulate sexual aggression in the apes requires additional research in which the factors are investigated independently.

Sexual Aggression as (Altered) Sexual Behavior

If sexual aggression is an extension of normal sexual behavior, then analysis can be subdivided further into those factors that may stimulate sexual arousal generally and those factors that relate to particular characteristics of each species' mating patterns.

General Stimulatory Influences

Three factors that may have general stimulating effects on sexual behavior are closely related to those that may stimulate aggression: (1) restriction of sexual activity to a relatively brief period of time, (2) novelty, and (3) close proximity. Limiting the male's opportunities for sexual interaction to a relatively short test may reduce the importance of discriminatory cues related to the sexual status of the female. Such an effect would be influenced by prior experience and learning in the test situation. The degree of novelty of one animal to another "can be of great importance to the sexual behavior of nonhuman animals" (Dewsbury, 1981, p. 464). The relationship between novelty and sexual behavior is comparable to the relationship between strangeness and aggressive behavior, considered earlier. Dewsbury points out, however, that the response to a novel animal varies with other factors, such as the species involved and the specific parameters of testing. No adequate data are available to evaluate whether periodic presentation of a female to a male, such as that used in testing the apes, constitutes novelty in a meaningful sense, or, if it is novelty, whether it accounts for increased male sexual initiative. Proximity of the test pair could stimulate male sexual initiative through a distortion of communicatory cues, both behavioral and nonbehavioral (Wallen, 1982). If a female normally communicates her readiness to copulate by establishing proximity with the male, the enforced proximity of the test cage could be interpreted by the male as readiness of the female to copulate. Nonbehavioral cues transmitted by the female could be more apparent in close proximity, moreover, than would be the case under more natural conditions. The latter suggestion could account for increased male initiative during the early follicular phase, but not during the luteal phase. As is the case for the possibilities suggested earlier for interpreting sexual aggression as aggressive behavior, these possibilities regarding sexual initiative have not been studied sufficiently to permit conclusions in the present context.

Specific stimulatory influences

Although the foregoing examples reflect factors with some broad species relevance that might stimulate sexual aggression, factors associated more directly with individual species-typical behavior are probably better sources of data for such an analysis. The approach taken in this section is to identify the components of behavior that are altered in the free-access tests, determine the context in which these components occur in the natural habitat, and then attempt to account for the alterations in terms of interactions between species-typical behavior and the test conditions.

Chimpanzees

Among the chimpanzees, a temporal difference was found between behavior recorded in the free-access laboratory tests and that in the field. Male sexual initiative in the tests occurred at times in the cycle during which copulation is uncommon in the wild, and the females responded to this initiative by presenting. One interpretation that accounts for such male behavior is that the introduction of the female to the male at the beginning of the test constitutes a reunion. Under natural conditions, reunion of chimpanzees after a period of separation is accompanied by an elaborate repertoire of greeting behavior that includes presenting and mounting (van Lawick-Goodall, 1968). Males, in addition, may perform ritualized patterns of aggressive behavior that include components described earlier in the laboratory studies as courtship (Yerkes & Elder, 1936). The increased male sexual initiative in the free-access pair tests, therefore, may be stimulated by an experimental manipulation (introduction of a female) that simulates a natural condition (reunion) in which such behavior is typical of the species. As noted, Tutin and McGinnis (1981) proposed that aggressive components of the male's courtship display create an approach-avoidance conflict for the female. Gradual resolution of such conflict is possible in the wild because spatial conditions permit the female to regulate her approach in a relatively unconstrained manner. The interpretation presented here for the success of male sexual initiative in the pair test is related to the enforced condition of proximity. The impact of the aggressive components of the male's display is exaggerated in close proximity and results in the female presenting as a subordinate gesture. Females can generally avoid the solicitations of males in the wild. Lack of female responsiveness accounted for 86% of 209 instances of unsuccessful male sexual initiative. Two instances in which avoidance was not possible were recorded, however, when the females involved were trapped high in trees. On both of these occasions, the copulations were described as "forced" (Tutin & McGinnis, 1981, p. 261). Apparently, therefore, male chimpanzees can mate with nonreceptive females even under natural conditions, when the female's options for refusal are limited.

Gorillas

Both quantitative and temporal differences between the wild and the free-access tests were found in male sexual initiative of gorillas. The males repeatedly performed more or less complete variants of the chest-beat display during the tests, a display performed much less frequently during sexual contacts in the wild that appears to be a component of male sexual initiative. The display was also more widely distributed in the cycle during these tests than in the wild and was followed by copulation dissociated from its temporal relationship to the presumptive time of ovulation. The interpretation proffered for this behavior is that the chest-beat display and variants thereof have multiple functions. They can serve as a cue for female presenting, as described, and they can also serve as a spacing mechanism. Gorillas in the wild spend most of their waking hours feeding in a spatially dispersed pattern. They come into proximity a few times a day during shorter periods of rest. The antecedent to dispersal and resumption of feeding that terminates the rest period is frequently the male chest-beat display, performed by the silverback. The significance of the display, therefore, may depend in part on the context in which it is performed.

Under natural conditions, cycling female gorillas are only in proximity to the silverbacked male with which they copulate during a few days of estrus (Harcourt, 1979). At other times, such females essentially avoid the male. Despite their limited association with the male, cycling females are charged and hit more often than females with dependent offspring that closely associate with the male on a regular basis. Therefore, the male's display is probably a natural response to the proximity of a cycling female. It is performed throughout the cycle in the free-access tests because the female is forced into proximity with the male on a daily basis. It is performed at high frequencies because the female is prevented from leaving the male. The presenting by the female under these conditions is viewed as a response to the sexual significance of the display (i.e., its function as a signal for such behavior).

Orangutans

Both qualitative and temporal differences between the free-access tests and the wild were found for orangutans. In the tests, males reverted to a pattern of forcible copulation, characteristic of an earlier stage of development in the wild, and initiated such copulation on a daily basis. Female proceptivity, which restricts copulation to 5–6 days per cycle in the natural habitat, was all but eliminated in these tests. The conditions under which forcible copulation occurs in the wild are when a (subadult) male meets a female for the first time or when it meets a familiar female following some *period of separation* (Rijksen, 1978). In the free-access tests, the female is introduced into the male's cage every day after a period of separation. The introduction of the female per se may thus constitute adequate stimulus for the male's aggressive sexual initiative in the tests. It may be relevant, in addition, that female orangutans in the wild are only in proximity to adult males at the time of estrus. The presence of a female in close proximity, therefore, may also stimulate male sexual initiative in this species. Adult males in the wild may learn to suppress their aggressive form of initiating copulation as a consequence of their increased size and reduced mobility. Because they are more mobile than fully adult males, adult females are probably able to thwart the aggressive behavior of such males in the wild. Data from the restricted-access tests in the laboratory support this conclusion. Over the course of testing with any given male, there was a gradual change in behavior. Initially the males tried to squeeze themselves through the restricted doorway, but within a short period of time they gave up. When the females first entered the males' cage voluntarily, the males tried, sometimes successfully, to mate with them as they had earlier in the free-access tests. As the females became more effective in escaping such attempts by retreating to their own cages, the males began blocking the doorway, once the female entered, to prevent escape. Ultimately, however, the females learned to avoid and escape the males, and the male's behavior changed dramatically to restrained waiting. In the laboratory, therefore, it was possible to promote species-typical behavior in adult males that had repeatedly displayed the aggressive form of sexual initiative, by providing females with options that are comparable to those they have in the wild.

Conclusions

The data reviewed emphasize that the behavioral dimorphisms in sexual initiative of the great apes are not immutable, but are, instead, influenced by environmental factors (i.e., the context in which the behavior takes place). The males of these species appear to be primarily responsible for the differences observed between the wild and the free-access tests. Several perspectives from which the data can be considered are presented. These include conceptualizing the increased male sexual initiative as aggression and as altered sexual behavior. When conceptualized as sexual behavior, the data can be subdivided further and examined in terms of possible stimulation of a general type (with some cross-species generality) and specific stimulation (i.e., stimulation specifically related to each species-typical pattern of sexual interaction). In the latter case, the resulting behavior is viewed as arising from an interaction between proximate (environmental) and ultimate causes. Although specific conclusions regarding the source(s) of stimulatory influences cannot be ascribed, the differences in behavior appear to be a function, principally, of male dominance over female and confinement of the pair in a limited spatial condition. These results on sexual behavior of the apes, moreover, reflect a general characteristic of many nonhuman primates, in that copulation in the free-access pair tests occurs more widely in the female cycle that it does under wild or semi-free-ranging conditions (Nadler, Herndon, & Wallis, 1986). Recognition of the pervasiveness of this phenomenon is likely to encourage further investigations into causal mechanisms. The present analysis, therefore, may have some heuristic value in making explicit several different ways in which the data can be approached. It may be gratuitous, in the absence of supporting data, to propose that the effects of (1) male dominance over females and (2) the compromise of female options regarding sexual initiation may have counterparts in human sexual relations. On the other hand, it may be relevant for investigators of human sexual behavior to be aware of such possibilities when designing and analyzing their research.

Acknowledgments

Preparation of this chapter was supported by U.S. Public Health Service Grant RR-00165 (Division of Research Resources, National Institutes of Health). Research by the author described herein was supported by NSF Grants No. GB-30757, BNS 75-06287, and BNS-7923015.

References

Bateman, A. J. (1948). Intra-sexual selection in *Drosophila*. *Heredity, 2,* 349–368.

Beach, F. A. (1978). Sociobiology and interspecific comparisons of behavior. In M. S. Gregory, A. Silvers, & D. Sutch (Eds.), *Sociobiology and human nature,* (pp. 116–135). San Francisco: Jossey-Bass.

Calhoun, J. B. (1962). A "behavioral sink." In E. L. Bliss (Ed.), *Roots of behavior* (pp. 295–315). New York: Harper and Brothers.

Darwin, C. (1871). *The descent of man, and selection in relation to sex.* London: John Murray.

Dewsbury, D. A. (1981). Effects of novelty on copulatory behavior: The Coolidge effect and related phenomena. *Psychol. Bull., 89,* 464–482.

Fossey, D. (1982). Reproduction among free-living mountain gorillas. *Am. J. Primatol. Suppl., 1,* 97–104.

Galdikas, B. M. F. (1981). Orangutan sexuality in the wild. In C. E. Graham (Ed.), *Reproductive biology of the great apes: Comparative and biomedical perspectives* (pp. 281–300). New York: Academic Press.

Goy, R. W., & Goldfoot, D. A. (1973). Hormonal influences on sexually dimorphic behavior. In R. O. Greep & E. B. Astwood (Eds.), *Handbook of physiology: Endocrinology* (Vol. 2, Part 1) (pp. 169–186). Baltimore: Williams & Wilkins.

Harcourt, A. H. (1979). Social relationships between adult male and female mountain gorilla in the wild. *Anim. Behav., 27,* 325–342.

Harcourt, A. H. (1981). Inter-male competition and the reproductive behavior of the great apes. In C. E. Graham (Ed.), *Reproductive biology of the great apes: Comparative and biomedical perspectives* (pp. 301–318). New York: Academic Press.

Harcourt, A. H., Stewart, K. J. & Fossey, D. (1981). Gorilla reproduction in the wild. In C. E. Graham (Ed.), *Reproductive biology of the great apes: Comparative and biomedical perspectives* (pp. 265–279). New York: Academic Press.

Horr, D. A. (1975). The Borneo orang-utan: Population structure and dynamics in relation to ecology and reproductive strategy. In L. A. Rosenblum (Ed.), *Primate behavior: Developments in field and laboratory research* (Vol. 4, No. 4, pp. 307–323). New York: Academic Press.

MacKinnon, J. R. (1971). The orang-utan in Sabah today. *Oryx, 11,* 141–191.

Maslow, A. H. (1936). The pole of dominance in the social and sexual behavior of infra-human primates: III. A theory of sexual behavior of infra-human primates. *J. Genet. Psychol., 48,* 310–338.

Moyer, K. E. (1974). Sex differences in aggression. In R. C. Friedman, R. M. Richart, & R. L. Vande Wiele (Eds.), *Sex differences in behavior* (pp. 335–372). New York: Wiley.

Nadler, R. D. (1976). Sexual behavior of captive lowland gorillas. *Arch. Sex. Behav., 5,* 487–502.

Nadler, R. D. (1977). Sexual behavior of captive orang-utans. *Arch. Sex. Behay., 6,* 457–475.

Nadler, R. D. 1981. Laboratory research on sexual behavior of the great apes. In C. E. Graham (Ed.), *Reproductive biology of the great apes: Comparative and biomedical perspectives* (pp. 191–238). New York: Academic Press.

Nadler, R. D. (1982). Laboratory research on sexual behavior and reproduction of gorillas and orang-utans. *Am. J. Primatol. Suppl., 1,* 57–66.

Nadler, R. D., Collins, D. C., Miller, L. C., & Graham, C. E., (1983). Menstrual cycle patterns of hormones and sexual behavior in gorillas. *Horm. Behav., 17,* 1-17.

Nadler, R. D., Herndon, J. G., & Wallis, J. (1986). Adult sexual behavior: Hormones and reproduction. In J. Erwin (Series Ed.), *Comparative primate biology* (G. Mitchell & J. Erwin, Eds., Vol. 2), *Behavior, conservation and ecology* (pp. 363–407). New York: Alan R. Liss.

Nadler, R. D., & Miller, L. C. (1982). Influence of male aggression on mating or gorillas in the laboratory. *Folia Primatol., 38,* 233–239.

Nishida, T. (1968). The social group of wild chimpanzees in the Mahali Mountains. *Primates, 9,* 167–224.

Rijksen, H. D. (1978). A field study on Sumatran orang-utans (Pongo pygmaeus abelii Lesson 1827). *Ecology, behavior and conservation*. Wageningen: H. Veenman and Zonen.

Rodman, P. S. (1973). Population composition and adaptive organization among orang-utans of the Kutai Reserve. In R. P. Michael & J. H. Crook (Eds.), *Comparative biology and behavior of primates* (pp. 181–209). London: Academic Press.

Schaller, G. B. (1963). *The mountain gorilla*. Chicago: University of Chicago Press.

Schürmann, C. L. (1981). Courtship and mating behavior of wild orang-utans in Sumatra. In A. B. Chiarelli & R. S. Corruccini (Eds.), *Primate behavior and sociobiology* (pp. 130–135). Berlin: Springer-Verlag.

Scott, J. P. (1966). Agonistic behavior of mice and rats: A review. *Am. Zool., 6,* 683–701.

Short, R. V. (1977). Sexual selection and the descent of man. In J. H. Calaby & C. H. Tyndale-Biscoe, Calaby (Eds.), *Reproduction and evolution* (pp. 3–19). Netley, South Australia: Griffin Press.

Southwick, C. H., Siddiqi, M. F., Farooqui, M. Y., & Pal, B.C. (1974). Xenophobia among free-ranging rhesus groups in India. In R. L. Halloway (Ed.), *Primate aggression, territoriality, and xenophobia* (pp. 185–209). New York: Academic Press.

Sugiyama, Y. (1969). Social behavior of chimpanzees in the Budongo Forest, Uganda. *Primates, 10,* 197–225.

Tinbergen, N. (1968). On war and peace in animals and man. *Science, 160,* 1411–1418.

Trivers, R. L. (1972). Parental investment and sexual selection. In B. Campbell (Ed.), *Sexual selection and the descent of man, 1871–1971* (pp. 136–179). Chicago: Aldine.

Tutin, C. E. G. (1979). Mating patterns and reproductive strategies in a community of wild chimpanzees *(Pan troglodytes schweinfurthii). Behav. Ecol. Sociobiol., 6,* 29–38.

Tutin, C. E. G., & McGinnis, R. P. (1981). Sexuality of the chimpanzee in the wild. In C. E. Graham (Ed.), *Reproductive biology of the great apes: Comparative and biomedical perspectives* (pp. 239–264). New York: Academic Press.

van Lawick-Goodall, J. (1968). The behaviour of free-living chimpanzees in the Gombe Stream Reserve. *Animal Behaviour Monographs, 1,* 161–311.

Wallen, K. (1982). Influence of female hormonal state on rhesus sexual behavior varies with space for social interaction. *Science, 217,* 375–377.

Yerkes, R. M. (1939). Sexual behavior in the chimpanzee. *Human Biology, 11,* 78–111.

Yerkes, R. M., & Elder, J. H. (1936). Oestrus, receptivity, and mating in the chimpanzee. *Comparative Psychology Monographs, 13,* 1–39.

Young, W. C., & Orbison, W. D. (1944). Changes in selected features of behavior in pairs of oppositely-sexed chimpanzees during the sexual cycle and after ovariectomy. *Journal of Comparative Psychology, 37,* 107–143.

9

The Study of Masculinity/Femininity from a Comparative Developmental Perspective

Leonard A. Rosenblum

The chapter by Nadler (Chapter 8) summarizes and illustrates a number of the most salient issues involved in the study of nominally male and female (i.e., masculine and feminine) behavior in nonhuman primates. Within the primate order, almost all elements of behavior that are generally or most frequently seen in one sex will, under certain circumstances, be seen in the other. Thus, it is commonly observed in various species that dominant females will mount and thrust against the presented perianal region of more subordinate partners, whether the latter are females or males. Indeed, one of the rare observations of apparent orgasm in a female monkey was observed in just such a female-female interaction (Goldfoot, Westerborg-van Loon, Groeneveld, Slob, 1980). Similarly, males of various primate species not only protect, but may also carry or even adopt orphaned or marooned infants of their troop. In the pair-bonding marmosets, the father generally carries its infants even more than does the mother after their first several weeks of life. Whether we focus on aggression, troop movements, parental care, or even sex itself, while sex-predominant behaviors are quite common, sex-exclusive behaviors are relatively rare.

Although we can reasonably expect that many sexually differentiated patterns are the primary product of genetic factors, the sources of normal sex predominance in various other behaviors may have their origins in sexually differentiated experiences encountered from early in life onward, including events preceding or immediately following birth. Thus, Sackett (1981) has reported that females carrying female fetuses were more likely to engage in fights (requiring treatment in the laboratory setting in which these data were gathered) than was the case when the fetus was male. To add to these early, potentially differentiating life events, the macaque mother, for instance, having given birth to a male infant, frequently turns her infant upside down and closely inspects, sniffs, and handles the infant's penis during the first several postpartum weeks (personal observation). If the species is of a gregarious nature and the mother seeks the close contact of other group members (as does the bonnet macaque), the infant newborn male will be handled by other troop members as well. Female infants receive but a few perfunctory inspections during the first few days of life. Coupled with reports in these species of higher rates of physical punishment of young males by their mothers, conceivably these early differences in handling stimulation might have long-term sequelae in adult sex-predominant patterns. At this point it would be foolish to speculate as to what these sequelae might be, but such potentially potent early experiential factors might serve to shift genetically equipotential patterns into those that are predominantly shown by one sex or the other.

As we consider additional questions that emerge from a developmental perspective, I would like to comment further on another of the points raised by Nadler—the significance of the stimulus partner and social context in influencing the form and appearance of potentially dimorphic behaviors. The same stimulus and contextual issues must be considered as we attempt to determine the ontogeny of various apects of behavior. In a study we conduct some years ago, Nadler and I (1971) tested a series of young bonnet macaque males for their sexual performance with either a young or a fully mature "receptive" female. The males were 2, 3, 4, and 5 years old (i.e., ranging from pre- to postpuberty). When tested with the older females, the 4- and 5-year-olds copulated readily. The 3-year-olds, often after considerable but constrained aggressive activity, also mated with the older females. The 2-year-olds never even approached these adult female partners.

This apparent disinterest on the part of the very young males seemed strange to us, because we had often seen even younger males, living with their mothers in social groups, frequently mount and thrust toward a number of apparently receptive adult females with whom they lived. Despite their reluctance with the older female, when tested with young (estrogenized) females in the paired testing study, all males mated readily, and the 2-year-old males had the highest rates of mounting and the highest frequency of orgasm (although apparently without ejaculation).

The two issues of social context and the role of the stimulus partner that must be considered when attempting to assess the sources of sexually differentiated behavior are neatly illustrated here. When tested with a large adult female, only the oldest males were sufficiently dominant over the female for her to permit this strange male to mount her. The 3-year-olds had difficulty coercing her into receptivity. The 2-year-olds knew better than to try. Clearly, if one only had observations of 2-year-olds with the more dominant adult females as partners, one might be led to conclude that 2-year-old male macaques lack the drive or the skills to perform the predominantly male sexual pattern. Such a conclusion, however, would be incorrect, because performance with a smaller and less dominant female was quite prodigious. But what of the fact that in the rearing group setting, even very young males mount adult females? The same issue of dominance appears to be operating here as well, only in this case the dominance of the male itself is not involved, but the dominance standing of the mother over the son's female partner. In a typical instance, the young male approaches an adult female of the group over which his mother is dominant, the female glances quickly toward the mother, and then permits or even encourages the young male to mount.

The impact of social context on emerging male and female patterns of behavior during development adds further to the complex task of determining which patterns are or are not the exclusive or predominant characteristic of one sex or the other and the sources of any such differentiation. The social context within which development occurs may influence patterns of partner choice, and, as Goldfoot and Neff (see Chapter 12) has demonstrated, early differences in sex composition of the rearing group may significantly alter ongoing and subsequent behavioral repertoires.

The often large social groups within which Peruvian squirrel monkeys live are generally segregated along sexual lines. The male and female subgroups remain near each another, within a "sphere of potential interaction," but contact and huddling groups are almost always unisexual. If one creates a group of unfamiliar 1-year-old males and females (2 to 3 years before puberty, but naturally weaned half a year earlier), no segregation takes place. The same occurs when the subjects are a year older and only a year or so from puberty.

If a group of adult males and females are added to the younger subgroups, a striking difference emerges between the 1- and 2-year-olds. The adults, as is typical, rapidly divide themselves into all-male and all-female subgroups. Within a week or two, the 2-year-olds increasingly have joined their same-sex adult subgroup, and by the third week, intersexual interaction among the 2-year-olds has been substantially reduced. Importantly, the functional segregation of the prepubertal subjects has not yet become an "idee fixe," because removal of the adult males and females almost immediately breaks down the segregation of the younger subjects. Lest we think that even when the patterns of sexual segregation in the adults become habitually present they are immutable, good evidence is available to show that when mature adults are gonadectomized, segregation disappears, largely as a result of the lowered rejection of the males by the females! Finally, in the 1-year-olds, no division of the males and females occurrs at all, suggesting that at this developmental stage, neither the characteristics of the partner nor the social context is sufficient to cause male and female differentiation.

Thus, as is the case in humans, in order to specify patterns that are sexually dimorphic and to speculate about their possible origins, we must (1) examine the patterns in a variety of settings, (2) determine whether the males and females are being confronted with functionally equivalent stimuli, and (3) at least examine the possibility that the behavior in question may not itself be sexually differentiated, but may emerge as the by-product of some other dimorphic pattern.

References

Goldfoot, D. A., Westerborg-van Loon, H., Groeneveld, W., & Slob, A. K. (1980). Behavioral and physiological evidence of sexual climax in the female stump-tailed macaque *(Macaca arctoides). Science, 208,* 1477–1479.

Rosenblum, L. A., & Nadler, R. D. (1971). Ontogeny of male sexual behavior in bonnet macaques. In D. Ford (Ed.), *Influence of Hormones on the Nervous System* (pp. 388–400). Basel: Kager.

Sackett, G. (1981). Receiving severe aggression correlates with fetal gender in pregnant pig-tail monkeys. *Developmental Psychobiology, 14,* 267–272.

IV

BEHAVIORAL GENETICS PERSPECTIVES

10

Sex Differences in Mental Retardation and Their Implications for Sex Differences in Ability

Steven G. Vandenberg

We have known for a long time that there are more male retardates than female, but the reasons are uncertain. Various causes have been proposed for this discrepancy. Because the incidence of stuttering, reading disability, delayed speech development, and even left-handedness is also higher in boys than in girls, a greater general vulnerability in boys than in girls has been suggested. This argument gains in plausibility when we remember that male infants have a higher mortality rate than do female infants.

Another theory advances the idea that the distribution of IQ is wider for males than for females, resulting not only in more retardates but also in more high-IQ individuals among the males. A special form of this theory ties this greater variability of the males to the X chromosomes. Because males receive only one X chromosome, while females receive two Xs, it has been proposed that even when a female receives an X chromosome that carries genes for low intelligence, she will be protected from the effects of those genes by the genes on her second X chromosome. On the other hand, this theory continues, genes for high intelligence on the one X chromosome will be tempered in their effects by genes on the second X chromosome. Only when receiving two X chromosomes carrying genes for high intelligence would a female be extremely intelligent, and that is the reason, according to this theory, for the scarcity of female genius.

Of course, most people have accepted a third explanation, which sees female social conditioning and the lower role expectations for girls held by parents and teachers, leading to more modest achievements of

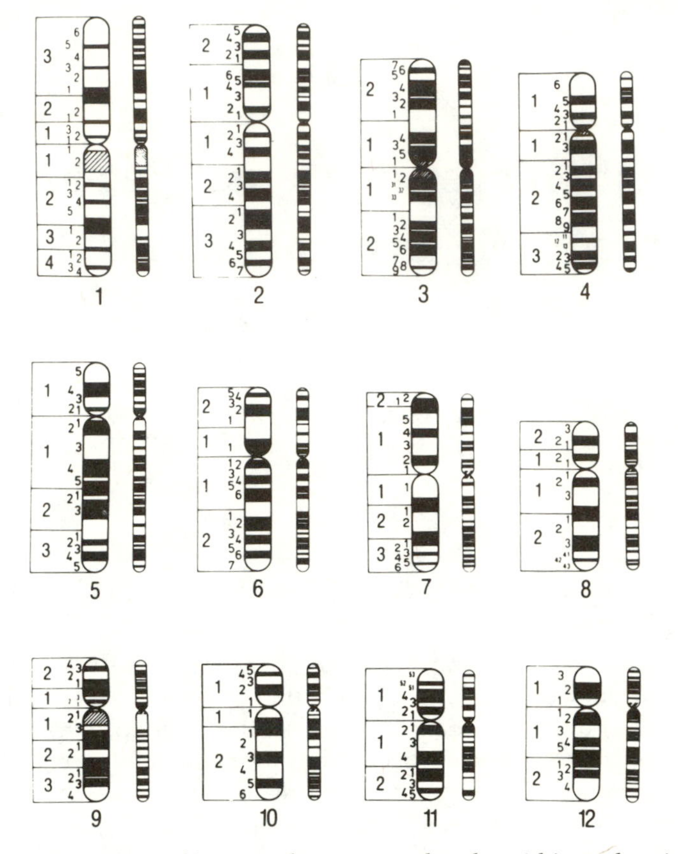

Figure 10–1. Diagrams of human chromosome bands; within each pair the metaphase and prometaphase bands are on the left and right, respectively.

girls, as the real cause of the discrepancy at the high end of the ability distribution. That theory does not, however, address the discrepancy at the low end. Because social factors are so important, early reports of pedigress showing an X-linked pattern of inheritance for mental retardation did not receive much attention, in part also because many authors suspected that the observed excess of male retardates might not be real but an artifact due to cultural factors such as an admission bias on the part of institutions as well as a greater tolerance by the family for low IQ in girls than in boys. In 1969, however, Davison noted that individuals with an IQ below 55 are male in disproportionate numbers and suggested that gene mutations carried on the X chromosome may contribute to such severe retardation. In 1974 Lehrke reviewed five published pedigrees (Allen, Herndon, & Dudley, 1944; Dunn, Renpenning, Gerrard, Miller, Tabata, & Federoff, 1963; Losowsky, 1961; Martin & Bell, 1943; Renpenning, Gerrard, Zaleski, & Tabata, 1962) and

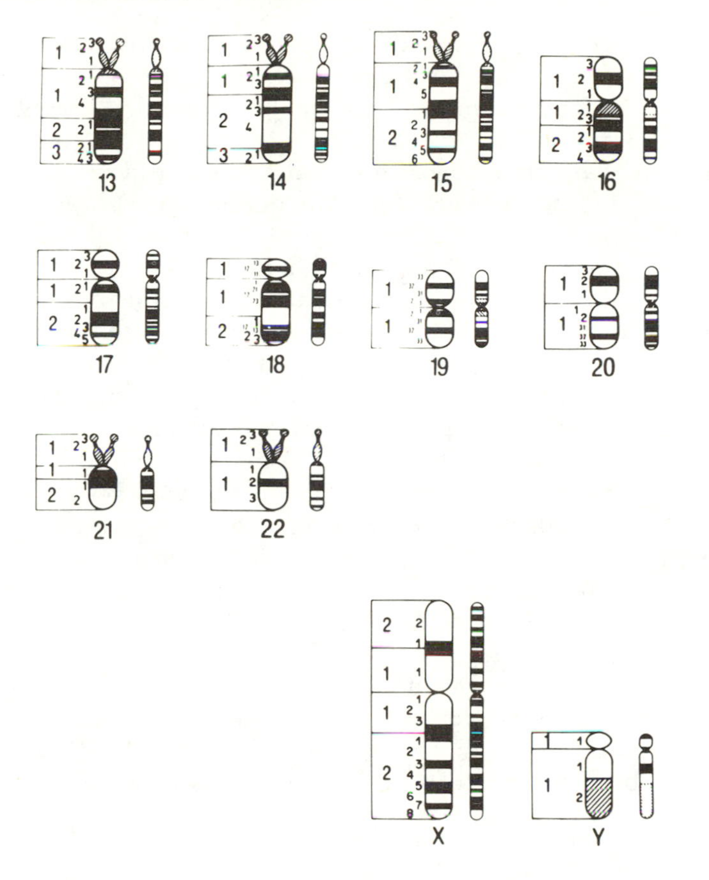

described five additional large pedigrees that showed no father-to-son transmission of mental retardation. In these pedigrees, there were 105 retardates out of 282 male offspring of mothers suspected of being carriers of genes leading to mental retardation, while very few female offspring were retarded. Wolff, Hameister, and Roper (1978) reported transmission of mental retardation by a *normal* male to 4 of his 12 daughters but to none of his 12 sons. Except for the absence of evidence that the mother was also not retarded, this is a classical pattern for X-linked recessive inheritance, wherein the father's X-linked condition was expressed in some of his daughters but none of his sons, who by virtue of being male, inherited a Y chromosome instead of an X chromosome from their father. One wonders what the intelligence of the father was, relative to that of his siblings and parents. The father may have been less intelligent than the other family members, although his IQ might still have been in the low normal range.

These descriptions of families that suggest X-linked inheritance of mental retardation may provide one reason for the considerable excess of males over females among the mentally retarded. As mentioned before, in the past this excess was frequently argued to be the result of a greater variability in intelligence in males, so that more males than females would be at *both* extremes of the distribution. In other words, in comparison with females, there would be more males among the severely retarded *and* among the highly gifted. Some authors, especially in the past, did not furnish any explanation for this discrepancy, but others attributed this sex effect to the fact that females have two X chromosomes and males one. Although it seems reasonable to expect possession of two X chromosomes to provide protection against a mutant allele leading to mental retardation, it is difficult to see how having two X chromosomes would reduce the incidence of very high intelligence, because that would require that a gene or genes for *high* intelligence also be located on the X chromosome. If that were the case, and if the genes for high intelligence were recessive, females would be highly gifted *only* if both X chromosomes carried these alleles—a much less likely event than for males to receive just one X chromosome with the recessive high IQ allele. This idea of a rare, recessive allele providing something beneficial to the individual is unusual. Rare recessive alleles that are harmful seem to be the rule, with the normal or beneficial allele dominant. Note, however, that a hypothetical X-linked spatial gene has been proposed to account for the sizable sex difference in spatial visualization ability in favor of males. It has also been suggested that this allele for good spatial ability has a frequency of .50 (Bock & Kolakowski, 1973), so we may just as sensibly regard the other allele, leading to poor spatial ability, as having a frequency of .50. But *it* would have to be the recessive allele.

In recent years methods have been developed for staining human chromosomes, in cells cultured usually from blood samples, so that various patterns of bands appear, as shown in Figure 10-1. Abnormalities have been described in these patterns, such as deletions, inversions, or insertions.

Fragile X Syndrome

In 1969 Lubs first described what has come to be called a fragile site on the X chromosome, based at band q27 near the end of the long arm, that may account for some cases of X-linked mental retardation, and a number of reports since then from England, Australia, Belgium, and the United States show an association between mental retardation and the "fragile X" abnormality (Daker,Chidiak, Frar, & Berry, 1981: Deroover, Frijns, Parloir, & Van Denberghe, 1977; Frijns & Van Den Berghe, 1982; Giraud, Ayme, Mattei, & Mattei, 1976; Harvey, Judge, & Wiener,

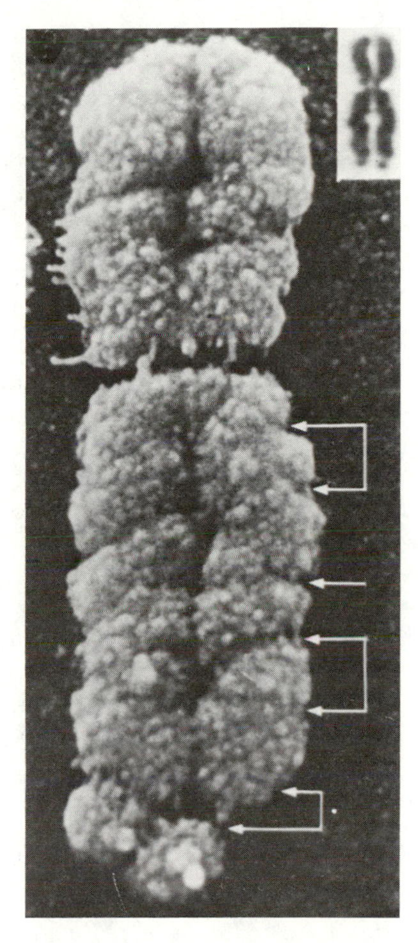

Figure 10–2. Scanning electron microscopy photo of an X chromosome, showing the large gap between the coils at the end of the long arm, which leads to the absence of staining that characterizes the "marker-X" or "fragile X" syndrome. (Courtesy of C. J. Harrison.)

1977; Nielsen, Tommerup, Poulsen, & Mikkelsen, 1981; Rhoads, Oglesby, Mayer, & Jacobs, 1982; Shapiro, Kuhr, Wilmot, Lilienthal, & Higgs, 1982; Sutherland & Ashforth, 1979; Van Den Berghe, Deroover, Parloir, & Frijns, 1978; Webb, Rogers, Pitt, Halliday, & Thesbald, 1981). A fragile site is an area on a mitotic methaphase or prometaphase chromosome that does not stain as this area normally does. Scanning electron microscopy reveals a defect in the quaternary coiling in this region so that the normal circumferential groove at this place is much wider and the region beyond resembles a satellite body (Harrison, Jack, Allen, & Harris, 1983), as shown in Figure 10-2. Why only a certain percent of the cells show this phenomenon is not understood, nor is the cause or the biochemical results, although Lejeune (1982) has made an interesting suggestion that monocarbons, which are essential to the transmethylase system, are involved. The transmethylases are necessary for the production of myelin, which insulates neurons. Lejeune (1982) has even suggested that in the infant early supplementation by folic acid and other donors of monocarbons may ameliorate the symptoms of fragile X and, in some cases, autism. Similar medication of the mother during pregnancy is another possibility, according to Lejeune. These suggestions are rather farfetched, because there may be no connection between the in-vitro condition and the patient's internal metabolism. No abnormality in folate metabolism has been reported. More may soon be known about the specific molecular changes producing the fragile X syndrome because restriction fragment polymorphisms are being used to construct a detailed map of the X chromosome (Drayne, Davies, Williamson, & White, 1983; Holden et al., 1983; Kunkel, Lalande, Aldridge, Flint, & Latt, 1983; Marsh, Yen, Mohandas, & Shapiro, 1983). A small beginning has been made toward the construction of such a map. Camerino, Mattei, Mattei, Jaye, and Mandel (1983) reported close linkage of the fragile X syndrome and hemophilia B, while Filippi et al. (1983) demonstrated linkage of the syndrome with the glucose 6-phosphate dehydrogenase (G6PD) and color-blindness cluster at Xq28. These two findings confirm that the locus is at Xq27, where the nonstaining area suggested it is located (see Fig. 10-3). The name *fragile X* suggests that breakage of the chromosome may occur at this point, but we have no evidence for this. Fragile sites have been reported on chromosomes 2, 10, 11, 16, and 20, but the fragile X has received the most attention, perhaps because it seems to be associated with a clear phenotype (i.e., severe mental retardation).

Sutherland (1977, 1979a, 1979b) has shown that fragile sites are more likely to be observed when a particular cell-culturing medium, commonly known as medium 199, that is deficient in folic acid is used. He proposed that higher pH values of the medium cause fragile sites to be revealed in a larger percentage of cells. This percentage varies widely among studies and also among individuals in a single study. Most, perhaps all, of the fragile sites seem to be genetic polymorphisms. Jenkins et al. (1983) found fragile X chromosomes in 0.6 percent of cells in five normal males and two normal females. The presence of fragile X in some cells of normal individuals was confirmed by Van Roy, De Smedt, Raes, Dumon, and Leroy (1983), making it necessary to define a cutoff value for the frequency at which the fragile X phenomenon is considered abnormal. Jacobs, Mayer, Matsuura, Rhoads, and Yee (1983) used 4% as this value. Turner and Turner (1974) have suggested that there might be a correlation between the number of cells showing a fragile X and the degree of mental retardation, but Sutherland's discovery of the effect of the culture medium suggests the possibility that any such correlations observed earlier were artifacts, or at least very unreliable. Nevertheless, gross differences in percentages of fragile X chro-

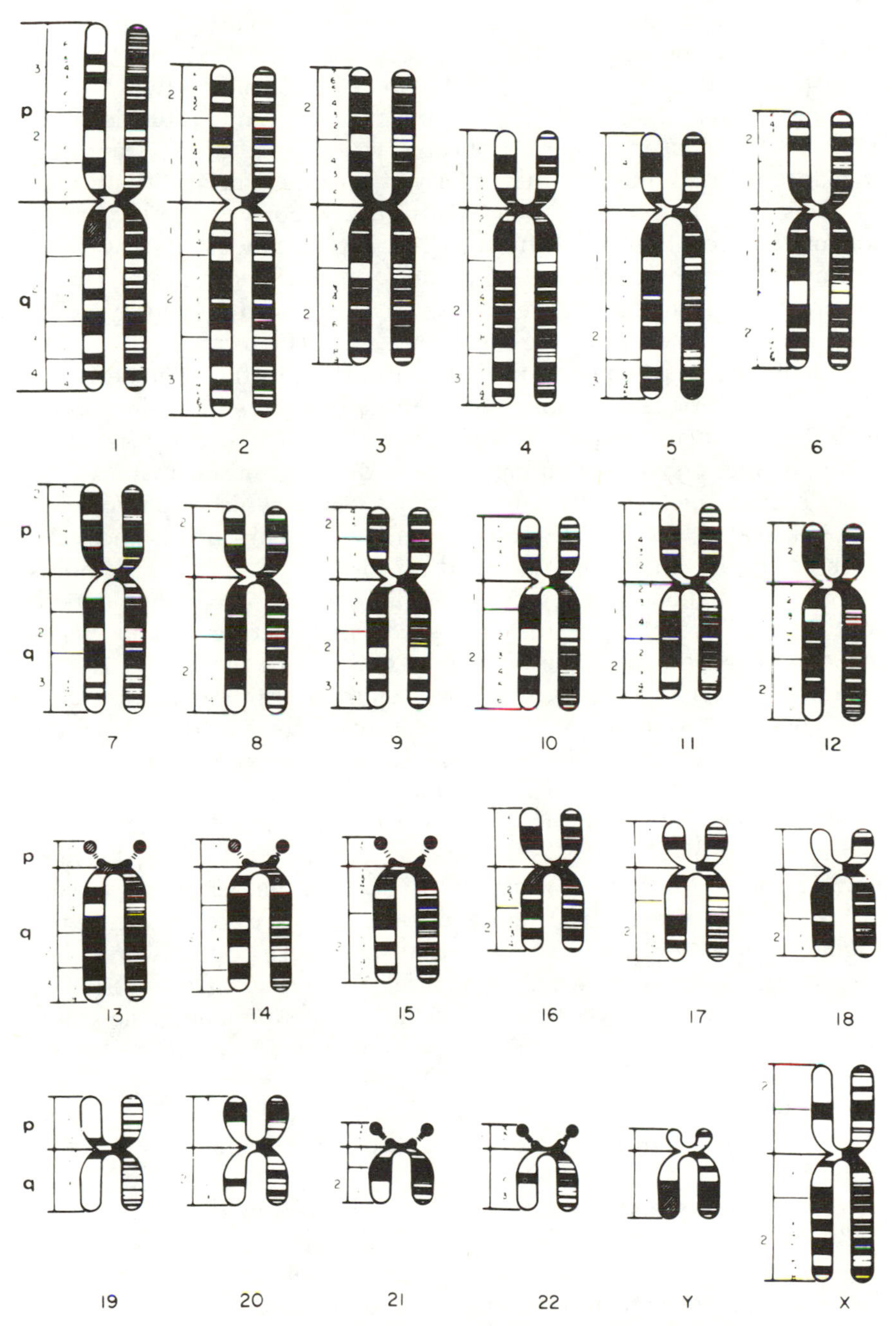

Figure 10–3. In each chromosome, the left chromatid shows the G-banding observed in mid-metaphase and the right that observed in late metaphase.

mosomes may be related to differences in mental ability. In the case of females, one also needs to take into account Lyon's (1961) hypothesis. Lyon proposed that in every cell of females, one of the two X chromosomes is inactivated; that this happens early in embryonic development, after only a small number of cell divisions; and that it is a matter of chance whether the maternally derived X or the paternal X is turned off. This hypothesis is consistent with observations of the incorporation of radioactive material during DNA replication in dividing cells, indicating that one of the two X chromosomes replicates late. This is the one thought to be inactive. Thus, according to the Lyon hypothesis, females are mosaic for the X chromosome. The term *mosaic* is used to describe the phenomenon in which a portion of the genome is not identical in every cell. Although such mosaicism was difficult to demonstrate in humans, it is now well established. The occurrence of inactivation after only a small number of cell divisions means that a whole organ system might be descended from a cell in which a particular X was inactivated, so that all the cells in that system (for instance, the retina) could have the same X turned off. If the Lyon hypothesis holds true for the fragile X, the degree of mental retardation could be dependent on the percentage of cells with a normal X chromosome that is active versus the percentage of cells in which a fragile X is active. In Uchida and Joyce's 1982 study of two retarded sisters who were heterozygous for the fragile X, the fragile X was active in 100 out of 126 cells in one sister and in 85 out of 120 cells in the other. For two controls with some fragile X chromosomes but normal intelligence, the numbers of cells in which the fragile X was active was lower: 40 out of 78 and 10 out of 32. The Wechsler scores for the two retarded sisters were 62 and 66 for the verbal part, 81 and 73 for the performance part, and 68 and 67 for total IQ. There had been earlier suggestions that the fragile X affects verbal ability more than performance test scores, but this pattern is common in many retardates. Eren and Disteche (1983) found no difference between the verbal and the performance scale scores of 10 fragile X males: verbal 57 ± 6, performance 55 ± 3. Three female carriers with fragile X had a mean IQ of 84, and four obligate carrier females without fragile X cells had a mean IQ of 118, suggesting a mild lowering effect on the IQ in females expressing the fragile X in some of their cells.

In 1983 Uchida and associates published data on three additional kindreds supporting the earlier paper that suggested that mental development was negatively affected in females heterozygous for fragile X whenever a substantial number of cells had the fragile X active (and the normal X inactive). In this second study a marked contrast was seen between the very low percentage of cells with fragile X in normal individuals and the percentage of such cells in retarded or borderline individuals. In addition, in the "slow learners," the percentage of fragile X that was active was much larger than the percentage that was inactive, whether or not either type of fragile X was common.

In another recent paper (McDermott, Walters, Howell, & Gardner, 1983) two females were heterozygous for fragile X: one retarded and one of normal intelligence. The retarded one has 74 percent of her fragile Xs active, while the normal one had 72 percent of her fragile Xs inactive.

The degree of mental retardation produced by the fragile X occasionally is mild enough that it may be overlooked unless there are other affected persons in the family. Perhaps the effect of the fragile X is particularly apt to be overlooked in more gifted families, where a drop of 30 points from an average family IQ of 125, for example, would still leave an IQ of 95 (Nielsen et al. 1981). Whether the apparently wide range of retardation among fragile X cases is entirely because of variation in average family intelligence, or whether the fragile X itself has a variable effect is not known. Such a variable effect could be because the number of cells showing the fragile X varies among patients, as does the percentage of those fragile Xs that are active in females. Another unusual feature of the condition is that retardation in females is found at a higher frequence than would be expected for a recessive trait. In the large kindred reported by Nielsen et al. (1981), four of the ten affected individuals were female. This proportion in females would suggest dominant transmission, but there are too many exceptions to this pattern to accept this hypothesis. If enough fragile X chromosomes are active, this may produce an effect resembling dominant inheritance.

The study of X-linked mental retardation is further complicated by reports suggesting genetic heterogeneity. We have already mentioned the remarkable family reported by Wolff et al. (1978) in which the father was seemingly unaffected. This was not the fragile X syndrome, and girls were affected instead of boys. Fragile X in males with normal intelligence has been reported (Daker et al., 1981; Frijns & Van Den Berghe, 1982; Nielsen et al., 1981; Rhoads et al., 1982; Van Roy et al., 1983; Webb et al., 1981).

An association between macroorchidism (enlarged testicles) and X-linked retardation has been reported in several papers (Cantú et al., 1976; Escalante, Grunspun, & Frota-Passoa, 1971; Richards, Sylvester, & Brooker, 1981; Shapiro, Summa, Wilmot, & Gloth, 1983; Turner, English, Lindsay, & Turner, 1972; Turner et al., 1975; Jacobs et al., 1983). To what extent the fragile X condition overlaps with macroorchidism is not known. Individuals in the early studies of X-linked retardation were not examined for either the fragile X or enlarge testicles (although extreme instances of the latter might have been mentioned in the published reports if they had been present).

In the pedigree described by Martin and Bell (1943) that was subsequently followed up by Richards et al. (1981), besides the long faces, prognathism, and large ears noted in the earlier report, the affected males also had macroorchidism and the fragile X chromosome. For that reason some British authors recommend the name *Martin-Bell syndrome* for the fragile X condition. Others use the expression *marker X.* Sometimes the name *Renpenning* is added, but the family described by Renpenning and associates (1962), although it displayed X-linked mental retardation, was found by Fox, Fox, and Gerrard (1980) not to have the fragile X chromosome.

Rubinstein (1964) suggested a type of mental retardation in which unusual facial features, broad thumbs and toes, and incomplete or delayed descent of tests was present, but no mention was made of sex linkage.

Juberg and Massidi (1980) have reported a male child and two maternal uncles with rudimentary scrotum, small penis, mental retardation, deafness, and certain characteristic facial features. They suggest that this is a unique type of X-linked retardation. Another form of X-linked retardation has been reported by Davis, Silverber, Williams, Spiro, and Shapiro (1982), who found nine retarded males with progressive spastic quadriparesis in three generations of one family. Finally, Fried and Sanger (1973) have suggested that there may be linkage between the blood group allele Xg and a gene causing mental retardation.

As Kaiser-McCaw, Hecht, Cadie, and Moore (1980) have emphasized, many unanswered questions remain concerning X-linked mental retardation. The frequency and distribution in the population of fragile X sites are perhaps the most interesting unknown factors. If X-linked retardation accounts for most of the excess males at the lower end of the IQ distribution, one would no longer expect to find an equivalent excess of males at the upper end of the distribution. A greater frequency of males than females with very high IQs has been suggested to account for the relative scarcity of female geniuses; if that explanation is no longer plausible, cultural explanations become more likely. Questions concerning the variable effect of the fragile X and the mode of transmission will not be answered until many more families have been studied.

Some preliminary figures for the incidence of fragile X were obtained by Shapiro et al. (1983). In 260 mentally retarded postpubertal males, 90 were found to have macroorchidism. Of those, 22 also had specific features of face and ears. Of 23 macroorchid males examined, only 10 were fragile X positive, and 11 had specific features of face and ear; of those 11 males, 8 were positive for fragile X. It is not clear from the abstract whether the 23 examined for fragile X were a random sample or not. If they were, 10 out of 23 or 43.48% might be the incidence of fragile X syndrome among institutionalized retardates who have macroorchidism. What percentage of retardates without macroorchidism may also have the fragile X is not known.

Herbst and Miller (1980) have calculated that the incidence in males of X-linked retardation is 1.83 per 1000 and of fragile X syndrome may be as high as 0.9 per 1000 births. If these estimates are correct, X-linked retardation indeed accounts for the excess of male retardates, and the fragile X syndrome accounts for half of the excess.

Jacobs et al. (1983) found five fragile X among 274 retardates in a community placement program in Hawaii. The incidence seemed roughly the same for Americans of European descent and for those of Japanese descent. This would tend to be an underestimate of frequency because severely retarded are not placed in the community placement program.

From Australia, Fishburn, Turner, Daniel, and Brookwell (1983) have estimated a frequency of 2.8 per 10,000 X-linked mental retardation with macroorchidism with or without the fragile X chromosome in males and 3.65 per 10,000 females heterozygous for fragile X.

In closing we note several reports of cases of autism with the fragile X (Brown et al., 1982; Gillberg, 1983; Jacobs et al., 1983; Meryash, Szymanski, & Gerald, 1982; Proops & Webb, 1981). Until now, no general screening of autistic children for fragile X has been performed.

In conclusion, an unexpected new discovery has opened up a whole new perspective on the phenomenon of excess male mental retardation, with implications for prenatal screening in affected families and, in the unlikely event that Lejeune (1982) is correct, for early treatment of affected babies and perhaps even of pregnant carrier females. In addition, this new cause of mental retardation strengthens the cultural explanation of fewer high-ability females by weakening the case for the "symmetric" argument that explains excess males in both the high and the low end of the distribution of IQ by the presence of one X in males but two in females.

References

Allan, W., Herndon, C. N., & Dudley, F. C. (1944). Some examples of the inheritance of mental deficiency: Apparently sex-linked idiocy and microcephaly. *American Journal of Mental Deficiency, 48,* 325–334.

Bock, R. D., & Kolakowski, D. (1973). Further evidence of sex-linked major gene influence in twins' spatial visualizing ability. *American Journal of Human Genetics, 25,* 1–14.

Bowen, P., Biederman, B., & Swallow, K. (1978). The X-linked syndrome of macro-orchidism and mental retardation. *American Journal of Medical Genetics, 2,* 409–414.

Brown, W. T., Friedman, E., Jenkins, E. C., Brooks, J., Wisniewski, K., Raguthu, S., & French, J. H. (1982). Association of fragile X syndrome with autism. *Lancet, 1,* 100.

Cantú, J. M., Saglia, H. E., Medina, M., Gonzáles-Didi, M., Morato, T., Moreno, M. E., & Pérez-Palacios, G. (1976). Inherited congenital normofunctional testicular hyperplasia and mental deficiency. *Human Genetics, 33,* 23–33.

Camerino, G., Mattei, M. G., Mattei, J. F., Jaye, M., & Mandel, J. L. (1983). Close linkage of fragile X-mental retardation syndrome to haemophilia B and transmission through a normal male. *Nature, 306,* 701-704.

Daker, M. G., Chidiac, P., Fear, C. N., & Berry, A. C. (1981). Fragile X in a normal male: A cautionary tale. *Lancet, 1,* 780.

Davis, J. G., Silverber, G., Williams, M. K., Spiro, A., & Shapiro, L. R. (1982). A new X-linked recessive mental retardation syndrome with progressive spastic quadriplegia. *American Journal of Human Genetics, 34,* 75A.

Davison, B. C. (1969). *Severe mental defect: The contribution by X-linked or sex-limited gene mutation.* M. D. thesis, The Queen's University of Belfast.

Deroover, J., Frijns, J. P., Parloir, C., & Van Den Berghe, (1977). X-linked recessively inherited non-specific mental retardation: Report of a large family. *Annals of Genetics, 20,* 236–268.

Drayne, D. T., Davies, K. E., Williamson, R., & White, R. L. (1983). Construction of a linkage map of the human X chromosome. *American Journal of Human Genetics, 35,* 171A.

Dunn, H. G., Renpenning, H., Gerrard, J. W., Miller, J. R., Tabata, T., & Federoff, S. (1963). Mental retardation as a sex-linked defect. *American Journal of Mental Deficiency, 67,* 827–848.

Eren, M., & Disteche, C. (1983) Behavioral phenotype of the fragile X syndrome: Relationship betweeen frequency of marker X, mental retardation and verbal ability, *American Journal of Human Genetics, 35,* 131A.

Escalante, J. A., Grunspun, H., & Fronta-Passoa, W. (1971). Severe sex-linked mental retardation. *Journal de Génétique Humaine, 19,* 137–140.

Filippi, G., Rinaldi, A., Archidiacono, N., Rocchi, M., Balazs, I., & Siniscalco, M. (1983). Brief report. Linkage between glucose 6 phosphate dehydrogenase and fragile X syndrome. *American Journal of Medical Genetics, 15,* 113–120.

Fishburn, J., Turner, G., Daniel, A., & Brookwell, R. (1983). The diagnosis and frequency of X-linked conditions in a cohort of moderately retarded males with affected brothers. *American Journal of Medical Genetics, 14,* 713–724.

Fox, P., Fox, D., & Gerrard, J. W. (1980). X-linked mental retardation: Renpenning revisited. *American Journal of Medical Genetics, 7,* 491–495.

Fried, K., & Sanger, R. (1973). Possible linkage between Xg and the locus for a gene causing mental retardation with or without hydrocephalus. *Journal of Medical Genetics, 10,* 17–18.

Frijns, J. P., & Van Den Berghe, N. (1982). Transmission of fragile (X) (q27) from normal male(s). *Human Genetics, 61,* 262–263.

Gillberg, C. (1983). Identical triplets with infantile autism and the fragile X syndrome. *British Journal of Psychiatry, 143,* 256–260.

Giraud, F., Ayme, S., Mattei, J. F., & Mattei, M. G. (1976). Constitutional chromosome breakage. *Human Genetics, 34,* 125–136.

Harrison, C. J., Jack, E. M., Allen, T. D., & Harris, R. (1983). The fragile X; a scanning electron microscope study. *Journal of Medical Genetics, 20,* 280–285.

Harvey, J., Judge, C., & Wiener, S. (1977). Familial X-linked mental retardation with an X chromosome abnormality. *Journal of Medical Genetics, 14,* 46–50.

Herbst, D. S., & Miller, J. R. (1980). Non-specific X-linked mental retardation. II. The frequency in British Columbia. *American Journal of Medical Genetics, 7,* 461–469.

Holden, J., Beckett, J., Mulligan, L., Phillips, A., Simpson, N., Partington, M., Hamerton, J., Wang, H-S., Donald, L., & White, B. (1983). A search for restriction fragment length polymorphisms (RFLPs) linked to the fragile X form of X-linked mental retardation. *American Journal of Human Genetics, 35,* 174A.

Jacobs, P. A., Mayer, M., Matsuura, J., Rhoads, F., & Yee, S. C. (1983). A cytogenetic study of a population of mentally retarded males with special reference to the marker (X) syndrome. *Human Genetics, 63,* 139–148.

Jenkins, E., Duncan, C., Brooks, J., Lele, K., Sanz, M., Nolin, S., & Brown,T. (1983). Low frequency fragile X chromosomes in cultures from normal people. *American Journal of Human Genetics, 35,* 136A.

Juberg, R. C., & Massidi, I. (1980). A new form of X-linked mental retardation with growth retardation, deafness and microgenitalism. *American Journal of Human Genetics, 32,* 714–722.

Kaiser-McCaw, B., Hecht, F., Cadien, J. D., & Morre, B. C. (1980). Letter to the editor: Fragile X-linked mental retardation. *American Journal of Medical Genetics, 7,* 503–506.

Kunkel, L., Lalande, M., Aldridge, J., Flint, A., & Latt, S. (1983). A large insert X chromosome specific library for the identification of RFLP haplotypes. *American Journal of Human Genetics, 35,* 176A.

Lehrke, R. G. (1974). *X-linked mental retardation and verbal disability*. New York: Intercontinental Medical Book.

Lejeune, J. (1982). Is the fragile X syndrome amenable to treatment? *Lancet, I,* 273–274.

Losowsky, M. S. (1961). Hereditary mental defect showing the pattern of sex influence. *Journal of Mental Deficiency Research, 48,* 60–62.

Lubs, H. A. (1969). A marker X chromosome. *American Journal Of Human Genetics, 21,* 231–244.

Lyon, M. F. (1961). Gene action in the X chromosome of the mouse *(Mus musculus L.). Nature, 190,* 372–373.

Marsh, B., Yen, P., Mohandas, T., & Shapiro, L. J. (1983). Isolation of human X chromosome DNA clones homologous to coding sequences. *American Journal of Human Genetics, 35,* 179A.

Martin, J. P., & Bell, J. (1943). A pedigree for mental defect showing sex linkage. *Journal of Neurology and Psychiatry N. S., 6,* 154–157.

McDermott, A., Walters, R., Howell, R. T., & Gardner, A. (1983). Fragile X chromosome: Clinical and cytogenetic studies on cases from seven families. *Journal of Medical Genetics, 20,* 169–178.

Meryash, D. L., Szymanski, L. S., & Gerald, P. S. (1982). Infantile autism associated with the fragile X syndrome. *Journal of Autism and Developmental Disorders,* 12, 295–301.

Nielsen, K. B., Tommerup, N., Polusen, H., & Mikkelsen, M. (1981). X-linked mental retardation with fragile X: A pedigree showing transmission by apparently unaffected males and partial expression in female carriers. *Human Genetics, 59,* 23–25.

Proops, R., & Webb, T. (1981). The "fragile" X chromosome in the Martin-Bell-Renpenning syndrome and in males with other forms of familial mental retardation. *Journal of Medical Genetics, 18,* 366–373.

Renpenning, H., Gerrard, J. W., Zaleski, W. A., & Tabate, T. (1962). Familiar sex-linked mental retardation. *Canadian Medical Association Journal, 87,* 954–956.

Rhoads, F. A., Oglesby, A. C., Mayer, M. & Jacobs, P. A. (1982). Marker X syndrome in an Oriental family with probable transmission by a normal male. *American Journal of Medical Genetics, 12,* 205–217.

Richards, B. W., Sylvester, P. E., & Brooker, C. (1981). Fragile X-linked mental retardation: The Martin-Bell syndrome. *Journal of Mental Deficiency Research, 25,* 253–256.

Rubinstein, J. H. (1964). A syndrome of mental retardation with abnormal facial features and broad thumbs and great toes. *Proceedings of the International Congress on the scientific study of mental retardation.* Copenhagen, 2, 812.

Ruvalcaba, R. H. A., Myhre, S. A., Roosen-Runge, E. C., & Beckwith, J. B. (1977). X-linked deficiency megalotestes syndrome. *Journal of the American Medical Association, 238,* 1646–1650.

Shapiro, L. R., Kuhr, M. D., Wilmot, P. L., Lilienthal, E. R., & Higgs, L. C. (1982). Multiple sibling mental retardation and the impact of the fragile X chromosome. *American Journal of Human Genetics, 34,* 112A.

Shapiro, L. R., Summa, G. M., Wilmot, P. L., & Gloth, E. (1983). Screening and detection of the fragile X syndrome. *American Journal of Human Genetics, 35,* 117A.

Sutherland, G. R. (1977). Fragile sites on human chromosomes: Demonstration of their dependence on the type of tissue culture medium. *Science, 197,* 265–266.

Sutherland, G. R. (1979a) Heritable fragile sites on human chromosomes. I. Effect of composition of culture medium on expression. *American Journal of Human Genetics, 31,* 125–135.

Sutherland, G. R. (1979b). Heritable fragile sites on human chromosomes, II. Distribution, phenotypic effects and cytogenetics. *American Journal of Human Genetics, 31,* 136–148.

Sutherland, G. R., & Ashforth, P. L. C., (1979). X-linked mental retardation with macro-orchidism and the fragile site at Xq27 or 28. *Human Genetics, 48,* 117–120.

Turner, G., Eastman, C., Casey, J., McLeavy, A., Procopi, P., & Turner, B. (1975). X-linked mental retardation associated with macro-orchidism. *Journal of Medical Genetics, 12,* 367–371.

Turner, G., English, B., Lindsay, D. G., & Turner, B. (1972), X-linked mental retardation without physical abnormality (Renpenning's syndrome) in sibs in an institution. *Journal of Medical Genetics, 9,* 324–330.

Turner, G., & Turner, B. (1974). X-linked mental retardation. *Journal of Medical Genetics, 11,* 109–113.

Uchida, I. A., Freeman, V. C., Jamro, H., Partington, M. W., & Soltan, H. C. (1983). Additional evidence for fragile X activity in heterozygous carriers. *American Journal of Human Genetics, 35,* 861–868.

Uchida, I. A., and Joyce, E. M. (1982). Activity in the fragile X in heterozygous carriers. *American Journal of Human Genetics, 34,* 286–293.

Van Den Berghe, H., Deroover, J., Parloir, C., & Frijns, J. P. (1978). X-linked non-specific mental retardation. *Clinical Genetics, 13,* 106.

Van Roy, B. C., De Smedt, M. C., Raes, R. A., Dumon, J. E., & Leroy, J. G. (1983). Fragile X trait in a large kindred: Transmission also through normal males. *Journal of Medical Genetics, 20,* 286–289.

Webb, G. C., Rogers, J. G., Pitt, O. B., Halliday, J., & Theobald, T. (1981). Transmission of fragile (X) (q27) site from a male. *Lancet, ii,* 1231–1232.

Wolff, G., Hameister, H., & Roper, H. H. (1978). X-linked mental retardation: Transmission of the trait by an apparently unaffected male. *American Journal of Medical Genetics, 2,* 217–224.

11

The Fragile X Chromosome Anomaly: A Comment

James R. Wilson

Before turning to Vandenberg's paper (Chapter 10), I would like to unburden myself of two complaints. I am struck by the lack of genetics in the research designs reviewed elsewhere at this symposium. It seems clear to me that the hypothesis of environmental transmission (e.g., social learning) of dominance status from mother to son in apes needs to be tested within a genetic design, perhaps with a cross-fostered group. Is it not possible that the traits leading to attainment and expression of dominance in male apes have been transmitted genetically by the mother and/or the father? And, in the development of psychopathologies in the offspring of masculine women, is it not possible that the psychopathology has at least some genetic basis?

My second complaint involves the statistical analyses reported in some of the papers in this volume. More than once we have been given an outline of analyses that involve computation of many—sometimes very many—*t* tests, without any indication of what procedure was employed to keep the alpha level constant during the serial application of these tests. I propose that this methodological weakness is at least partially an explanation for the many failures to replicate specific findings. That is, the statistical significance of a given "finding" actually may never have been assessed, if the alpha level was not kept constant during computation of many *t* tests on the same data set by using a procedure that requires ever larger *t* values for significance during computation of many *t* tests on the same data set. A somewhat different but related matter is the use of univariate statistics to assess significance in what apparently is a multivariate design. Here again, one or two "significant" finding are all too likely to emerge, not as a function of what is really the truth, but because many univariate tests were applied without a prior multivariate indication of significance within the data set.

Enough of complaints! Let me turn now to Vandenberg's paper. My most general comment is that Vandenberg's contribution goes a long way toward achieving genetic balance in this volume. His review of the fragile X syndrome also addresses the symposium theme of sex differences quite directly, in the sense of reviewing the evidence for and against a currently "hot" hypothesis concerning the almost universal finding of more frequent mental retardation in males than in females.

Because males are hemizygous for the X chromosome, any gene on the chromosome is likely to be expressed, as the Y chromosome is not a homolog in the usual sense. This, of course, is the basis for the greatly increased frequency in males of X-linked traits such as certain forms of color-blindness, certain hemophilias, and over 100 other known traits. Females, having two X chromosomes, can be heterozygous carriers of the same recessive gene, and pass the gene on to half of their offspring, on average, but be unaffected themselves. Likewise, any anomaly in the macrostructure of the X is likely to be expressed much more frequently in males. As reviewed by Vandenberg, a specific anomaly, an extreme narrowing at a specific location on the long arm of the X, has been shown to be heritable and, in many cases, to be associated with mental retardation and macroorchidism.

The fragile X [fra(X)] syndrome, as this anomaly has come to be called, may not be ideally named. Fragility would seem to imply the likelihood of breakage at the "weak" point, with the distal portion then being lost during mitosis or meiosis because it lacks a centromere for attachment of the spindle fibers. Loss of the segment would, however, be termed a deletion. Because the severely constricted fra(X) can be followed in pedigrees, its fragility must be more apparent than real. Vandenberg pointed out this nomenclature problem; I only wish to emphasize here that the developmental anomalies associated with fra(X) are not associated with loss of part of chromosome, but with some difficulty in transcribing genetic information at, and perhaps beyond, the constriction. The use of *fragile* is, however, consistent with previous use in naming aberrant constriction sites on some of the autosomes (e.g., Sutherland, Jacky & Baker, 1984).

Whether the transcription difficulty is only *at* the fragile site, or includes all loci beyond the site, is not yet known, but will probably be critical information for attempts at amelioration. If only one or a few genes are involved (for example, a gene or genes on the biosynthetic pathway of myelin), devising a treatment may prove somewhat easier; if several hundred genes are involved, prospects for amelioration may be dimmer.

As ably reviewed by Vandenberg, the fra(X) syndrome is somewhat variable in expression. In some affected persons, cognitive ability seems to be in the normal range; in most, however, considerable deficit is measurable. One reason for this variation may be that some affected individuals are otherwise genetically endowed with abilities considerably above average, and the interaction of this endowment with the fra(X) syndrome results in near-average intelligence. This hypothesis could probably be examined further by testing nuclear family members who are not affected by fra(X), on the grounds that intellectual functioning tends to be somewhat similar within families. Another reason for the variation may be genetic mosaicism. In most cases studied so far, the fra(X) is not expressed in all cells, but only in some fraction of them. On the whole, a correlation does seem to exist between the proportion of cells in which fra(X) is expressed and cognitive deficit. The source of this mosaicism could be somatic mutation (i.e., during mitosis), with some cell lines descending from the fra(X) mutation and others descending from unmutated cells. Other sources could be outlined: perhaps the fra(X) is "repaired" in some cell lines, or perhaps two Xs initially were present, and one or the other was lost during early cell divisions.

Vandenberg's review is certainly neoteric. I counted 6 references dated 1982, and 12 dated 1983 (many of the latter are in symposium reports). In addition to being up to date, his review is timely, in that increasing attention is being paid to the fra(X) syndrome, and information should accumulate rapidly. A report by Camerino, Mattei, Mattei, Jaye, and Mandel appeared late in 1983, and part of their summary is quoted here.

> This syndrome has recently been recognized as one of the major causes of genetically determined mental retardation, and as one of the most important X-linked diseases with respect to its frequency (analogous to that of Duchenne muscular dystrophy or of haemophilia A) and severity. In the absence of treatment, genetic screening for this disease would seem to be particularly important. Prenatal diagnosis is now feasible although difficult and detection of heterozygous carriers is only possible in 50% of cases. The recent demonstration of genetic linkage between the glucose 6-phosphate dehydrogenase (G6PD)-colour blindness cluster (at Xq28) and the fragile X locus has suggested that the fragile site is indeed the site of the mutation. We show here that fragile X and haemophilia B loci are closely linked, using as genetic marker a

polymorphism of the coagulating factor IX gene [authors' citations omitted].

This quick establishment of linkage for fra(X) is unusual in that just what gene or genes may be involved is not yet known. Because of its hemizygous condition in males, the X chromosome has long had the best linkage map. This, together with the visible nature of the fra(X) constriction at Xq27, apparently has aided molecular geneticists in their work to ascertain linkage. The linkage with hemophilia B and coagulating factor IX is close, and the presence or absence of the latter two genes may prove to be helpful in genetic screening for fra(X), as pointed out by Camerino et al. (1983).

In summary, Vandenberg has brought to our attention a recently discovered genetic anomaly that may "account" for a substantial proportion of the greater number of males who are mentally retarded. Although it is a truism, we must still say that the relationship between fra(X) and mental retardation is not fully established, and more studies will be needed. More research will also be needed to learn just what the genetic difficulty may be. Insofar as the relationship is confirmed, we will have a first-order explanation for the sex difference at the low end of the ability scale. Studies of the fra(X) syndrome will probably not lead directly to a better understanding of the apparent excess of males at the high end of the ability distribution. Such studies will possibly do so indirectly, though, by increasing knowledge about the X chromosome and the genes it carries.

References

Camerino, G., Mattei, M. G., Mattei, J. F., Jaye, M., & Mandel, J. L. (1983). Close linkage of fragile X-mental retardation syndrome to haemophilia B and transmission through a normal male. *Nature, 306*, 701–704.

Sutherland, G. R., Jacky, P. B., & Baker, E. G. (1984). Heritable fragile sites on human chromosomes. XI. Factors affecting expression of fragile sites at 10q25, 16q22, and 17p12. *American Journal of Human Genetics, 36*, 110–122.

V

DEVELOPMENTAL PERSPECTIVES

12

Assessment of Behavioral Sex Differences in Social Contexts: Perspectives from Primatology

David A. Goldfoot and Deborah A. Neff

This chapter reviews strategies used in the scientific measurement of behavioral sex differences. It is part "how to do it" and part "what to make of it" in nature, because our aim is to point to logical and technical restrictions associated with studies of sex differences as well as to present data. Moreover, the chapter focuses on results from nonhuman primate species instead of human beings. The use of animal studies to learn something about human beings is a controversial and complex topic, to say the least, but when applied to research in behavioral sex differences, the issue risks becoming political and even more confusing than for other behavioral categories, demanding explicit and careful justification. Because relatively few of us in animal behavior disciplines have written specifically about the role our data might play in understanding sex differences in humans, it seems appropriate in this forum to precede the main theme of this chapter with a short consideration of current thought about that practice.

The Assessment of Behavioral Sex Differences

Definition

Few behavioral responses are dimorphic in an absolute sense. In many cases both sexes display a given behavior pattern. It is the relative difference in frequency, intensity, or context associated with the display of a particular behavior pattern that is termed a *sex difference.* A behavioral dimorphism is therefore a relational construct, an expression of statistical comparison, and not a directly observable phenomenon.

Conditions of Assessment

Individual Testing

If the goal of a study is to go beyond the description of a dimorphically expressed behavior and to investigate possible etiological and causal factors determining its display, then experimental design calls for standardized, identical conditions of testing for both sexes. The least complicated way to accomplish this from a statistical point of view is to study individuals of each sex, one at a time, in a standardized, nonsocial environment. The reason for turning to a nonsocial environment is to eliminate all concurrent social factors that might influence response tendencies differentially, and to have unambiguous environmental constancy for evaluations of both sexes. In animal studies such tests can be given using activity wheels, emergence tests, open-field arenas, T mazes, etc. In human studies, tests of standarized vocabulary, mathematics, spatial perception, dexterity, pupil dilation, etc., are analogous situations, provided that the tester is not directly interacting with the subject. The advantage of testing subjects individually is that statistically "clean" experiments are possible: subjects can be given identical task presentations under exactly the same conditions; concurrent social variables are eliminated; and such things as test-retest reliability can be easily managed. Moreover, if a statistically reliable sex difference is found under such conditions, it increases the chances somewhat that some underlying biological sex difference might be identified that mediates the response. This is by no means a safe assumption, however, because earlier social history, learning, etc., might be responsible for the difference and would need to be investigated.

The disadvantage of testing subjects in nonsocial environments is that the most interesting behavioral characteristics of individuals are usually seen in social contexts (e.g., sex, aggression, parenting, group dynamics).

Dyadic Testing

In dyadic testing, two individuals are observed together, and their individual and interactive behavior is recorded, usually by designating one individual as the "experimental" and the other as the "stimulus" subject. Such a strategy has been used in hundreds of animal studies to investigate sexual, aggressive, and parenting sex differences. An immediate complication, however, is that behavioral and nonbehavioral characteristics (appearance, size, odor, etc.) of each member of the dyad will influence the behavior of the other. Moreover, the "test" is not just a momentary response by the experimental subject toward the stimulus subject, but involves complicated interactive sequences between the two subjects, with each influencing the other. For example, let us say that we want to know if there is a sex difference in the display of presenting behavior in adult rhesus monkeys. We construct a testing arena and test both male and female experimental animals, one at a time, with a stimulus male. We might also adjust concurrent hormonal conditions by gonadectomy and exogenous replacement to have circulating hormones equivalent for the two sexes, if that is related to our hypothesis. It would seem that we have achieved stimulus equivalence with this strategy. In fact we have not, because even though both the experimental males and females are tested with a male, that stimulus male in each case sees quite different types of partners, and his responses might be quite different, depending on the sex of the experimental partner with whom he is paired. In turn, because his responses are different in each case, the ultimate responses of our male and female "experimental" animals are clearly measured under nonidentical conditions. Goy and Goldfoot (1973) emphasized the need for social stimulus equivalency when pursuing biological hypotheses related to sex differences, and suggested that each sex be tested with identical stimulus partners. We can see now that the strategy has some real limitations. How does one maintain identical conditions of testing when each sex might induce differing patterns of response in partners of each sex? Larssen (1973) raised this issue some time ago and cautioned that because such tests are interactive and dynamic, they are not entirely within the control of the experimenter.

No completely satisfactory answer to this problem exists, but several strategies have been developed which address this issue methodologically.

Use of "neutral" stimulus animals

Investigators have sometimes attempted to modify behavioral responses of the stimulus animal, either by training (Brain & Nowell, 1970) or by sensory impairment (Denenberg, Gaulin-Kremer, Gandelman, & Zarrow, 1973), to achieve a relatively undifferentiated response to experimental animals of either sex. Others have employed dummies (Seitz, 1940), photos, movies, or videotapes to elicit responses from both sexes from a nonreactive social stimulus. Reducing the complexity of the stimulus further to a salient component of the natural stimulus set (e.g., presentation of an odor, a sound, a tactile stimulus) has also been attempted (Goy & Young, 1957).

Use of restricted social stimulus options

Investigators have sometimes used normal stimulus animals, but have arranged the testing situation to give the experimental animal relatively more control over the social options within the test. For example, Nadler (1982) has used a testing cage in which the female can approach or retreat from a stimulus male by squeezing through a portal too small for the male to enter. Pomerantz and Goy (1983) have used a tether to restrict the stimulus male's ability to approach or pursue his partners. Several investigators have used operant tasks in which the experimental animal may gain access to the stimulus partner or not, depending on whether it presses a lever (Keverne, 1976). All of these strategies are designed to reduce some of the interactive aspects of dyadic testing that can influence the behavior under consideration. In one sense, they are artificial and do not reflect the multiply determined nature of social interactions. In another sense, they can be misleading, because they reduce, but certainly do not eliminate, differential responding (a tethered male can act in a variety of different ways to male or female partners, although he cannot contact them if they stay beyond his reach). Nonetheless, studies of this type can help to unravel and identify components of social interactions in order to clarify specific roles contributed to by hormones, by prior experiences, etc.; from this point of view, these manipulations are extremely helpful.

Use of data from the pair instead of from the individual

Patterson and Moore (1979) have suggested that the most appropriate unit of analysis is sometimes the pair instead of the individual. Thus, if we applied the suggestion to studies of sex difference, we would not ask whether males differ from females, but would instead measure how behavior seen in male-male testing differs from that seen in male-female and female-female testing. Attributes differing for each sex could then be deduced from comparisons of their qualities as initiators of specific behavioral events, recipients of specific behavioral events, and patterns of response unique to each dyadic type, instead of in terms of individual sex differences per se. This approach has strong appeal for those interested in specific manifestations of sex differences as a function of social environment (e.g., alliance formation, reconciliation, etc.), but the data would be hard to use if the goal of the project were to identify specific physiological principles operating to generate the differences seen.

Group Testing

Three or more animals tested simultaneously in the same environment constitute a definition of "group." Group testing is often desirable when attempting to add layers of social complexity to behavioral situations believed to be modulated by social factors. Concepts such as social competition or alliances simply do not manifest themselves without group contexts. In addition, many behavioral characteristics of social species, including those potentially sexually dimorphic, might require a complex social environment to be expressed. Limiting testing to dyadic interactions might result in missing behavioral aspects crucial for interpretation in natural social circumstances. Recent studies by McClintock (1981), McClintock, Anisko, and Adler (1982), Keverne, Meller, and Eberhart (1981), and an older study by Goldfoot (1971) illustrate how aspects of sexual behavior more complex than represented by pair testing can be studied in laboratory environments to yield important new information. Not only do words of caution relevant to pair testing extend to group testing, but the difficulties increase geometrically with each additional group member. In addition, entirely new phenomena associated with social grouping not encountered in pair tests raise additional issues in which careful interpretation of results is essential.

Sex ratio effects

Studies that focus on animals living in groups in order to measure behavioral sex differences usually include comparisons of the frequencies of a variety of behavior patterns displayed, such as mounting, presenting, and rough-and-tumble play. Harking back to the previous discussion, one might immediately wonder if the number of animals of each sex available as partners might not influence behavioral measures or, more important, might change the social dynamics of the group. Preliminary analyses from our own work suggest that the relative number of males and females in small groups (four to six members) does indeed influence overall frequencies of display of targeted behaviors. As an example, let us consider rough-and-tumble play of juvenile rhesus females living in five-member groups. In all social situations, males show this form of play more frequently than females. We have found that this form of play is displayed at very low frequencies by females living in all-female groups, but it is considerably higher in females living in one-male, four-female groups, and low again in females from two-male, three-female groups. We think what occurs is that males initiate this particular response more than females (a "predisposition" hypothesis), and that males tend to select male partners for this activity. Thus, when no male is present, we see low levels of the behavior among females. When two males are present, they play with each other and do not involve the females very much, so again rough play is of low frequency among the females. In one-male groups, however, the male has only female partners, and so he engages them in rough play, and they reciprocate enough to generate higher scores than in the other conditions. As it turns out, we see males displaying higher rates of rough play in each situation in which they are present, but quite a bit of the variance seen in the scores of both sexes can be traced to the specific sex ratio within the group. Several other forms of behavior are influenced by sex ratio in our preliminary analyses of the data, including presenting and mounting.

One statistical approach to studies in which sex ratios between groups cannot be held constant is to divide each animal's scores by the number of partners of each sex available. Thus, in a group of three males and two females, each female has three male partners and one female partner. Her total female-to-male frequency could be divided by three, and her female-to-female frequency could be divided by one to reflect the fact that by chance more of one partner type than another was available for interactions. Chi squares can also be calculated based on these proportions in order to test for significant nonrandom distributions of interactions. We have used these statistical manipulations in some studies, but believe they have major drawbacks, because they do not really tap the social dynamics at work within a group. Quite often animals form and rarely display similar levels or types of behavior to all members of a given sex. Under ideal circumstances, when one is concerned that sex ratio is influencing data, one is better advised to conduct an experiment to test that hypothesis specifically rather than to try to use statistical approaches to "cope" with the problem.

Secondary effects of the presence of other animals

In addition to combinations of pairwise activities within groups, a major social phenomenon in many social species is that both direct and indirect interactions occur among three or more animals at a time. A clear example of an indirect interaction can be found from work by Perachio (1978), where a male rhesus monkey implanted with electrodes in his hypothalamus would mount a female partner whenever an electric signal was delivered by the experimenter. When placed alone with the female, the male mounted with each and every electrical stimulation. When the male was placed with the female and another male that was dominant to the experimental male, however, the electrical stimulation was ineffective in driving the mounting behavior. Apparently the male assessed the social situation and inhibited his mounting when near a dangerous competitor, even though direct aggression with the second male did not occur.

Because indirect influences are often at work in social situations in which sex differences are measured, we offer a second example of the phenomenon, this time taken from Gibber (1981), who was interested in determining whether there were sex differences in forms of parenting behavior in rhesus monkeys. Gibber devised a test in which an adult monkey was placed in a cage adjacent to a "stranded" neonate. A passageway led to the infant, and Gibber observed whether males showed lesser or greater tendencies to contact the infant than females. She found that both adult males and females looked at, approached, and picked up stranded infants with similar latencies and frequencies. She then modified the test by placing an adult male and female simultaneously in the apparatus. With these new conditions of testing, females displayed virtually all of the parenting behavior observed. No male picked up or approached an infant. Many of the males who showed care-giving responses when tested alone were used under these new testing conditions, and so the results unambiguously demonstrated a strong social inhibition of infant-directed responses to stressed infants by males when in the presence of an adult female. If Gibber had used only the latter paradigm to assess sex differences, she might have concluded that males have far less proclivity to attend to infants than females. By using more than one condition of testing, and by evaluating meaningful social interactions, she has a far more interesting and far more realistic set of findings on which to base further studies. Lamb (1979) provides specific examples and has additional suggestions for research designs that include effects of third parties on dyadic relationships.

Covariance, confoundment, and unclear causality problems

Sometimes a sex difference measured in a study originates from factors correlated with, but not logically dependent on, the genetic sex of the individuals in the study. These noncausal relationships are extremely troublesome because they can serve as a major source of confusion among researchers, as well as among critics of sexual dimorphism research. In fact, we are not absolutely confident that our own thinking about this issue is devoid of confusion, but we present our views here as a point of discussion. Basically, the problem for us is as follows: Let us consider some attribute such as physical size or strength that covaries with genetic sex in the sense of having overlapping but statistically distinct distributions for each sex. Now, if a measured behavioral sex difference is entirely accounted for by a size or strength factor, then it follows that how big or strong the organism is, and not what sex it is, actually controls the sex difference. Thus, animals of both sexes matched exactly for these causal variables should not reveal the behavioral sex differences found in a random sample of males and females for that response.

In one sense, we would be simply describing a legitimate sex difference in which genetic sex or strength is the causal explanation for a measured sex difference. In another sense, however, we are tapping an area of sex differences where the emphasis on sex could be considered to be misleading, because the controlling mechanism is not an inherent aspect of sexuality, but instead is an associated phenomenon of relative size/strength differences.

Factors of relative social dominance in group situations could work like this as well. For example, in our heterosexual monkey groups of preadolescents, comprised of four to six individuals, we usually find that males mount at much higher frequency and in much greater proportion (98% versus 25%) than do females. We also have noted that males usually occupy higher positions of social dominance in our groups than do females. Finally, we have noted that high-and middle-ranking dominant males mount at higher frequencies than do low-ranking males, and that almost all of the females that do mount are high ranking (Goldfoot, 1978). We suspect, therefore, that some social factor related to the distribution of dominance rank has some influence over the display of mounting by low-ranking individuals of either sex. When we study groups in which there are no males (Goldfoot, Wallen, Neff, McBrair, & Goy, 1984), we see statistically higher percentages of females mounting than we do in mixed-sex groups. We see further reductions in percentages of animals mounting in lower social positions when we manipulate rearing conditions such that dominance contests are more severe than in our standard conditions of rearing (Goy & Goldfoot, 1974). Under these conditions, the probability of a female mounting is near zero. These observations all support the notion that some social factor related to dominance rank is influencing the probability of display of our target behavior, and that the factor influences both sexes.

Because of much work with endocrinological variables, we know that mounting is strongly influenced by the prenatal endocrine environment (e.g., Goy, 1978), and so sex differences in mounting could never be considered to be determined entirely by social factors. Now we suspect that mounting is not *exclusively* determined by hormonal factors; in fact, we must understand the social conditions influencing our target behavior in order to interpret accurately effects of hormonal manipulations.

Many other investigators have observed the way in which social rank influences the frequency of response of many sociosexual responses. For example, Keverne et al. (1981), discussing their work with group-housed talapoin monkeys, write as follows:

> [Our] experiments show that in a social group of talapoin monkeys a number of behaviours are not shown equally by different individuals but are related to the status of that individual in the group's aggressive hierarchy. Thus, high ranking individuals take part in more sexual interactions, are monitored more closely by others, receive little aggression and display aggressive behaviour to others. In marked contrast, the lowest extreme of the hierarchy is characterised by monkeys which take no part in sexual activity, show high levels of visual monitoring while being monitored little, receive relatively high levels of aggressive behaviour, but are themselves not aggressive to others. (p. 90)

Keverne et al. were not concerned with sex differences in their study, and in fact were discussing only males in this particular quotation, but their observations, as well as those from many others (Simpson, 1973; Nishida, 1970; Goldfoot, 1971; Ploog, Blitz, & Ploog, 1963), are in agreement that one's position of dominance influences the probability of display of a variety of social behaviors. Our point, of course, is that many times position of dominance and sex of the individual are confounded, usually with males having dominance advantages. In some respects, then, what is interpreted as a sex difference might in fact be mediated at some level of control by the social position the individual occupies. Other kinds of dimorphic behavior (e.g., rough-and-tumble play) do not necessarily show the same sensitivity to these kinds of social factors. In our own data, display of rough-and-tumble play does not seem to be related to position of dominance and is not shown at elevated frequencies by females in our all-female groups, even by those females that do show elevations in mounting (Goldfoot et al., 1984; Goldfoot & Wallen, 1979). Distinct kinds of behavioral sex differences seem to exist, with some resulting, at least in part, from identifiable physical or social attributes that co-vary with, but are not necessarily aspects of, one's sex.

Animal Studies of Behavioral Sex Differences

The major reason animals are used to study mechanisms governing expression of behavioral patterns is control. Simple accessible environments can be created in which behavior can be observed. Past experience can be controlled or at least documented. Subjects can be exposed to conditions of physical or social environmental stress and deprivation that are simply not possible with human subjects (Schwartz, 1978) In our own field, virilizing agents are administered transplacentally or directly to the fetus; glands are removed; steroids, etc., are administered exogenously and/or measured in response to various physiological or environmental conditions; lesions are made to specific areas of the brain.

For a comprehensive review of the work accomplished in this field, the reader is referred to Goy and McEwen (1980), in which the findings of a neurosciences research program devoted to the biology of sexual differentiation are summarized. In spite of the obvious advantage afforded by the use of animals for research of this nature, problems have been raised that potentially restrict the interpretation of results. The first caution comes from the evolutionary principle that each species has had its own unique set of survival pressures that have influenced its current morphology and behavior in its present natural environment. Phenotypic expression of behavioral patterns at very simple levels can be dependent on the physical and social parameters of the environment in which they evolved. Even when attempts are made to provide the animals with environmental stimulus characteristics relevant to the behavioral system of study, there is room for disagreement between the experimenter's view of the situation and the animals. As control procedures attempt to parcel variables that may influence expression of behavior patterns, cues from the environment become more and more arbitrary. Possibly the more arbitrary the experimental environment, the more arbitrary is the behavioral response of the animal.

The second problem comes from the psychological/ethological principle that all behavior studied is related to the environmental context in which it occurs, even behavior strongly influenced by genetic, hormonal, or other concurrent physiological conditions (Bleier, 1976, 1979; Hubbard & Lowe, 1979; Tobach & Rosoff, 1978). Some argue that if the animal species studied has a social organization divergent from that of human beings, the relevance of sex-typed behavior to human situations would be questionable, and hence the finding would become vulnerable to misunderstanding. Most people would agree that human beings exist in the most complex social network known to science. Possibly major social variables playing important roles in determining human sex differences are absent from our animal models, which admittedly have varied but simpler social environments. Thus, potentially important controlling dimensions of the social environment salient to human behavior are likely to be overlooked or deemphasized in animal studies. This objection is currently being overcome with a greater emphasis on naturalistic conditions of experimentation, but it is far from a trivial objection.

The third problem of interpreting animal data for relevance to human activities constitutes a type of interaction of the first two problems. Because one common goal of animal research is to discover endocrine or other physiological mechanisms that account for a behavioral sex difference, and because scientists have indeed been quite successful in identifying such mechanisms throughout the animal kingdom, we tend to assume that a similar or identical biological mechanism underlies an analogous or homologous sex difference in humans (Bleier, 1976). This conclusion is in fact premature and of course needs additional documentation within our own species before it can be accepted.

> There are, of course, measurable biological and measurable psychological and behavioral differences (viewed statistically) between men and women. The serious logical fallacy often made, however, is to assume that there is a necessary *causal* relationship between these differences; for example between androgen and

> "aggressiveness." For this assumption there is no convincing evidence whatsoever in humans, and to use animal data as evidence that particular psychological or behavioral sex differences in humans are caused by particular *biological* sex differences is to ignore a fact that in other contexts is recognized to be of supreme importance: that humans are *qualitatively* different from all other animals. They are different precisely because of their brain and their culture which is the unique product of that brain. (Bleier, 1976, p. 68)

We find ourselves in agreement with part but not all of the position quoted. On the one hand, we acknowledge that biological principles derived from animal studies need not fit the human case or represent the causal factor(s) responsible for an observed sex difference. Moreover, we would add that when the animal species studied is a primate, the tendency to assume analogous mechanisms is even greater (Leibowitz, 1979; Weisstein, 1982) because of the genetic relatedness of these species to ourselves. Given the diversity of genera within the primate order and the lack of cross-species similarity of behavioral characteristics within them (Goldfoot, 1977; Mitchell, 1979), such tendencies need to be recognized as potentially inaccurate or misleading. Nonetheless, selected species of nonhuman primates do share with humans many physiological, endocrine, and anatomical similarities, and in addition manifest several behavioral characteristics including complex and wide-ranging sensitivities to social conditions (Goldfoot et al., 1984; Epple, Alveario, & Katz, 1982; Keverne, 1982; Bernstein, Gordon, & Peterson, 1979; Dixson & Herbert, 1977; Goldfoot, 1977; Missakian, 1972; de Waal, 1982; Stephenson, 1976; Carpenter, 1942). Studies of nonhuman primates, then, *might* contribute to the understanding of human behavior, as comparative data suggestive of research directions, not as demonstrations of findings directly applicable to humans.

We disagree, therefore, with the position quoted that comparative animal research is irrelevant to hypotheses concerning human behavior. Comparative data have revealed biological mechanisms broadly expressed throughout vertebrates that are strongly associated with the development and/or expression of dimorphic behavior patterns, and we would be mistaken to ignore the possibility of biological involvement in behavioral sex differences of human beings (Adkins-Regan, 1981; Goy & McEwen, 1980). The evidence of behavioral heredity, and hence evolutionary history, is simply too extensive to be dismissed as irrelevant for human beings. The comparative data cannot tell us what determines human behavior, but they are invaluable in guiding the construction of hypotheses concerning the participation of underlying mechanisms influencing human responses. Provided that we do not prematurely conclude similarity of mechanism from similarity of appearance, comparative data, far from being ignored, should be even more thoroughly considered to help us understand the evolutionary role and biological significance of sex differences in *Homo sapiens*.

We now turn in this discussion to a strange set of circumstances. Although exclusively detailed comparative data from a variety of laboratory species exist for hormonal (Adkins-Regan, 1981), neuroanatomical (Goy & McEwen, 1980), and biochemical (Feder, 1981; Crowley & Zemlan, 1981; McEwen, 1981) phenomena associated with behavioral sex differences, relatively little information from animal studies is available about social factors that might be involved in the development or maintenance of these response differences. Yet, we have nearly the opposite situation in the case of data regarding humans: much is known about social influences, and little is known about endocrine mediation (Money & Ehrhardt, 1976; Maccoby & Jacklin, 1974). With two entirely different data bases as well as with many species differences, no wonder the tired story of blind "wise men" feeling different parts of the elephant to understand its nature comes to mind. Money (1981) has discussed this situation extensively, and argues for more complete theoretical and experimental syntheses of biological and social determinants. We agree, and emphasize that the admonition applies not only to those who study humans, where more biological information is obviously necessary, but also to those who study animals, where more social information would probably provide considerably more clarity to these issues.

As we see it, then, studies investigating a possible role of social factors influencing the development and/or expression of behavioral sex differences are needed within a comparative framework and should be pursued in animal species. Moreover, such studies need to incorporate, when possible, hypotheses that would be sensitive to the interaction of social and physiological events. We know of few scientists who disagree with the view that sex differences are multiply determined. The challenge now is to design experiments that can study biological and social factors simultaneously to assess directly the interaction of experiential and biological variables that are hypothesized to be involved in the expression of sex differences.

Conclusion

This chapter began by discussing three sources of criticism and potential misunderstanding associated with studies of sex differences. The issues of inappropriate generalization and inappropriate social context were both examined and related to the central question of how animal studies might be used to understand human behavior. We acknowledged that data from animal studies need very careful and measured interpretations that reflect both the intent of the study and the inherent limitations of the experimental design that was utilized. In most cases, investigations of sex differences have focused on one or two operative factors, and have not attempted to incorporate all variables that might account for a given dimorphically expressed behavior under natural conditions. The approach of most investigators concerned with hormonal and other physiological mechanisms, in fact, has usually been to eliminate carefully, or at least to hold constant, socially mediated variables that affect the target behavior under consideration. This is an appropriate scientific strategy, provided that scientist and reader are both entirely aware that behavioral dimorphisms can be expected to be determined by multiple factors, and that both the description of the dimorphism and the conclusion related to the variable under study are limited to the exact conditions of evaluation used in the study.

The second part of this chapter dealt with specific kinds of approaches that can be taken to study behavioral sex differences. Part of the discussion emphasized limitations associated with designs that involved more than one subject observed at a time in the same environment. We suggested that studies involving measurements of social behavior be considered dynamic, and not entirely within the control of the investigator. We listed such factors as partner sex, sex ratio of the group, dominance influences, and group phenomena such as coalition formation that can all influence aspects of responding dimorphically. We also emphasized that choice of measurement (e.g., frequency of response or percentage of individuals classified as responders) can give different impressions of the nature of a sex difference. This realization is critical in evaluating just what sex difference is being reported and what possible underlying mechanisms might be hypothesized.

In a review of dimorphic behavior emphasizing biological mechanisms, Goy and Goldfoot (1973) described three categories of behavioral sex differences: (1) responses influenced by the prenatal endocrine environment and not dependent on concurrent hormones, (2) responses influenced by concurrent hormones and not dependent on prenatal endocrine conditions, and (3) responses dependent on both prenatal and concurrent endocrine factors. The differentiation of types of sex differences into these categories is important for eventually elucidating biological mechanisms responding to hormonal influence and resulting in manifestations of dimorphic behavior. We now suggest that social processes and physical attributes be identified and added to the list of moderator variables that affect dimorphic responding. We see these factors as interactive with endocrine-mediated events, and believe that most, if not all, behavioral sex differences reflect complicated interactions of endocrine, physical, and social variables. Our position, in fact, is that manipulation of social variables is as essential to biologically oriented studies as it is to socially oriented studies, because obvious misinterpretations of the effects of biological variables can occur under inappropriate or restrictive testing conditions. Also, to avoid the "assumed biological mediation" error, including relevant social and environmental factors in research designs is essential. Clearly, therefore, for a more comprehensive evaluation of an individual's potential to display a given behavior, a strategy of "comparative socialization" could be employed in which parameters of social condition as well as physiological factors are manipulated within the same experiment. One could then gauge more accurately the degree to which social factors regulate a dimorphic response, and at the same time, identify the extent to which the biological variable under study interacts with differing environmental situations. As we have already seen, sex difference studies rely heavily on concepts such as "predisposition" and "probability of response." We know that prenatal hormones can affect measures of these concepts, and yet we know that a female monkey's probability of mounting, for example, ranges from near 0% to 67%, depending on the social conditions of rearing and/or testing, quite independent of hormonal considerations. We see no need to revise the concepts of predisposition or response probability, nor to deny any of the demonstrated endocrine mediation of dimorphic behavior. We suggest, however, that the concepts of predisposition and behavioral dimorphism should only be defined relative to the social and environmental factors in which they are estimated. As we have attempted to show, they can be highly misleading concepts, unless adequate measures of cross-situational stability are demonstrated, or unless the specific environmental and social conditions necessary for their manifestation are clarified.

Acknowledgements

This chapter, publication 25-028 of the Wisconsin Regional Primate Research Center, was supported in part by grants from NIH (RR-00167) and NIMH (MH-21312). Special thanks are extended to R. W. Goy for years of discussion and exchange of opinion.

References

Adkins-Regan, E. (1981). Early organizational effects of hormones: An evolutionary perspective. In N. T. Adler (Ed.), *Neuroendocrinology of reproduction* (pp. 159–228). New York: Plenum Press.

Bernstein, I. S., Gordon, T. P., & Peterson, M. (1979). Role behavior of an agonadal alpha male rhesus monkey in a heterosexual group. *Folia Primatologica, 32*, 263–267.

Bleier, R. (1979). Social and political bias in science: An examination of animal studies and their generalizations to human behavior and evolution. In

R. E. Hubbard & M. Lowe (Eds.), *Genes and gender: II. Pitfalls in research on sex and gender* (pp. 49–70). New York: Gordian Press.

Bleier, R. H. (1976). Brain, body, and behavior. In J. I. Roberts (Ed.), *Beyond intellectual sexism: A new woman, a new reality.* New York: David McKay.

Brain, P. F. & Nowell, N. W. (1970). Some observations on intermale aggression testing in albino mice. *Comm. Behav. Biol., 5,* 7–17.

Carpenter, C. R. (1942). Sexual behavior of free-ranging rhesus monkeys *(Macaca mulatta).* II. Periodicity of oestrus, homosexual, autoerotic and nonconformist behavior. *Journal of Comparative Psychology, 33,* 143–162.

Crowley, W. R., & Zemlan, F. P. (1981). The neurochemical control of mating behavior. In N. T. Adler (Ed.), *Neuroendocrinology of reproduction* (pp. 451–476). New York: Plenum Press.

Denenberg, V. H., Gaulin-Kremer, E., Gandelman, R., & Zarrow, M. X. (1973). The development of standard stimulus animals for mouse *(Mus musculus)* aggression testing by means of olfactory bulbectomy. *An. Behav., 21,* 590–598.

Dixson, A. F., & Herbert, J. (1977). Testosterone, aggressive behavior and dominance rank in captive adult male talapoin monkeys *(Miophithecus talapoin). Physiology and Behavior, 18,* 539–543.

Epple, G., Alveario, M. C., & Katz, Y. (1982). The role of chemical communication in aggressive behavior and its gonadal control in the tamarin *(Saguinus fusciocollis).* In C. T. Snowdon, C. H. Brown, & M. R. Petersen (Eds.), *Primate communication* (pp. 279–302). London: Cambridge University Press.

Feder, H. H. (1981). Perinatal hormones and their role in the development of sexually dimorphic behaviors. In N. T. Adler (Ed.), *Neuroendocrinology of reproduction* (pp. 127–158). New York: Plenum Press.

Gibber, J. R. (1981). Infant-directed behaviors in male and female rhesus monkeys. Unpublished doctoral dissertation, University of Wisconsin–Madison, Dept. of Psychology.

Goldfoot, D. A. (1977). Sociosexual behavior of nonhuman primates during development and maturity: Social and hormonal relationships. In A. M. Schrier (Ed.), *Behavioral primatology: Advances in research and theory: Vol. I* (pp. 139–184). Hillsdale, NJ: Lawrence Erlbaum Assoc.

Goldfoot, D. A., & Wallen, K. (1978). Development of gender role behaviors in heterosexual and isosexual groups of infant rhesus monkeys. In D. J. Chivers & J. Herbert (Eds.), *Recent advances in primatology: Vol. 1: Behavior. Proceedings of the Sixth Congress of the International Primatological Society.* (pp. 155–159). New York: Academic Press.

Goldfoot, D. A., Wallen, K., Neff, D. A., McBrair, M. C., & Goy, R. W. (1984). Social influences upon the display of sexually dimorphic behavior in rhesus monkeys: Isosexual rearing. *Archives of Sexual Behavior, 13* (5), 395–412.

Goy, R. W., & Goldfoot, D. A. (1973). Hormonal influences on sexually dimorphic behavior. In R. O. Greep & E. B. Astwood (Eds.), *Handbook of physiology: Endocrinology: Vol. 2, Part 1* (pp. 169–186). Baltimore: Williams and Wilkins.

Goy, R. W., & Goldfoot, D. A. (1974). Experiential and hormonal factors influencing development of sexual behavior in the male rhesus monkey. In F. O. Schmidt & F. G. Worden (Eds.), *The neurosciences; Third study program* (pp. 571–581). Cambridge: MIT Press.

Goy, R. W., & McEwen, B. S. (1980). *Sexual differentiation of the brain* (pp. 1–63, 102–211). Cambridge: MIT Press.

Goy, R. W., & Young, W. C. (1957). Somatic basis of sexual behavior patterns in guinea pigs: Factors involved in the determination of the character of the soma of the female. *Psychosom. Med., 19,* 114–151.

Hubbard, R., & Lowe, M. (Eds.). (1979). *Genes and Genders: II Pitfalls in Research on Sex and Gender.* New York: Gordian Press.

Kass-Simon, G. (1976). Female strategies: Animal adaptations and adaptive significance. In J. I. Roberts (Ed.), *Beyond intellectual sexism: A new woman, A new reality.* New York: David McKay.

Keverne, E. B. (1976). Sexual receptivity and attractiveness in female rhesus monkeys. *Advan. Study Behav., 7,* 155–200.

Keverne, E. B. (1982). Olfaction and the reproductive behavior of nonhuman primates. In C. T. Snowdon, C. H. Brown, & M. R. Petersen (Eds.), *Primate communication.* London: Cambridge University Press.

Keverne, E. B., Meller, R. E., Eberhart, A. (1982). Dominance and subordination: Concepts or physiological states? In A. B. Chiarelli & R. S. Corruccini (Eds.), *Advanced views in primate biology:* Main Lectures of the VIIIth Congress of the International Primatological Society, Florence, July 7–12, 1980 (pp. 81–94). New York: Springer-Verlag.

Lamb, Michael E. (1979). The effects of the social context on dyadic social interaction. In M. E. Lamb, S. J. Suomi, & G. R. Stephenson (Eds.), *Social interaction analysis: Methodological issues* (pp. 253–268). Madison: University of Wisconsin Press.

Larsson, K. (1973). Sexual behavior: The result of an interaction. In J. Zubin & J. Money (Eds.), *Contemporary sexual behavior: Critical issues in the 1970s* (pp. 33–51). Baltimore: Johns Hopkins University Press.

Leibowitz, L. (1979). "Universals" and male dominance among primates: A critical examination. In R. Hubbard & M. Lowe (Eds.), *Genes and gender: II. Pitfalls in research on sex and gender.* New York: Gordian Press.

Leshner, A. E. (1978). *An introduction to behavioral endocrinology.* New York: Oxford University Press.

Maccoby, E. E., & Jacklin, C. N. (1974). *The psychology of sex differences* (pp. 277–376). Stanford: Stanford University Press.

McClintock, M., Anisko, J. J., & Adler, N. T. (1982). Group mating among Norway rats. II. The social dynamics of copulation: Competition, cooperation and mate choice. *Anim. Behav., 30,* 410–425.

McClintock, M. K. (1981) Simplicity from complexity: A naturalistic approach to behavior and neuroendocrine function. In I. Silverman (Ed.), *Laboratory and life: New directions for methodology of social and behavioral research:* No. 8. San Francisco: Jossey-Bass.

McEwen, B. S. (1981). Cellular biochemistry of hormone action in brain and pituitary. In N. T. Adler (Ed.), *Neuroendocrinology of reproduction* (pp. 485–518). New York: Plenum Press.

Missakian, E. A. (1972). Genealogical and cross-genealogical dominance relations in a group of free-ranging rhesus monkeys *(Macaca mulatta)* on Cayo Santiago. *Primates, 13,* 169–180.

Mitchell, G. (1979). *Behavioral sex differences in nonhuman primates.* New York: Van Nostrand Reinhold.

Money, J. (1981). The development of sexuality and eroticism in humankind. *Quart. Rev. Biol., 56,* 379–404.

Money, J., & Ehrhardt, A. A. (1972). *Man & woman, boy & girl.* Baltimore: Johns Hopkins University Press.

Nadler, R. D. (1982). Laboratory research on sexual behavior and reproduction of gorillas and orangutans. *Amer. J. Primatol., Suppl. 1,* 57–66.

Nishida, T. (1970). Social behavior and relationship among wild chimpanzees of the Maliali mountains. *Primates, 111,* 47–87.

Patterson, G. R., & Moore, D. (1979). Interactive patterns as units of behavior. In M. E. Lamb, S. J. Suomi, & G. R. Stephensen (Eds.), *Social interaction analysis: Methodological issues.* Madison: University of Wisconsin Press.

Perachio, A. A. (1978). Hypothalamic regulation of behavioral and hormonal aspects of aggression and sexual performance. In D. J. Chivers & J. Herbert (Eds.), *Recent advances in primatology, Vol. 1: Behavior. Proceedings of the Sixth Congress of the International Primatological Society.* New York: Academic Press.

Ploog, D. W., Blitz, J., & Ploog, F. (1963). Studies on social and sexual behaviour of the squirrel monkey (simiri sciurens). *Folia Primatol., 1,* 29–66.

Pomerantz, S. M., & Goy, R. W. (1983). Proceptive behavior of female rhesus monkeys during tests with tethered males. *Hormones and Behavior, 17,* 237–248.

Schwartz, B. (1978). *Psychology of learning & behavior,* New York: Norton.

Seitz, A. (1940). Die Paarbildung bei einigen Cichliden. I. Die Paarbildung bei *Astatotilapia* strigigena). *Zeit. f. Tierpsychol., 4,* 40–84.

Sherman, J. A. (1976). Some psychological "facts" about women: Will the real Ms. please stand up? In J. I. Roberts (Ed.), *Beyond intellectual sexism: A new woman, a new reality.* New York: David McKay.

Simpson, M. J. A. (1973). The social grooming of male chimpanzees. In R. P. Michael & J. H. Crook (Eds.), *The comparative ecology and behaviour of primates.* New York: Academic Press.

Stephensen, G. R. (1973). Testing for group specific communication patterns in Japanese macaques. In E. W. Menzel (Ed.), *Symposia of the Fourth International Congress of Primatology: Vol. 1.* Basel: Karger.

Susman, R. L. (1984). *The pygmy chimpanzee: Evolutionary biology and behavior.* New York: Plenum Press.

Tobach, E., & Rosoff, B. (Eds.). (1978). *Genes and gender.* New York: Gordian Press.

Waal, F. de. (1982). *Chimpanzee politics.* London: Jonathan Cape.

Weisstein, N. (1985, November). Tired of arguing about biological inferiority? *Ms,* 41–46.

13

Assessment of Behavioral Sex Differences: A Commentary

R. W. Goy

The notion is not strange that behavior bears, or can come to bear, a predictable relationship to the specific conditions under which it occurs or by which it is elicited. Such a notion has underlain all of stimulus-response (S-R) theorizing in psychology from the earliest associationisms to modern learning theory. In ethological theories as well, the notion of an optimal stimulus with precisely identifiable critical elements figures strongly in an accounting of response tendencies, and deviations from the optimum stimulus can be quantitatively related to variations in frequency or intensity of responding. Perhaps because this notion of the determining influence of specific conditions is so ubiquitously represented in theories of behavior, readers will find the message from Goldfoot and Neff (Chapter 12) easy, even reassuring, to accept. Yet the caveat they articulate so clearly is sufficiently profound as applied to sex differences that it merits more than easy acceptance, and some expatiation of this theme is warranted because of its essential role in systematization. An appropriate way to begin this further elaboration of their caveat is with its repetition: "A definition of a behavioral sex difference must be statistically defensible, and must include reference to the environment and the conditions of assessment in which it is measured."

A noteworthy departure of the Goldfoot and Neff statement from the basic tenets of S-R theories is that it attempts to deal not with a response, but with a *difference* between responses from two individuals or classes. Our ability to account for the difference between two individuals' responses (or between two classes' responses) to the same stimuli has been much the same within or without the framework of S-R theory. Such differences, when not due to chance alone, are attributed speculatively, but not without reason, to organismic variables such as genetic endowment, past experience, current state, motivational level, or perceptual, emotional, or cognitive processes. These variables, as psychologists know very well, are not totally independent of one another. For one example, a stimulus that in one individual produces an initial interfering emotion can be "neutralized" by appropriate repetitive experience. For another, attentional processes can be altered quantitatively and selectively by corresponding manipulations of motivational level. The interdependencies and interactive nature of these variables render more difficult than one might wish the pure analysis of the causes of *differences* between individuals or classes in their response to specific stimuli. Costly and large-scale experimental designs have to be employed to determine if the difference in response as measured might be because of interactions of these underlying variables that are unique to each class. In a theoretical framework, the discovery of unique interactions is not necessarily trivial. Suppose, for instance, that on a given task females performed at higher rates than males and that prior relevant experience interacted positively with motivation in females, whereas current state interacted negatively with motivation in males. Such findings, given an adequate experimental design, might be the kind that would improve our ability to interpret a sex difference in observable events at the level of causation by psychological process or mechanism.

It is not worthwhile here to indicate the scope of the experimental design(s) that would be required to permit statistically valid inferences regarding underlying mechanisms of behavioral sex differences. The specific designs would vary considerably, depending on the specific stimuli (social and nonsocial) used in assessment, because these stimuli would dictate our estimates of which organismic variables would need to be held constant and which would need to be varied systematically. The main point to be taken from this discourse is that studies of sex-related social behavior rarely, if ever, have been carried out on a scope permitting such analyses and inferences. Even at the simplest level of dyadic testing, designs of the required scope have been largely ignored, and investigators have chosen single-variable experiments over more complex routes of inquiry. The generally adopted approach, then, has been inefficiently slow as well as contributory to major gaps in our knowledge.

Confronted with such obvious deficiencies in our state of knowledge, one could welcome Goldfoot and Neff's appeal for increased systematic variation of social variables as, indeed, a step forward. The nature of their appeal is not for multivariate designs per se, but to evaluate sex differences in performance of specific behavior in the dyadic situation and in groups varying in size, sex ratio, age-class ratio, etc.—in short, in varying situations and social contexts. They have shown with examples from their own work that large-scale multivariate designs are not essential to the discovery of underlying causal factors. We have no disagreement about the need for or value of such an increased scope of investigation, but it seems important to point out previously unmentioned hazards of interpretation that might arise when comparing results obtained under different assessment conditions. Goldfoot and Neff (Chapter 12) have already described experimental results showing that more juvenile female monkeys mount when housed only with other females than when housed in groups containing both males and females. In the latter housing situation, many more males show mounting than females. They offer evidence from other single-variable studies to suggest that the situational difference in female mounting performance is attributable to a differing access to higher dominance positions; that is, in heterosexual groups males more commonly occupy positions of higher social dominance, but in all-female groups females necessarily occupy high-ranking positions. The description of these results is a testimony to the power of systematic variation of social variables to uncover an underlying psychosocial mechanism that contributes to a behavioral sex difference. As these results are described, however, one can infer from this information that there is no sex difference in mounting behavior, but only a sex difference in the tendency or ability to win or assume high dominance status. Such an inference can be shown to be correct only for the particular measure of mounting behavior used, however—namely, proportion of animals showing the behavior regardless of the frequency with which it is displayed. When the frequency of mounting, which is also dimorphic, is considered, isosexually housed females do not mount more often than those females that show mounting in heterosexual housing conditions; they do not mount as often as males housed either heterosexually or isosexually, and they display a form of mounting response that is rarely, if ever, used by males. Furthermore, additional analysis has shown that in heterosexual housing conditions, dominant females do not mount as frequently as males who are equal in dominance status.

The point of articulating the limitation of the influence of dominance status on the sex difference in mounting is to emphasize, or reemphasize, that not only are specific conditions of testing crucial to the definition of a sex difference, but so is the specific aspect that is measured of the behavioral response. As the analysis provided by Goldfoot and Neff of mounting by juvenile monkeys has shown, dominance strongly influences the sex difference in proportion of each sex showing mounting, but it has much less influence over the sex difference in frequency of mounting as well as the type of mounting. Hence the conclusion that one aspect of mounting is socially mediated and its dimorphic expression is situationally influenced should not be generalized to mean that all dimorphic aspects of mounting are similarly disposed at the causal level. After all, it would be somewhat unconvincing to assert that there is no sex difference in playing baseball. Within the women's league women may attain batting averages as good as men have in the men's league. Batting averages, however, are not the only measure of baseball playing, and the styles of play have totally different characteristics. Furthermore, when a team of women plays against a team of men, even batting averages become disparate. In the case of baseball, we can comfortably "explain" the sex difference in terms of differential physical attributes even without assessment of the difference in varying situations.

Surely a part of the value of the type of multiple-assessment approach that Goldfoot and Neff are advocating is that it reorients our thinking about and conceptualization of behavioral sex differences. For example, in studying the behavior of juvenile monkeys living in heterosexual groups, Goldfoot and Neff's position could be interpreted to mean that the question of why more males than females display mounting has been transformed to the new question of why more males than females win or assume higher dominance positions. Let us hope, casting cynicism aside, that the pursuit of this new question will not lead us back along the circle to the conclusion that the male monkeys generally have higher dominance because they show more frequent mounting than females. Despite the circumstance that years of study have shown little or nothing about the factors that confer high social dominance on one individual as opposed to another, it seems entirely reasonable to anticipate that Goldfoot and Neff's analysis of monkey mounting behavior will lead to new experimental attacks on the problems of social dominance within the context of sex differences. If the example provided by Goldfoot and Neff were used as a model for investigation of other behavioral sex differences, we might reasonably expect that future work would be as clearly redirected toward an examination of variables more fundamental to issues of masculinity/femininity for those behavioral sex differences as well.

Is there a paradoxical message in the paper by Goldfoot and Neff? Some implication in their article insinuates itself into the thought that no behavioral sex difference should be accepted for incorporation into theory construction until it has been shown not to exist—that is, until evaluated in enough situations and at enough levels of complexity sufficient to its dissolution. Are they saying, in effect, that there is no sex difference until there is no sex difference? Neither I nor they would stress this implication to the point of a maxim, but the thought itself is worth more than passing mention if for no other reason than their frequent use of examples of sex differences conspicuous to one testing situation that become minimal or disappear altogether in another testing situation. In the broader context of their writing, they clearly believe that a multiple-assessment approach, even when it uncovers situations that dissolve the sex difference, leads to greater precision in defining the sex difference or to a potential discovery of underlying mechanisms. One cannot argue for the value of such objectives, or for the use of multiple approaches to achieve them, more persuasively than they have done already. One can, however, see clearly that the pursuit of variations in testing conditions until a manifest sex difference no longer exists could be a totally idle effort in some instances. For example, would the field of reproductive endocrinology really advance if investigators continued developing hormone assays to the point where a sex difference in circulating levels of testosterone could no longer be measured? I believe the only lesson that could be learned from such an exercise is one of how to measure. Nothing would or could be learned about the sex difference. In fact, the history of chemical assays for reproductive hormones provides an example of the multiple-assessment approach employed in the opposite direction from the examples provided by Goldfoot and Neff: namely, to magnify instead of diminish the measurable sex difference.

At this point any erroneous impression I have created about what Goldfoot and Neff are advocating by their contrary examples of the ephemeral nature of sex differences in behavior should be corrected. Their purpose in so elaborately demonstrating this property of behavioral sex differences is to convince all of us that behavioral dimorphisms are determined by multiple factors, and each description of the dimorphism as well as any conclusion related to the variable under study is limited to the exact conditions of assessment used. The role and extent of influence of these factors, or even their nature, cannot be realized in a single-variable method of evaluation or assessment of the behavioral trait under study, because investigators tend to focus on the influence of the single factor that interests them most. No one should expect, however, that every investigator will undertake multiple evaluations of a behavioral response in order to demonstrate the possible influences of all factors. The message from Goldfoot and Neff can go a long way to help us all understand why individual investigators, each working blindly on a team of elephant feelers, can come to different conclusions about the factor that influences a given behavioral response, and even different conclusions about whether or not behavior is sexually dimorphic.

I would not like to end this commentary by conveying the impression to readers that only behavioral dimorphisms are subject to modification and/or extensive influence by environmental variables. On the contrary, any trait that is expressed as a result of neural integration of environmental input and that is also ordinarily sexually dimorphic could be expected to be manifested in ways that would reflect alterations in the environmental input. I am confident that Goldfoot and Neff would agree with this premise as asserted. A concrete example illustrating the premise can be found in the pattern of pituitary gonadotropin secretory activity in rats. This pattern can be described as sexually dimorphic in the sense that it is tonic in the male and episodic or cyclical in the female rat. Moreover, in the female, but not so conspicuously in the male, the cyclical character of gonadotropin release is heavily regulated by brain structures that integrate environmental information about the circadian changes in illumination. If an adult female rat is placed in constant light (i.e., if the input of environmental information is drastically altered), the pituitary pattern of gonadotropin secretion is converted from one of cyclicity to the tonic pattern characteristic of the male. This should not be interpreted to mean that male rats normally go around with the sun and the stars in their eyes, but it almost certainly means that the male rat brain does not integrate and make use of environmental information about circadian changes in illumination in the same way as the female rat brain. One cannot help but wonder—indeed hope—that the type of social/environmental manipulations advocated by Goldfoot and Neff will lead to similar significant discoveries about dimorphisms in brain function and the expression of behavioral sex differences.

14

Early Sex Role Behavior and School Age Adjustment

Michael Lewis

The study of human sexuality is complex and varied, ranging from the study of the biological to the study of the psychological. To define masculinity and femininity, we should specify many features that are normally used to understand these terms. We will briefly consider three of them before turning our attention to the more specific task of determining whether a failure in any one of these aspects has significant impact on the organism's subsequent development. Thus, this essay will move from a general discussion of the development of gender identity, sex role behavior, and sexual preference, and the models underlying their development, to a more specific focus on the role of early sex role behavior on subsequent adjustment.

Gender Identity, Sex Role, and Sexual Preference

Gender Identity

Gender identity is the belief that one is either a male or female. Such a belief system is more than a simple labeling phenomenon because not only does the organism have cognitive schema (knowledge about males and females and knowledge of one's self as either one or the other), but, even more important, the organism has a strong affective commitment to that label and schema. This affective commitment can be easily seen when a young child of 2 to 4 years of age is challenged by an experimenter who says, "What a nice little (girl or boy) you are" using the incorrect gender label. In general, children of this age will respond to the challenge by becoming upset over the observer's mislabeling their gender. The affective response of the children informs us (1) that they have a gender identity, and (2) that their schema has an affective as well as cognitive component.

The developmental research on gender identity is not considerable. Much of the existing work follows Kohlberg's (1966) argument that identity is not established until sometime around 6 years of age. For theorists committed to a cognitive/developmental approach, the proof of the establishment of gender identity is that organisms younger than 6 years of age are unable to maintain gender consistency in the face of experimental manipulation of this concept. Thus, for example, if the experimenter puts a skirt and long hair on a male, a child younger than 6 years may say that the figure is no longer a man but a woman (DeVries, 1969; Marcus & Overton, 1978). This lack of conservation is used as evidence to support the belief that gender identity is not well established until particular cognitive features have emerged.

Contrary arguments have been proposed. Part of the confusion, as suggested by Emmerich, Goldman, Kirsh, and Sharabany (1977), is that gender identity has been linked with the understanding of gender and all it implies. Gender identity can be said to exist at a variety of *levels of knowledge.* Experimental evidence would appear to indicate gender identity at considerably younger ages than those proposed by Kohlberg. Money and Erhardt (1972) have argued that overandrogynized females (those who have male external genitalia) are reluctant to give up the male gender label they have been assigned since birth if sex reassignment is performed later than 18 to 24 months of age. Likewise, Lewis (Lewis & Brooks-Gunn, 1979; Lewis & Weinraub, 1979; Lewis, 1981) has reviewed the infancy data and has shown that gender identity, as part of self-identity, is established as both an affective and a cognitive schema sometime in the second year of life. Moreover, increased focus on gender identity, measured by children's classification behavior, indicates that children 2–3 years of age show gender identity (Weinraub & Brown, 1983; Weinraub, Clemons, Sockloff, Eldridge, Gravely & Myers, 1984). These and other studies (Kuhn, Nash & Brucken, 1978; Thompson, 1975) suggest that gender identity should be considered as having both an affective and a cognitive component and that aspects of identity can be shown to exist by 2 years of age.

Sexual Preference

The second feature of human sexuality, sexual preference, has received almost no attention in the developmental literature. The understanding of sexual attraction and the choice of sexual partners cannot readily be studied prior to puberty and the emergence of a strong sex drive. Thus, relatively little research information is related to the development of sex-object preference. Individual differences related to homosexuality, heterosexuality, bisexuality, and fetishes have gone relatively unexplored. Only one aspect of sex-object preference has been touched on developmentally: that related to young children's dressing up in opposite or same-sex adult clothes. Green (1975) has argued that early socialization practices seem to be related, in part, to later cross-sex dressing. Even though research in this area is limited, it would be of some interest to explore developmentally children's understanding and knowledge of dating, mating, and marriage practices as they exist in their particular cultures. Moreover, young children's playmate structure and subsequent sexual preference should be explored.

Sex Role Behavior

Sex role behavior commonly refers to the organism's knowledge and behavior as it relates to other conspecifics having the same sex. Sex role behavior is recognized to be in part culturally related. Thus, for example, for U. S. men, dressing in a skirt is inappropriate, whereas Scottish men do wear a skirt (or kilt). Wearing a skirt as a sex-role-appropriate behavior is culturally defined. The debate over whether sex role behavior is biologically determined or mostly culturally determined we will leave for others (Money & Erhardt, 1972). The argument is heated, interesting, informative, and yet not settled. One would be hard put to imagine that sex role behavior did not have some biological connection; in like fashion, to excuse the culture from exerting an enormous sex role effect or definition would be foolhardy.

The question of sex role behavior and its acquisition has engaged the interest of developmental scientists. Although there are several literature reviews of early sex role behavior differences, many of them appear flawed. For example, Block (1977) has pointed out the omissions and technical errors to be found in one highly used review, Maccoby and Jacklin (1974). Nevertheless, there appears to be general agreement that by 2 to 3 years of age, children's play behavior with peers, their toy behavior, and their knowledge about sex roles and activities indicate that they possess considerable sex role knowledge (Lewis & Weinraub, 1979; Serbin, 1980; Weinraub & Brown, 1983, for reviews).

How early can sex role behaviors and knowledge be demonstrated? Not all studies present the same findings, but sex role behavior in the first and second year of life have been reported, both in terms of parental behavior toward the child and in terms of children's behavior. Parental sex role behavior has been reported for: (1) giving children sex-appropriate toys (Rheingold & Cook, 1975; Lewis & Michalson, 1982); (2) types of clothes children wear (Brooks & Lewis, 1974); (3) parent-child interactions (for example, Lewis, 1972; Lewis & Cherry, 1977; Moss, 1974; Parke & O'Leary, 1975; Smith & Daglish, 1977); and (4) adult attitudes (Rubin, Provenzano, & Luria, 1974). In turn, early sex role behavior in children has been observed by the end of the first year and on into the second and third. These include: (1) play behavior (for example, Goldberg & Lewis, 1969; Feiring & Lewis, 1979; Fagot, 1974; Smith & Daglish, 1977; Serbin, 1980; Connor & Serbin, 1977; Rikers, 1975; Serbin, Connor, Burchardt, & Citron, 1979; Fagot & Patterson, 1969); and (2) knowledge of sex role behavior (Lewis & Brooks-Gunn, 1979; Weinraub & Brown, 1983; Weinraub et al., 1984; Kuhn et al., 1978). There is also some support for differential social networks by 3 years with children playing more with the same-sex than with opposite-sex playmates (Feiring & Lewis, 1979; Jacklin & Maccoby, 1978).

From a developmental perspective, at between 2 and 3 years of age, children clearly have established gender identity and show (both through behavior and their knowledge) that they know of some of the sex-role-appropriate behavior of their culture. Doubtless these skills are tied to the children's developmental level, and with increased information and mental ability their demonstration of gender identity and sex role knowledge will increase. That they will be able to conserve gender identity under experimenter-induced variations and that they will become better able at labeling, classifying, and articulating sex-role-stereotyped behaviors should not subtract from our willingness to describe even young children as possessing these skills. It is reasonable to assume that these skills, present by the end of the second year of life, exhibit an influence on subsequent development. There is, however, little data to demonstrate how these early skills relate to subsequent development and adjustment, and we now turn toward that issue.

Sex Role Play Behavior and Development

Before considering the developmental issues pertaining to individual differences in early sex role play, we should consider both the context of sex role toy play and the possible mechanisms underlying its occurrence. Parental concern for children's appropriate sex role toy play remains quite strong in spite of changing values. Parents in general are concerned with the types of toys with which their children play. Such concern manifests itself in the toys they purchase for their children (Rheingold & Cook, 1975) and the types of toys the children have in their homes (Lewis & Michalson, 1982). It also is manifested in their behavior toward their children when they play with toys inappropriate for their sex role. The negative sanctions that accompany their children's inappropriate sex role toy play acts both as a negative reinforcement toward that play and as a cognitive guide to what behaviors and what objects are sex role appropriate (see Lewis & Weinraub, 1979).

In return, young children's toy-play behavior reflects sex-role-appropriate behavior and the socialization of that behavior. Although claim for a biological basis for toy-play behavior might be made, especially on the basis of toys that require gross motor activity, we have little reason to assume that biological imperatives play an active part in influencing sex role toy behavior per se. Instead, socialization practices appear to account for the differences observed in children's play. Lewis (1980) articulated two distinct processes, although more are possible. From the point of view of the child, the first set of processes can be considered passive—that is, they act *on* the child. The second set is active, in that they require the child's interaction and construction of reality. In the first set are simple reinforcement arguments that suggest that positive and negative sanctions shape the young child's toy-play behavior. These sanctions can be delivered by parents, relatives, other adults, siblings, and friends (Serbin, 1980). Young children's toy-play behavior is thus influenced by these external direct pressures. Similarly, less direct external pressures influence the child's behavior, and while not falling into a reinforcement paradigm per se, can be said to passively determine sex-role-appropriate toy play. Instead of reinforcing the child's action, they determine that action by limiting possible choices. Parent and adult selection of the toys children play with acts in this fashion. The parent will have little need to reinforce negatively a boy's playing with dolls if no dolls are bought for the child in the first place. Such a pattern of behavior control can be brought about in other situations beside the home (for example, at school) and by socialization agents other than the parents (for example, relatives, teachers, and peers).

For example, Lewis & Michalson (1982) asked 128 mothers to report which of a set of 10 toys were likely to be found in their children's home. Figure 14-1 presents the percent of parents who report their 2-year-old child as having a specific toy in their home. There were approximately equal numbers of mothers of boy and girl children (62 males, 68 females).

Notice that male children have significantly more toy guns, blocks, and trucks in their homes, while female children have significantly more dolls and toy pianos. That children tend to play with the toys they are given by their parents should come as no surprise.

In the second set of processes, we need to consider those active processes that usually consist of cognitive structure facilitators—that is, those events likely to lead the child into creating the rules that govern its sex role behavior. For these processes we can assume that the critical factor is in the child's *own* creation of the rule. Thus, passive forms of control can operate in an active manner by supplying the information the child needs in order to create a cognitive structure. This cognitive structure allows both understanding of the action of those di-

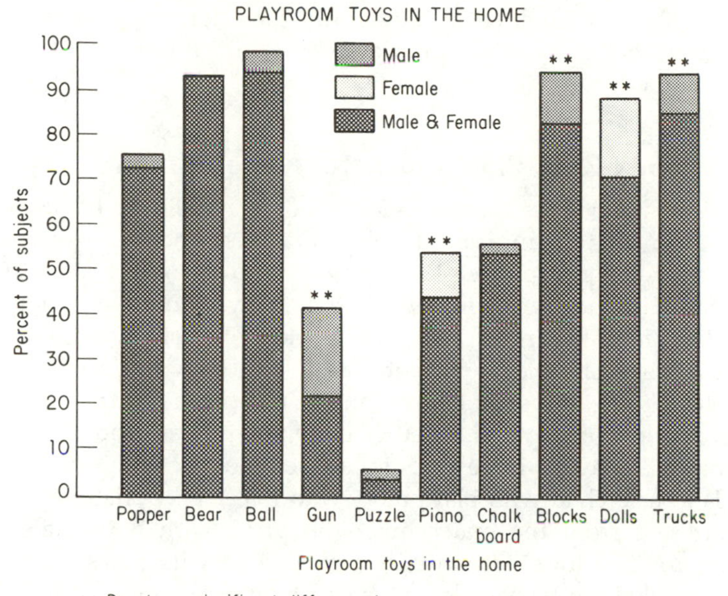

Figure 14–1. Playroom toys in the home.

recting the child and the creation of a rule of action for the child. Such rule generation allows for the creation of appropriate behavior in situations in which external reinforcement or control is not present. Social learning theory (Mischel, 1966) attempted to bridge this gap; what is needed, however, is a stronger cognitive orientation. Such an orientation relates sex role behavior to gender identity (see Money in this volume, Chapter 2; Lewis, 1980). In this view the child's knowledge of its own gender identity as well as its knowledge of the social actions of others is critical in determining sex-role-appropriate behavior.

Regardless of the mechanisms we choose to evoke to account for sex-role-appropriate play, we have considerable data to support the belief that by 2–3 years of life children have knowledge of what toys are sex role appropriate (see Serbin, 1980; Weinraub et al., 1984), and they tend to play with the appropriate toys (Parten, 1933; Goldberg & Lewis, 1969; Fagot & Patterson, 1969). Moreover, considerable evidence indicates that adults in our society, in general, are in agreement as to what constitutes sex-role-appropriate toys or roles (see Serbin, 1980, for a review). For example, Connor and Serbin (1977) asked a group of undergraduates to classify 35 classroom play activities. The authors report that the undergraduates were aware of the cultural expectations: boys play with blocks and trucks; girls paint and play with dolls. Likewise, the Rheingold and Cook (1975) study and the data reported in Figure 14-1 indicate that parents are providing toys in the home that are sex role stereotyped.

Sex Role Play and Later Development

We have much evidence that by 2–3 years of age, children have knowledge of their sex-role-appropriate play, that they increasingly engage in more sex-role-appropriate play behavior, and that there is considerable overlap in the toys with which they play. Moreover, there are large individual differences, some play almost exclusively with sex-appropriate toys, some play with both, and some play more with opposite-sex toys. One important question, from both a diagnostic and a theoretical point of view, is whether early sex-role-appropriate (or inappropriate) behavior is related to subsequent developmental outcomes. Is a male child who plays more with male-appropriate toys likely to be different from the male child who plays with both male- and female-appropriate toys? Likewise, is a male child who plays more with female toys likely to be at risk for psychosexual maladjustment? Given the concerns by parents and society (and the ease with which members of society of all ages can label toys as sex role appropriate), we should determine whether early sex-role-appropriate toy-play behavior is related to subsequent development.

Developmental Models

Three models are often used in development to examine the effects of early behavior on outcomes later in life. The first model, a trait model, states that the status of the child at time one (t_1) will predict the status of the child at time two (t_2).

$$St_1 \longrightarrow St_2$$

In the present example this model suggests that the play behavior of the child at t_1 would predict adjustment or psychopathology at t_2.

The second model, an environmental model, states that the environment at t_1 will predict the outcome of the child at t_2.

$$E_1 \longrightarrow St_2$$

Although the environment includes at least parents, siblings, and peers, we restrict our analysis to parental values regarding sex-role-linked behavior. Using model two, it is not the child's earlier behavior that is important in predicting later developmental outcomes but the parental attitudes that determine the child's later adjustment and psychopathology.

Finally, the third model, an interactional model, states that outcome

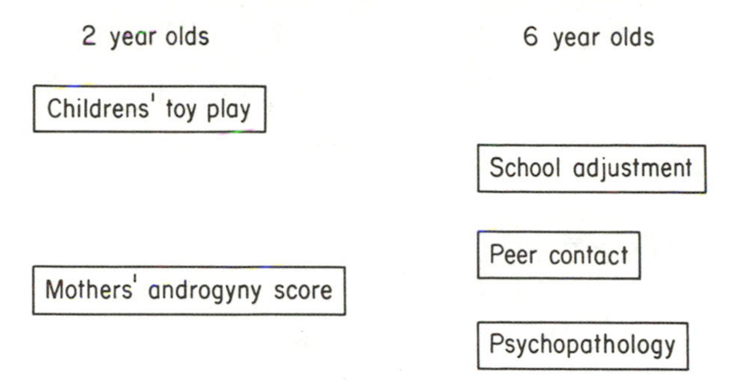

Figure 14–2. Sources for predicting 6-year outcomes for 2-year data.

can only be understood as the interaction between the child's status or behavior and the parental values.

$$(S \times E)\ t_1 \longrightarrow St_2$$

Given these three models of development, we can address the question as to individual differences in sex role play behavior and subsequent development—in particular, the relationship between early sex-role-appropriate play behavior and subsequent social adjustment.

To address this developmental question using each of the three models, five different data sources will be considered. Figure 14-2 presents these data schematically. The data include both 2- and 6-year-old data. When the children were 2 years old we observed their toy play and their mother's attitudes toward sex role orientation or what Bem (1984) has called their gender schema. At 6 years of age the data include children's school adjustment, peer contact, and psychopathology scores. Figures 14-3 presents each model as proposed, using the avail-

Figure 14–3. Models of development: model 1, a trait theory; model 2, an environmental model; model 3, an interactional model.

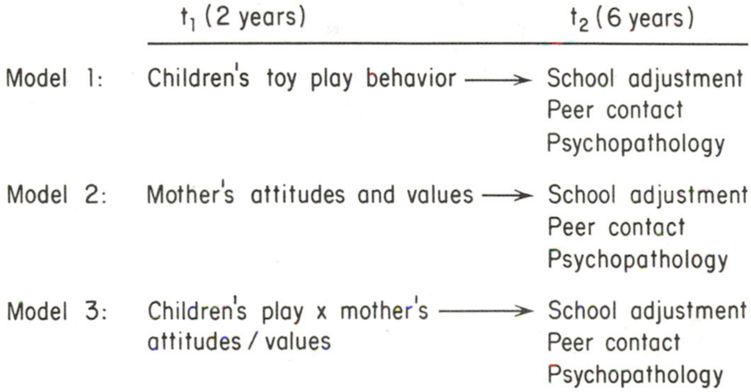

	t_1 (2 years)		t_2 (6 years)
Model 1:	Children's toy play behavior	⟶	School adjustment Peer contact Psychopathology
Model 2:	Mother's attitudes and values	⟶	School adjustment Peer contact Psychopathology
Model 3:	Children's play x mother's attitudes / values	⟶	School adjustment Peer contact Psychopathology

able data. Given that model 3 best describes the developmental processes across a range of data (see Lewis, 1972; Sameroff & Chandler, 1975), our working hypothesis is that children's behavior and maternal gender schema *both* play a role in determining subsequent adjustment and pathology.

Model 1—Children's Characteristics

At 2 years of age, 128 subjects (62 males and 66 females) were seen in a playroom setting where their toy-play behavior was observed. Such a setting is a standard procedure for obtaining data on sex-role-appropriate toy-play behavior. The toys in the playroom were rated independently by a group of observers on the basis of each toy's sex role appropriateness. There was high agreement for the 25 judges ($p<.001$). The toys judged to be appropriate for boys included toy guns, trucks, blocks, and balls, The female toys included dolls, teddy bear, toy piano, and a blackboard with chalk. Observation of the children's toy play occurred over a 15-minute period. Each subject's play behavior was scored according to the amount of time spent playing with the male and female toys. Subjects were given a sex ratio score, calculated by dividing the time spent playing with female toys by the time spent playing with male plus female toys ($F/M+F$).

While such a measure will be used to reflect the child's toy-play behavior, there are several questions which might limit the usefulness of the measure. For example, how stable is such a measure? The question can be addressed by looking at a subject's score for the first half of the play session and comparing it to the toy play in the second half of the session. Such a procedure resulted in toy-play split reliability of .93, indicating at least limited stability. We also would be interested to know whether such stability would be found over a longer time frame; unfortunately, no such data are available. Thus, the subject's toy ratio may not represent a stable individual characteristic. Nevertheless, in order to test our models we will assume that these individual differences represent some stable characteristics.

The mean data show a significant sex of child by toy interaction. Males played more with male toys (136.50 seconds) than with female toys (118.20 seconds), while females played more with female toys (139.65 seconds) than with male toys (84.00 seconds), $F(1, 122)=6.64$. $p<.02$. Such a finding supports sex differences in toy play and lends at least some support to the individual subject data.

More support can be found in the maternal report of number and type of toys available in the home. Recall that mothers were asked to fill out a questionnaire asking which of the toys in the playroom the child had at home. Mothers report that their boy children have on the average 3.26 of the male toys at home while they only have 2.66

of the female toys. Likewise, females have at home on the average 2.92 female toys and 2.89 male toys. The sex of child by toys available in the home interaction was significant [$F(1, 122) = 4.09$, $p < .05$] and indicates that parents provide their children with sex-role-appropriate toys. This finding is consistent with Rheingold & Cook's (1975) findings about toys found in the home. We should note that the sex role differentiation in terms of toys is greater for boys than for girls, a finding suggested by much of the work on sex role socialization in the young.

At 6 years three different measures were obtained; data were available for fewer subjects than seen earlier.[1] These include a rating of the child's school adjustment, a teacher rating of the child's school adjustment, the Classroom Behavior Inventory (Schaefer, Edgerton, & Aaronson, 1979). This latter measure was obtained from the teachers of the children when they were in first grade. The scale consists of nine subscales for the child: (1) creativity, (2) verbal intelligence, (3) extroversion, (4) independence, (5) considerateness, (6) task orientation, (7) social adjustment, (8) general competence, and (9) overall adjustment.

The Child Behavior Profile (Achenbach & Edelbrock, 1978), a parental report reflecting the child's psychopathology, was also obtained when the children were 6 years old. This measure of general adjustment or psychopathology has been found to be related to clinical judgments of poor childhood adjustment (Achenbach & Edelbrock, 1981). Parents were asked to fill out a 113-item questionnaire, and three general scores were available for each child: (1) an internalizing score reflecting the degree of psychopathology associated with problems reflected chiefly by depression and withdrawal, (2) an externalizing score associated with problems reflected chiefly by aggression and acting out, and, (3) an overall total score.

The final measure obtained when the children were 6 years was an estimate of the number of male and female friends each child had. Children were questioned about the number and gender of children they played with and who they would invite to their birthday party. Their mothers were also questioned about the children's social network, and a significant correlation was obtained between parent and child report ($r = .67$, $p < .001$).

In order to test model 1, the children's 2-year-old toy-play behavior was compared to their 6-year-old school adjustment scores, psychopathology scores, and number of friends. Recall that each child's toy-play behavior was converted into a ratio score; amount of play time

1. The numbers of subjects who we were able to follow decreased over the 6 years due mostly to changes in location. Lewis et al. (1984) found no differences at 1 year between subjects who moved and were no longer available for study and subjects who were seen at 6 years.

Table 14-1
Correlation Between Toy Play at 2 Years and Teachers' Ratings at 6 Years

	Female	*Male*
Creativity	.05	−.20
Verbal intelligence	.17	−.20
Extroversion	.10	.12
Independence	−.06	.05
Considerateness	.01	.02
Task orientation	−.03	.09
Social adjustment	.09	.10
General competence	.04	−.08
Overall adjustment	.05	.01

$^{*}p<.05$.

with female stereotyped toys divided by amount of play time with female plus male stereotyped toys.

The results for two of the three measures were the same; there was little relationship between sex-typed toy play at 2 years and subsequent behavior at 6 years. The psychopathology scores are best viewed as an "either/or" indication of psychopathology. Children who score above a particular level are considered to have adjustment problems, while those under that level are considered to be normal. A correlational approach was therefore not appropriate. Looking at those subjects with scores in the maladjusted range and comparing them in terms of their female/female + male toy ratio did not reveal any significant relationship. The correlation matrix between toy play at 2 years and teacher ratings and peer network at 6 years is presented in Table 14-1.

Note that there is no relationship between toy play at 2 years and any of the school adjustment measures at 6 years. No significant correlation between amount of female toy play and later peer contact was seen when each sex was considered separately, but some relationship does appear when the sample is combined. For the total sample there is a correlation ($r = -.19$, $p<.10$) between number of male peers at 6 years and female toy play at 2 years and a correlation of $r = +.29$, $p<.004$, between number of female peers and female toy play at 2 years. Thus, apparently more female toy play at 2 years is related to less male and more female peer contact at 6 years. As expected, 6-year-old boys reported significantly more male than female friends (4.00♂ vs. 1.43♀, $p<.001$) and girls significantly more female than male friends (4.09♀ vs. 1.36♂, $p<.001$). At 2 years boys play more with male than female toys and girls more with female than male toys, and at 6 years there was a sex-of-subject by sex-of-peer interaction. Individual differences in early sex role play were somewhat related to later peer contact; on the other hand, early toy preference did not appear related to later school adjustment or psychopathology.

These findings make clear that subject characteristics may be related to some outcomes but unrelated to others. Such differential results suggest that each of the three developmental models proposed offers some value in understanding the developmental process. Alternatively, one might argue that such findings are related to measurement problems. Perhaps the measurement of individual children's toy preference as a function of sex role behavior was insufficient, and a better measure would show a relationship between toy play and adjustment. Obtaining a more stable and more representative measure of sex role toy behavior would certainly be better. Nevertheless, that individual differences in toy play at 2 years were related to later peer contact makes such a need less likely. Perhaps better predictive findings could be found when only extreme cases of toy play were examined. In this regard we observed both male and female extreme cases of opposite-sex toy play and could find little more predictive value when only such subjects were examined.

Although better measurement of individual characteristics is desirable, the relationship between early play and later peer contact would suggest that the toy-preference measure may be sufficiently sensitive. The results indicate that different developmental outcomes require different models of development. In the case of school adjustment and psychopathology, subject characteristics alone do not show any relationship, while for peer contact early toy-play patterns are related. Such findings should alert us to the fact that *development is a complex transformational process wherein some outcomes are more readily observed from early behavior than others.* As we will see, school adjustment is related to early behavior, but only when a more complex model is employed. For some outcomes subject characteristics alone are insufficient for predictive purposes. For example, the work of Broman, Nichols, and Kennedy (1975), among others, has pointed out that infant characteristics, however measured, usually provide only limited support for observing some outcomes. Such problems may be improved with upgraded measures of the individual, but they will never disappear without the appreciation that some developmental outcomes require that we measure *both* the individual and the environment to which the individual must adapt.

Model 2—Maternal Attitudes and Values as Environment

The second model related to developmental change holds that the environment of the child may be a better predictor of subsequent characteristics than the child's own earlier characteristics. To explore this model we will look at one aspect of the child's environment that should be highly related to the child's sex role toy behavior.

Maternal gender schema was observed using the Bem Sex Role Inventory (Bem, 1985). The BSRI was "designed to assess the extent to which the culture's definitions of gender appropriateness are incorporated into an individual's self-description." Put differently, the BSRI presents "an individual with a heterogeneous collection of attributes and then [assesses] the extent to which he or she clusters the collection into the two categories designated by the culture as masculine and feminine" (p. 21). Such an instrument should be able to sort individuals who are inclined to organize information on the basis of gender. Thus, there should be two groups of individuals—those who show no or little gender schema (these are called androgynous) and those who do. We may consider those who do show a strong gender schema to contain both *sex typed* (those who have a strong preference for like sex roles) and *cross-sex typed* (those who have a strong preference for opposite sex roles). Several studies have related performance on the BSRI and sex-role-appropriate behavior (Bem & Lenney, 1976; Helmreich, Spence & Holahan, 1979). In general, sex-typed individuals were more likely to choose behaviors that are consistent with their own gender than androgynous or cross-sex-typed individuals. More significantly, they avoided gender-inappropriate activities (Bem, Martyna, & Watson, 1976).

Mothers of the children were asked to fill out the BSRI when their children were 2 years old. Using the standard procedure (see Bem, 1984, for a review of this procedure), of the 128 mothers who filled out the form, 47 or 37% were found to be androgynous, 68 (51%) were sex typed or had traditional gender schema, and 13 or 12% were cross-sex typed, that is, they had a masculine-like gender schema.

Although the present model seeks to relate early maternal attitudes and values to later childhood adjustment, we divert the discussion for a moment to look at the relationship of the maternal gender schema to a child's 2-year-old toy-play behavior.

Table 14-2 presents the child's toy-play behavior as a function of maternal gender schema. The results reveal a significant sex-by-toy interaction but no effect for maternal gender schema. Because the number of cross-sex-typed mothers is so small, we focus on those mothers who showed either an androgynous or feminine sex-typed gender schema. No significant differences were seen between both boy and girl children in terms of their toy play as a function of maternal gender schema.[2] In both the androgynous and feminine sex-typed mothers, both boy

2. Interestingly, the cross-sex-typed mothers had children (both male and female) who did not show sex-role-appropriate behavior. Although not significant, girls of cross-sex-typed mothers played more with male toys than female toys, while boys of cross-sex-typed mothers played equally with both types of toys.

Table 14-2
Mean Number of Periods (15 Seconds) of Toy Play by Sex and Maternal Androgyny Scores

	Female Toys	*Male Toys*	*F/M+F*
Males			
Androgynous	6.94	9.44	.43
Feminine	8.48	9.22	.49
Females			
Androgynous	10.15	5.31	.66
Feminine	9.07	5.46	.64

and girl children played more with sex-role-appropriate toys. Children's play behavior at 2 years did not seem to differ as a function of maternal attitudes. This lack of relationship is also visible in the maternal report concerning toys the children have in their homes. Overall, boys are reported to have more sex-role-appropriate toys in their homes, but there were no significant differences between androgynous and feminine sex-typed mothers in the toys they give their male children. Such findings allow us to conclude that: (1) males and females show sex-role-appropriate play behavior by 2 years, (2) mothers show both strong and weak gender schema, and (3) there is little relationship at 2 years between the children's play behavior and their mother's gender schema.

We now return to the relation between maternal gender schema when the child was 2 and the child's adjustment at 6 years of age. Table 14-3 presents the two maternal gender schema groups as well as the children's school adjustment and peer relationships at 6 years. Higher scores indicate more of the particular ability or problem. Although, as already reported, there are sex-of-subject by sex-of-peer interactions, few other findings are of significance. In general, then, the mother's gender schema

Table 14-3
Maternal Androgyny Scores at 2 Years and Teachers' Ratings and Peer Control at 6 Years

	Creat.	*Verb. IQ*	*Ext.*	*Ind.*	*Consid.*	*Task Or.*	*Adj.*	*Comp.*	*Total*	*No. of Peers Males*	*No. of Peers Females*
Males											
Feminine mothers (23)	49.4	49.7	49.3	50.3	48.2	48.1	48.8	49.4	48.7	3.63	1.50
Androgynous mothers (18)	53.5	52.0	47.7	49.7	45.9	44.8	46.8	50.0	47.7	4.76	1.12
Females											
Feminine mothers (25)	51.2	51.4	50.5	50.3	52.5	53.9	51.5	51.7	52.6	1.45	4.36
Androgynous mothers (17)	52.9	53.6	51.7	51.1	54.8	51.8	53.2	52.3	53.8	1.22	3.94

Table 14-4
Maternal Androgyny at 2 Years and 6-Year-Old Psychopathology Scores

	Internal	*External*	*Total Pathology*
Males			
Androgynous (18)	47.0	49.9	48.1
Feminine (23)	48.0	46.8	47.3
Females			
Androgynous (17)	47.4	46.1	46.1
Feminine (25)	48.5	48.6	48.4

does not appear to be related to the child's subsequent school performance. Thus, androgynous or sex-typed mothers have equally well-adjusted children as seen in teacher ratings.

Table 14-4 presents maternal gender schemata at 2 years and psychopathology scores as obtained at 6 years using the Child Behavior Profile inventory. No significant differences in level of psychopathology at 6 years were found as a function of maternal gender attitudes. The findings reveal that the child's environment, measured in terms of maternal gender schema, is not particularly related to subsequent adjustment at school age. From the point of view of the specific data, traditional maternal sex typing and less traditional androgynous behavior appear to be equal in terms of their consequence on subsequent development. In other words, mothers do no harm by holding to either traditional or androgynous behavior.

Model 2 indicates that examining the environment alone is insufficient to predict subsequent dysfunction. One possible explanation for this lack of relationship is the measurement of environmental factors. Perhaps a measure of the attitudes of both father and mother is necessary to observe a relationship over time, or perhaps we should use some other measure of maternal attitudes. Although such measurement issues need to be addressed, we wish to argue instead that the difficulty of measurement is not what results in poor predictive outcomes. Any measurement system, no matter how improved, will to some degree suffer from poor prediction if it does not include *both* measures of the child and measures of the environment. We will discuss this interactive model next.

Model 3—Interaction, Transaction, and Goodness of Fit

In our considerations to this point, we have looked at two simple but useful models of development, each focusing on important aspects of the process of continuity and change. Few would deny the importance of either the child's early characteristics or the nature of the environment in which the child lives, but clearly the most accurate predictor of developmental outcomes, in general, is found in the relationship of both these factors. The relationship of the child's status and environment has been considered from an interactional (Lewis, 1972), transactional (Sameroff & Chandler, 1975), and, more recently, goodness-of-fit model (Lerner, 1983). These models of the interdependence of subject's current status with current environment have much in common, and we will consider them similar for the purpose of looking at early sex role behavior and later psychopathology. For the argument we wish to make, it may be better to consider our data as fitting a *goodness-of-fit model.* For us, such a model holds that when a subject characteristic fits an environmental demand, the developmental outcome will be favorable. In the absence of such a fit, however, we can expect unfavorable developmental outcomes. For example, consider the variable of temperament. A particular child is physically quite active. If the child were raised in a household where such activity was considered favorably, we would expect a positive developmental outcome. On the other hand, if the same child were raised in a household where such activity was frowned on, we would expect conflict between the characteristics of the child and environmental demands, and this conflict should lead to a less positive developmental outcome than in the first example. Thus, a goodness-of-fit model would propose that outcomes are a function of the fit between subject characteristics and environmental demands. Parenthetically, such a model can be used to discuss either the development of subject characteristics or environmental demands and as such is interactional or transactional in nature. For example, the developmental outcome in terms of fit at an earlier point in time can be used to describe the subject's characteristics at a later point in time. Such a model of fit is equally relevant for our data because we have both the child's sex role toy play and the mother's gender schema.

Using a goodness-of-fit model we would argue that those children whose toy-play behavior fits their mothers' preference for sex-typing or androgynous behavior will have better developmental outcomes than those children whose behavior does not fit their mothers' gender schema. In particular we would predict that boys who play more with girl toys—that is, their $F/M+F$ ratio is high—will have negative outcomes if their mothers have a sex-typed schema; in other words, their mothers have a traditional belief that boys should play more with boy toys than girl toys. Boys who play more with girl toys will have positive outcomes if their mothers have an androgynous schema; that is, their mothers do not object to boys playing with girl toys. Likewise, girls who play more with girl toys will have negative outcomes if their mothers have androgynous schema but positive outcomes if their mothers have sex-typed schema.

Table 14-5 presents the data for boys and girls separately and divides those subjects with maternal sex-typed or androgynous scores. The

Table 14-5
Correlation Between Toy Play at 2 Years and Teachers' Ratings at 6 Years as a Function of Maternal Androgyny

	Creat.	*Verb. IQ*	*Ext.*	*Ind.*	*Consid.*	*Task Or.*	*Adj.*	*Comp.*	*Total*	*No. of Peers Males*	*No. of Peers Females*
Males											
Feminine mothers (25)	−.36†	−.51*	.20	−.32*	.05	−.53†	−.13	−.54†	−.37†	.04	.05
Androgynous mothers (18)	.56*	.59*	.04	.58*	.02	.53*	.04	.63*	.51*	−.09	+.22
Females											
Feminine mothers (23)	−.08	.31	.08	.00	.26	.14	.25	.12	.19	−.29	−.24
Androgynous mothers (17)	−.13	−.17‡	−.04	−.21	.00	−.17	−.03	−.21	−.17	.09	−.09

$^{*}p<.01$.
$^{\dagger}p<.05$.
$^{\ddagger}p<.10$.

correlations presented are between the amount of girl toy play ($F/M+F$) and school adjustment at 6 years. As can be seen from the table, high female toy play for boys is associated with positive outcomes *only* for boys whose mothers have an androgynous gender schema but *not* for boys whose mothers have a traditional sex-typed gender schema. Likewise, high female toy play for girls is associated with negative outcomes *only* for girls whose mothers have an androgynous gender schema but *not* for girls whose mothers have a traditional sex-typed gender schema. Because these findings support the goodness-of-fit model, we have collapsed the sex differences and observe the relationship of goodness of fit independent of sex of subject.

Figure 14-4 presents the correlations for these groups of positive and negative goodness of fit as they relate to both 6-year-olds' school adjustment and peer behavior. Consider first the children's school adjustment at 6 years. Subjects whose sex-role toy-play behavior fits the maternal gender schema show, in general, more positive school performance than children whose sex-role toy-play behavior does not fit the maternal gender schema. This holds for verbal intelligence ($p<.01$ for the difference between the two correlations), independence ($p<.05$), task orientation ($p<.01$), school adjustment ($p<.01$), competence ($p<.01$), and total school adjustment score ($p<.01$). Such findings support the goodness-of-fit model and suggest that both the child's sex role behavior and the mother's gender schema allow us to observe some developmental outcomes. Moreover, the failure of either the child's characteristics or the mother's schema alone to predict school adjustment outcomes cannot be judged to be a function of inadequate measurement.

The data show that both developmental models help predict the 6-year outcomes. Because these behaviors, child status, and maternal en-

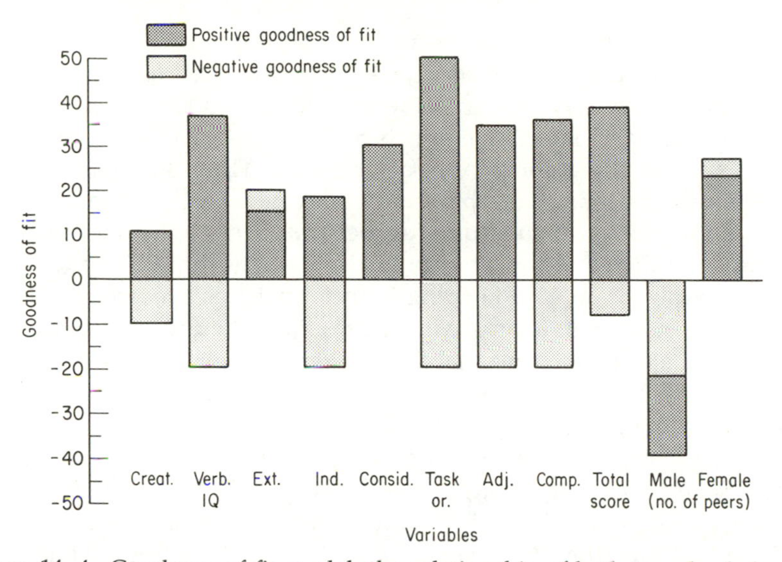

Figure 14–4. Goodness-of-fit model: the relationship of both toy play behavior and maternal attitudes on subsequent school adjustment.

vironment in combination predict later behavior, these results allow us to conclude that the measurement issues have not negated the opportunity to find developmental outcomes and that developmental outcomes can be revealed when a model of interaction between organism and environment is employed.

Figure 14-4 also presents the data on the number of male and female peers reported as friends when the children were 6 years old. As we can see, there was no significant difference as a function of goodness of fit. The data do reveal, however, a difference in male and female peers that we observed early; that is, the more the subjects play with female toys at 2 years, the more female and less male peer contact reported at 6 years. The findings of Figure 14-4 again show that the interaction between environment and child characteristics is relatively unimportant in predicting subsequent peer contact. What appears important, although only moderately so, is the amount of toy play at 2 years.[3]

3. The relationship between playing with a higher ratio of female to male toys and having more female friends holds for both sexes. The data indicate that males and females who play more with female toys at 2 years do not play exclusively with females as peers at 6 years, but the relative amount of peer contact is more than for the males who at 2 years do not play as much with female toys. The ratio of male to female peers at 6 years is about 4 to 1 for males in general; in the case of males who play more with female toys, the ratio is closer to 3 to 2, with still more male peers.

Two possible reasons can be used to account for such results: (1) both early toy choice and later peer structure are related to a third factor, or (2) early toy play in some manner *causes* later peer behavior. Both toy play and later peer structure could be related to some third factor, such as the amount of sex hormones. Thus, for example, a female fetus exposed to an abnormally elevated level of the male hormone androgen may subsequently become a girl who shows higher amounts of rough-and-tumble play in childhood (Erhardt & Baker, 1973). Such an explanation is in keeping with model 1 as presented. Alternatively, we could argue that playing with femalelike toys makes the child feel like or feel more comfortable with females, or because of less experience with male toys makes the child feel less competent with males. Even some combination of both is likely. In this view the play with female toys at 2 years also has a causal effect on peer behavior at 6 years. Such a view can be supported by some recent work (La Freniere, Strayer, & Gauthier, 1984). The latter view supports model 3 in that children's status (their toy-play behavior) interacts with the behavior of others to determine the peer-structure outcome. Thus, even within a single finding, multiple models are possible. In order to address which view is most correct, we would have to trace the play choice back into earlier childhood to see whether differences that have a biological cause can be seen at early ages.

Developmental Principles and Sex Role Behavior

At the beginning of this essay, we raised the question of defining masculinity and femininity, and we stated that no definition would be complete that did not take into account gender identity, sex role behavior, and sexual preference. In observing only sex role behavior development, we have come to face a wide set of issues. As in any discussion of development, we have had to raise the issue of the model(s) of development best able to reveal the course of change.

Although we have argued for an interactive model, other models of development cannot be rejected. For some developmental outcomes, more simple models of development may be sufficient to understand change. Social class as an environmental factor exhibits strong influence on many developmental outcomes and is quite useful in predicting the course of development (Golden & Birns, 1983). Likewise, subject characteristics, such as sex (e.g., Weinraub & Brown, 1983), temperament (e.g., Carey, 1981), and even genetic abnormalities (Hanson, 1981) are predictive of later outcomes. Our data on early toy preference and later peer contact show this. Our major premise is that a more complex model (given the same level of measurement ability) usually will provide the best prediction. This assumption requires, however, that we know what subject characteristics and what environmental influences should be measured. In the case of school adjustment as outcome, subject characteristics as measured and maternal gender schema are both useful in predicting development, at least until the first grade of school. For peer contact, early subject characteristics alone appear related to subsequent behavior. Psychopathology, however, does not appear related to either model. No single course of development can be used to predict all outcomes.

Thus, the first development proposition we wish to state is the following:

Proposition 1. Development is a complex process, and there are at least three possible and reasonable paths:

a. Individual differences in outcome are related to earlier individual differences.

b. Individual differences in outcome are caused by differences in earlier environments.

c. Individual differences in outcome are caused by differences in both earlier environments and earlier status.

The issue of outcome, in particular the outcome of masculinity/femininity, suggests problems not only of developmental models but also of the nature of the outcome. A single measure of masculinity/femininity would be convenient, but it is impossible. Even if we could choose among gender identity, sex role behavior, and sexual preference for our definition, the problem that each of these constructs has a variety of outcomes would still be present. Consider sex role as an example; play behavior, peer contact, and social behavior are all used as measures. Not only are there multiple outcomes; these outcomes are often unrelated to each other. Thus, the exposition of the course of development leading to one outcome may tell us little about the course of development leading to another. It is therefore possible, indeed probable, that different outcomes are unrelated or only moderately related to one another and that they may have different developmental paths. This requires that some caution be used when we move from one particular gender-linked behavior to the more broad construct of masculinity/femininity. In like fashion, in nonhuman animals, one sex role behavior may not be highly correlated with another. Thus, the study of the relationship of brain functioning to behavior *A* may tell us little about how behavior *B* functions when both *A* and *B* are each recognized as sex-role-appropriate behaviors. The second developmental proposition then becomes:

Proposition 2. Behavioral outcomes that appear to form a coherent whole are often unrelated;

and as a corollary

2a. The developmental paths of partially related behaviors (making up what appear to be a coherent whole) may be quite different.

These two general propositions have general reference and are also related to the specific topic of sex role behavior. The data on school adjustment inform us about one of our concerns, namely the health outcome of playing with sex-role-related toys. Recall parental concern as voiced around the question of whether a male child will be maladjusted if he plays at times with female toys such as dolls. Our answer to such parental inquiry rests on our findings that play alone does not affect school adjustment, but the interaction of parental value and toy play does. Thus, if the parent does not object to the child's choice of toys, the choice itself will have little effect on subsequent health. Although we cannot specify children's health outcomes past 6 years, school adjustment, at least as seen from the teacher's perspective, does not appear to be impaired by early toy-play choice. This leads to Proposition 3.

Proposition 3. Being less masculine or feminine (defined at least in terms of toy play) does not lead to psychopathology if the environment does not define the play as such.

The data at 2 years indicate both environmental pressure for appropriate sex role behavior—in terms of toys bought by parents—and societal commitment to sex-role-appropriate behavior. Although there is large overlap, girls and boys do play with sex-role-appropriate toys by 2 years and by 3 to 6 years show clear sex-role peer-play preferences. These data strongly support the more current view that sex role behavior is learned early and that by school age, before 6 years, children have clear sex role preferences including social activities and play behavior. Thus, by this age gender identity is established, and many important and central sex role behaviors are in place.

Our final proposition speaks to the early emergence of sex-role-appropriate behavior in our species.

Proposition 4. Without social support many sex-role-related behaviors would not appear differentially in the child's behavioral repertoire. With social support and encouragement they emerge within the first 2 years of life.

As many have argued, much of sex role behavior is learned and as such is early influenced by a variety of environmental factors. If we allow boys and girls to engage in behavior that is not sex role linked (for example, playing with toys), children will engage in less sex role behavior. If the adult (and child) culture does not define sex-role-appropriate behavior, far less will appear. Thus, for example, if boys and girls are allowed to play with both boy and girl toys, they will do so. Moreover, if allowed to have contact with male and female peers, they will do so far more than if they are restricted by social norms.

In summary, no definition of masculinity/femininity for the human species can be defined independently of gender identity, sex role behavior, and sexual preference. This being the case, we can see that the definition of masculinity/femininity cannot avoid a developmental consideration. Because the early part of life witnesses the emergence of gender identity and sex role learning, but not until puberty do we see the development of sexual preference, a developmental perspective is necessary; the first portions appear in the first year or two of life, while the final aspects do not emerge until later.

Acknowledgments

This research was supported in part from a grant from the William T. Grant Foundation. Special consideration is to be given to John Jaskir for data analysis and to Dr. Candice Feiring for data collection.

References

Achenbach, T. M., & Edelbrock, C. S. (1978). The classification of child psychopathology: A review and analysis of empirical effects. *Psychological Bulletin, 85*(6), 1275–1301.

Achenbach, T. M., & Edelbrock, C. S. (1981). Behavioral problems and competencies reported by parents of normal and disturbed children aged 4 through 16. *Monographs of the Society for Research in Child Development, 46*(1, Serial No. 188).

Bem, S. L. (1985). Gender schema theory. In T. B. Sonderegger (Ed.), *Nebraska Symposium on Motivation 1984: Psychology and Gender* (Vol. 32). Lincoln: University of Nebraska Press.

Bem, S. L. & Lenney, E. (1976). Sex typing and the avoidance of cross-sex behavior. *Journal of Personality and Social Psychology, 33,* 48–54.

Bem, S. L., Martyna, W., & Watson, C. (1976). Sex typing and androgyny: Further exploration of the expressive domain. *Journal of Personality and Social Psychology, 34,* 1016–1023.

Block, J. H. (1977). Assessing sex differences: Issues, problems and pitfalls. *Merrill-Palmer Quarterly,* 140–147.

Broman, S. H., Nichols, P. L., & Kennedy, W. A. (1975). *Preschool I.Q.: Prenatal and early development correlates*. New York: Wiley.

Brooks, J., & Lewis, M. (1974). Attachment behavior in thirteen-month-old, opposite sex twins. *Child Development, 45*, 243–247.

Carey, W. B. (1981). The importance of temperament-environment interaction for child health and development. In M. Lewis and L. A. Rosenblum (Eds.), *The uncommon child*. New York: Plenum Press.

Connor, J. M., & Serbin, L. A. (1977). Behaviorally-based masculine and feminine activity preference scales for preschoolers: Correlates with other classroom behaviors and cognitive tests. *Child Development, 48*, 1411–1416.

DeVries, R. (1969). Constancy of generic identity in the years three to six. *Monographs of the Society for Research in Child Development, 34*(3), Serial No. 127.

Emmerich, W., Goldman, K. S., Kirsh, G., & Sharabany, R. (1977). Evidence for a transitional phase in the development of gender constancy. *Child Development, 48*, 930–936.

Erhardt, A. A., & Baker, S. W. (1973). *Hormonal aberrations and their implications for the understanding of normal sex differentiation*. Paper presented at the meetings of the Society for Research in Child Development, Philadelphia.

Fagot, B. I. (1974). Sex differences in toddlers' behavior and parental reaction. *Developmental Psychology, 10*, 554–558.

Fagot, B. I., & Patterson, G. R. (1969). An in vivo analysis of reinforcing contingencies for sex-role behaviors in the preschool child. *Developmental Psychology, 1*, 563–568.

Feiring, C., & Lewis, M. (1979). Sex and age differences in young children's reactions to frustration: A further look at the Goldberg and Lewis (1969) subjects. *Child Development, 50*, 848–853.

Goldberg, S., & Lewis, M. (1969). Play behavior in the year-old infant: Early sex differences. *Child Development, 40*, 21–31.

Golden, M., & Birns, B. (1983). Social class and infant intelligence. In M. Lewis (Ed.) *Origins of intelligence* (2nd ed,) pp. 347–398. New York: Plenum Press.

Green, R. (1974). *Sexual identity conflict in children and adults*. New York: Basic Books.

Hanson, M. J. (1981). Down's syndrome children: Characteristics and intervention research. In M. Lewis & L. A. Rosenblum (Eds.), *The uncommon child* (pp. 83–114). New York: Plenum Press.

Helmreich, R. L., Spence, J. T., & Holahan, C. K. (1979). Psychological androgyny and sex-role flexibility: A test of two hypotheses. *Journal of Personality and Social Psychology, 37*, 1631–1644.

Jacklin, C. N., & Maccoby, E. E. (1978). Social behavior at thirty-three months in same-sex and mixed-sex dyads. *Child Development, 49*, 557–569.

Kohlberg, L. A. (1966). A cognitive-developmental analysis of children's sex-role concepts and attitudes. In E. E. Maccoby (Ed.), *The development of sex differences* (pp. 82–173). Stanford: Stanford University Press.

Kuhn, D., Nash, S., & Brucken, L. (1978). Sex role concepts of two-and three-year olds. *Child Development, 49*(2), 445–451.

La Freniere, P., Strayer, F. F., & Gauthier, R. (1984). The emergence of same sex affiliative preferences among preschool peers: A developmental/ethological perspective. *Child Development, 55*, 1958–1965.

Lerner, J. V. (1983). A goodness of fit model of the role of temperament in psychosocial adaptation in early adolescence. *Journal of Genetic Psychology, 143*, 149–157.

Lewis, M. (1972). State as an infant-environment interaction: An analysis of mother-infant interaction as a function of sex. *Merrill-Palmer Quarterly, 18*, 95–121.

Lewis, M. (1980). Self-knowledge: A social-cognitive perspective on gender identity and sex role development. In M. E. Lamb & L. R. Sherrod (Eds.), *Infant social cognition: Empirical and theoretical considerations* (pp. 395–414). Hillsdale, NJ: Erlbaum.

Lewis M., & Brooks-Gunn, J. (1979). *Social cognition and the acquisition of self.* New York: Plenum.

Lewis, M., & Cherry, L. (1977). Social behavior and language acquisition. In M. Lewis & L. Rosenblum (Eds.), *Interaction, conversation and the development of language: The origins of behavior* (Vol. 5, pp. 227–245). New York: Wiley.

Lewis, M., Fiering, C., & Kotsonis, M. (1984). The social network of the young child: A developmental perspective. In M. Lewis (Ed.), *Beyond the Dyad: The Genesis of Behavior* (Vol. 4). New York: Plenum.

Lewis, M., Feiring, C., McGuffog, C., & Jaskir, J. (1984). Predicting psychopathology in six year olds from early social relations. *Child Development, 55*, 123–136.

Lewis, M., Feiring, C., & Weinraub, M. (1981). The father as a member of the child's social network. In M. Lamb (Ed.), *The role of the father in child development* (2nd ed., pp. 259–294). New York: Wiley.

Lewis, M. & Michalson, L. (1982). *Sex differences in children's toy play behavior.* Presented at Eastern Psychological Association Meeting, New York.

Lewis, M., & Weinraub, M. (1979). Origins of early sex-role development. *Sex Roles, 5*(2), 135–153.

Maccoby, E. E., & Jacklin, C. N. (1974). *The psychology of sex differences.* Stanford: Stanford University Press.

Marcus, D. E., & Overton, W. F. (1978). The development of cognitive gender constancy and sex role preferences. *Child Development, 49*(2), 434–444.

Mischel, W. (1966). A social-learning view of sex differences in behaviour. In E.E. Maccoby (Ed.), *The development of sex differences* (pp. 56–81). Stanford: Stanford University Press.

Money, J., & Erhardt, A. *Man and woman, boy and girl.* Baltimore, The Johns Hopkins University Press, 1972.

Moss, H. A. (1974). Early sex differences and mother-infant interaction. In R. C. Friedman, R. M. Richart, & R. L. Vande Wiele (Eds.), Sex differences in behavior. New York: Wiley.

Parke, R. D., & O'Leary, S. (1975). Father-mother-infant interaction in the newborn period: Some findings, some observations, and unresolved issues. In K. Riegel & J. Meachan (Eds.), *The developing individual in a changing world, Vol. II, Social and environmental issues.* The Hague: Mouton.

Parten, M. B. (1933). Social play among preschool children. *Journal of Abnormal and Social Psychology, 28*, 136–147.

Rikers, G. A. (1975). Stimulus control over sex-typed play in cross-gender identified boys. *Journal of Experimental Child Psychology, 20*, 136–148.

Rheingold, H. L., & Cook, K. V. (1975). The contents of boys' and girls' rooms as an index of parents' behavior. *Child Development, 46*, 459–463.

Rubin, J., Provenzano, F., & Luria, Z. (1974). The eye of the beholder: Parents' views on sex of newborns. *American Journal of Orthopsychiatry, 44,* 512–519.

Sameroff, A. J., & Chandler, M. J. (1975). Reproductive risk and the continuum of caretaking casualty. In F.P. Horowitz (Ed.), *Review of child development research* (Vol. 4). Chicago: University of Chicago Press.

Schaefer, E. S., Edgerton, M., & Aaronson, M. (1979). *Classroom behavior inventory: A teacher behavior check list.* Chapel Hill: Frank Porter Graham Child Development Center.

Serbin, L. A. (1980). Sex-role socialization: A field in transition. In B. B. Lahey & A. E. Kazdin (Eds.), *Advances in clinical child psychology.* New York: Plenum Press.

Serbin, L. A., Connor, J. M., Burchardt, C. J., & Citron, C. C. (1979). Effects of peer presence on sex-typing of children's play behavior. *Journal of Experimental Child Psychology, 27,* 303–309.

Smith, P., & Daglish, L. (1977). Sex differences in parent and infant behavior in the home. *Child Development, 48,* 1250–1254.

Thompson, S. K. (1975). Gender labels and early sex role development. *Child Development, 46,* 339–347.

Weinraub, M., & Brown, L. (1983). Crushing realities: Development of children's sex role knowledge. In V. Franks & E. Rothblum (Eds.), *The Stereotyping of women: Its effects on mental health.* New York: Springer.

Weinraub, M., Clemens, L. P., Sockloff, A., Eldridge, T., Gravely, E., & Myers, G. (1984). The development of sex role stereotypes in the third year: Relationships to gender labeling, gender identity, sex-typed toy preference, and family characteristics. *Child Development, 55,* 1493–1503.

15

The Varied Meanings of "Masculine" and "Feminine"

Eleanor E. Maccoby

In the spirit of this volume, I want to begin this chapter by considering the several meanings of "masculinity" and "femininity." I can see a rough triparite division in the way these concepts have been employed.

1. A masculine or feminine person is one who exemplifies those characteristics that have been shown to differentiate the sexes. Thus a 4-year-old boy would be considered masculine if he enjoyed (and frequently engaged in) rough-and-tumble play; if he preferred to play with blocks and trucks; and if, during free-play periods at nursery school, he tended to play outdoors in the company of other boys. A 4-year-old girl would be seen as feminine if she liked to wear dresses, played with dolls and art materials, and didn't get into fights. At age 10, a masculine boy would be one who engaged in active sports, avoided girls, and wasn't particularly diligent about his schoolwork; a feminine girl would be one who had one or two close girlfriends, did not try to join boys' sports play groups, paid attention to the teacher in class, did not brag, liked to baby-sit, and preferred romantic television shows. At age 15, a masculine boy would be one who excelled in spatial-visual tasks, liked and did well at math, was interested in cars and machinery, and knew how to repair mechanical gadgets; a feminine girl would be more interested in English and history than math or science. These lists of differential characteristics are not exhaustive, but they give the flavor of what the replicably sex-differentiated characteristics are.

2. A masculine or feminine person is one who displays the charac-

teristics prescribed by male and female sex roles. Here I am using the term *role* in the sense of a social position or status for which certain behaviors are socially expected or required. Just as the role of clerk at a grocery checkout counter carries prescriptions for role behavior (to charge all customers the same price for the same item, to deal with customers in the order of their place in line, to collect money before packaging and handing over the groceries), so it can be said that an individual's gender carries with it certain social prescriptions for behavior. Sex as a role is different from occupational roles in that it is more diffuse. Some writers on role have suggested, in fact, that gender should not be thought of as a role per se—at least not in the sense that occupations are roles—but that gender can more accurately be thought of as infusing occupational and status roles to varying degrees. Thus, the roles of police officer or nurse or bank president are highly gender infused, while the roles of grocery clerk or college student are not, at least not in present-day American culture. In certain social situations, however, gender does emerge as a role in the traditional sense—that is, an individual in a given situation is expected to behave in one way if male, another way if female. Behavior on a first date fits this pattern reasonably well. Societies differ greatly with respect to how gender-infused their social roles are. In present-day Iran, gender determines almost all aspects of an individual's life: the number of public places in which the person is allowed to be, the educational experiences the person is allowed to have, the occupations that can be entered, the company that is kept. In other societies, social expectations and demands are linked to gender in a more limited set of situations and with respect to fewer aspects of behavior. The key to the idea of masculinity and femininity as referring to an individual's conformity to sex roles, however, is that in any society social expectations and prescriptions specify how a male or female person is to act, and what social functions that person is allowed or expected to perform. This concept of masculinity or femininity implies that sanctions, of varying degrees of severity, are imposed on individuals who deviate too far from prescribed sex roles; sex role training in childhood is seen as pressing children (through reward, punishment, and example) toward adopting those behaviors that are thought to be appropriate for their sex and avoiding those that are not.

3. A third meaning of masculinity and femininity has to do with the attraction between the sexes. In this sense, a feminine young woman is one who is attractive to men. She knows, among other things, how to flirt gracefully and not too openly; she chooses clothes and hairdos that bring out the best aspects of her face and figure, and she doesn't allow herself to get too fat; she knows how to be a "good listener" and otherwise flatter the male ego. What it takes to be a masculine man in

terms of attractiveness to women is not so clear. Perhaps it includes an element of being dominant among men. A rugged body type (as distinct from a delicate one) or a deep voice may be among the ingredients; of course, paying court to women and showing attraction toward them figures strongly. Probably the most essential ingredient in this definition of masculinity or femininity is that the person should not be, or seem to be, homosexual. Men known to be homosexual may be seen as effeminate even if they do not use feminine gestures, dress, or speech patterns.

Researchers have not given much attention to the various ways in which human males and females signal their sexual interest in each other, nor the variations in the signals to which potential partners are responsive. At present, we must assume that the variation in both signal and response is probably very great. To some degree, these things are culture specific, and even within cultures, human beings are surely much more variable in these respects than lower animals. The range of sexual fetishes testifies to this variety. The implication is that there are many ways to be masculine or feminine under definition 3.

We must note several implications of the differences in these three definitions. The first is that they are not always isomorphic. Thus, boys are more likely to be involved in rough-and-tumble play than girls, fitting definition 1, and being interested in such play is part of the cultural expectations associated with gender. Thus, rough-and-tumble play is a sexually dimorphic behavioral element that fits into both definitions 1 and 2. On the other hand, the replicable sex difference in spatial ability fits into definition 1 but not 2. A teenaged boy need not have a high score on tests of spatial ability to be seen as masculine. Indeed, there is some evidence that teenaged males with the most androgenized bodies—heavy upper-body muscles, ample facial and body hair—and who therefore *look* highly masculine are likely to have a feminine pattern of intellectual abilities, in that they are low in spatial ability relative to other intellectual domains (Peterson, 1973). Fifth-grade males with high spatial skills have a less firmly established masculine identity, are less aggressive, and are less likely to be seen as masculine by their age-mates than boys with relatively poor spatial ability (Ferguson & Maccoby, 1966). Thus, a characteristic that empirically distinguishes boys from girls is not part of the cluster of attributes that contribute to the social definition of masculinity. In fact, in this case spatial ability would have to be coded as a masculine trait under definition 1 and as a feminine trait or (at least correlated with feminine traits) under definition 2. The point of these examples is that not all characteristics that empirically differentiate the sexes (definition 1) are demanded or expected by society as part of the social definition of how a male or female person ought to behave (definition 2). Conversely, social expectations do not always match social reality. That is, in some cases the sexes do not actually differ in ways that social demands and expectations would prescribe. For these reasons I do not feel we can comfortably adopt Money's usage of the term *gender role behavior* for all behaviors that are sexually dimorphic (Chapter 2). We need terms that will distinguish those dimorphic behaviors that are socially expected or demanded from those that are not, and the term *role,* in its most widely accepted usage, is such a term, specifying definition 2 instead of 1.

To continue the argument, the third definition of masculinity and femininity is not always consistent with the other two. In all cultures the type 2 definition of femininity embodies the idea of nurturance, and a consistently dimorphic behavior pattern is found (definition 1). That is, over a wide range of ages, and in many societies, females show more positive response to infants than do males, and spend more time caring for infants and young children (Feldman & Nash, 1977, 1978; Edwards & Whiting, 1977). Concomitantly, such behavior is seen as part of the female role, and there is more social pressure on females than males to behave in nurturant ways. Little girls are more likely to be given dolls to play with, and when old enough, are more likely to be assigned baby-sitting duties. In adulthood, mothers are expected to be more involved in the care of their own infants than the fathers. Although the nurturant or maternal element figures strongly in the first and second definitions of femininity, it does not in the third definition. Indeed, it may even be an antithetical element. Teenaged girls, at the time they become most actively interested in being sexually attractive to males, show a *drop* in their responsiveness to babies, compared to their interest during the preadolescent period (Feldman & Nash, unpublished data reported in Maccoby, 1980). Furthermore, we have no reason to believe that the sexiest women become the best mothers. Indeed, the reverse may be true. Mate attraction and mate selection seem to constitute one phase of the reproductive cycle and care of the young a distinctly different phase. Although both are aspects of being an adequately functioning female adult, they may yield somewhat incompatible definitions of adult femininity. In giving dolls to girls but not boys, a culture signals its inclusion of nurturance in its definition of a little girl's femininity; but aspects of type 3 femininity are also present and encouraged. Little girls are given jewelry; they are flattered (especially by their fathers) when they dress up in ribbons and pretty clothes and have their hair nicely arranged. A father may say, with affection and approval, that his daughter is a "little flirt" or even "She's going to be quite a sexpot," while any signs of type 3 femininity in young sons are especially disturbing to many fathers.

The multifaceted nature of the meaning of femininity in childhood is illustrated by some findings from the Stanford Longitudinal Study (Maccoby & Jacklin, in press). When one cohort of children was 45 months old, several measures of sex typing were taken. One set of questions (separately asked of mother and father) inquired about whether their daughters were flirtatious, liked frilly clothes, cared how their hair looked, and liked to wear jewelry. These items formed a femininity scale. A second scale inquired about how rough and noisy a girl was in playing with her age-mates; whether she preferred to play with boys, girls, or both equally; whether she ever got into fights; and what her favorite "pretend" roles were (mostly taken from TV characters). On the basis of these items, a "tomboy" scale was constructed, representing a girl's willingness to play with boys, engage in rough and noisy play, fight occasionally, and enact masculine fantasy roles. To our surprise, the girls who were most feminine on the first scale were also somewhat more likely to be tomboys ($r = .38$, $p < .05$). This finding alerts us to several points. Clearly, individuals may display some elements but not others among the set of stereotyped "masculine" or "feminine" attributes. Also, different aspects of the cluster may take on more or less importance at different ages. The two attributes (sexually attractive and gentle-kind-considerate-nurturant) are no doubt involved in the concept of femininity at all ages, but the sexual aspect would appear to become dominant during adolescence, whereas the second component may emerge especially strongly during the ages when most women are involved in the care of children. In fact, in certain societies at certain historical periods, women are expected to subordinate their sexual attractiveness when they become mothers. We should not overlook the symbolism embodied in the Christian doctrine that Christ's mother is virginal.

We might speculate about possible age changes in the salience of different aspects of "masculinity." Being good at sports, for example, may be more important in defining a boy as masculine in middle childhood than it will be when he is an adult. These matters have hardly been thought about from a life-span perspective, however.

Deaux (Chapter 18) has referred to the concepts of masculinity and femininity as "fuzzy sets." I think this is an accurate usage that has some important implications. The notion of fuzzy sets has emerged in psychology in response to discoveries about the nature of prototypes or stereotypes. For many of our concepts—abstract ones such as "justice," or even one as concrete as "chair"—we can agree quite well on a concept's central meaning and can reliably identify an instance when we see one, but we cannot list a set of attributes or elements that every exemplar must have. In fact, no single instance is likely to be a complete exemplar of the prototype. If "masculinity" and "femininity" are truly fuzzy sets—and I believe they are—we will not have much luck in trying to list exactly what attributes are necessary and sufficient to determine whether a person (or even a given bit of behavior) is to be called masculine or feminine. The fact that such a list cannot be devised does not mean that the concepts are without meaning, however, nor even that they are more fuzzy than other concepts with which we deal quite comfortably all the time. Given that the concepts of masculinity and femininity do represent fairly loose clusters of attributes, not all of which are needed for any given exemplar, clearly there is almost infinite variety in the ways and degrees of an individual's being masculine or feminine. Nevertheless, in some instances all observers will agree that an instance is *not* a member of the class. A boy playing with dolls or wearing a dress in a case in point. We know that there are many things a boy cannot do if he is to be considered adequately masculine, but few things are explicitly proscribed for an adequately feminine girl. Neither parents nor age-mates show much disapproval of a girl if she wears masculine clothing or engages in masculine activities, but a boy is seen as effeminate if he crosses the gender line in a variety of ways. Males not only *receive* more social pressure for gender conformity; they also *exert* more sex-typing pressure. Studies of parent-child interaction have shown that fathers are more likely than mothers to treat children of the two sexes differently (Langlois & Downs, 1980; see also previously unpublished data from Langlois & Downs in Maccoby, 1980, p. 241; Jacklin, DiPietro, & Maccoby, 1984). Luria and Herzog (1985), after extensive observations in school playgrounds, refers to play settings as "gender school," and points out that boys teach each other the required masculine behavior and enforce it quite strictly. Girls also pass on the female culture (e.g., distinctively female games such as jump rope or jacks) and congregate mostly with each other, but individual "tomboy" girls can join boys' activities without losing their status in the girls' groups; the reverse is not true for boys. We can ask whether the greater sex-typing pressure exerted on and by males is a cross-culturally widespread phenomenon. Whiting and colleagues (Whiting, Kluckbohn & Anthony, 1958) have argued that it is, and that it is rooted in the requirement for societies to counteract the feminine identifications stemming from boys' early close ties to their mothers. Others argue that the attachment bonds of infancy have little or nothing to do with the establishment of gender identity or male and female sex typing (Kohlberg, 1966; Maccoby & Jacklin, 1974).

However this issue may be resolved, some important questions remain concerning the ontogenesis of sexually dimorphic forms of behavior. Money (Chapter 2) has distinguished among attributes that are sex irreducible, sex derivative, sex adjunctive, and sex arbitrary. To my mind these distinctions are useful in pointing to a dimension that has to do with the degree of flexibility societies have in assigning roles to the two sexes. No society can assign males the function of bearing children. On the other hand, societies have complete freedom to make assignments of distinctive hairstyles or modes of dress and body adornment. The deepest controversies lie in the intermediate area between these two extremes. Money rightly points out that one cannot think of any characteristics in the human that are completely sexually dimorphic, and indeed that there is large overlap between the sexes with respect to all aspects of behavior and motivation. The dimorphism lies, he says, in the threshold for activation of a particular kind of behavior, and these thresholds have varying degrees of direct linkage with biological sex. Money lists the following as being fairly directly derived from biological sex: the standing versus the sitting position in urination; readiness to engage in parenting behavior, including nest building, monitoring and controlling the whereabouts of the young, defending the young, and forming close attachment bonds to them (cuddling, carrying, clinging); high kinetic-energy expenditure (activity level); roaming and territorial marking; competitive rivalry and the formation of dominance hierarchies; fighting off predators; sexual arousal by vision versus touch. Some students of gender differentiation in human childhood would accept this list, and others would argue that some of these dimorphisms in humans are actually more the product of arbitrary cultural assignment than of any derivative linkage with biological sex.

The dimorphism in parenting is a good case in point. Readiness for parenting in the two sexes is an important issue in the current zeitgeist, when men are being urged to take a more equal role in child care in the interests of freeing women for more equal participation in the nondomestic sphere. The care of infants and young children has clearly been much more a female than male function in all human societies—and among lower primates as well. Adult males, however, are also clearly capable of carrying out all the needed parenting functions (other than lactation) with adequate skill if exposed to the young in situations where child caretaking is required. Furthermore, in human families a strong attachment bond between young children and their fathers is usually formed whether the father has been centrally involved in child care or not. There is some controversy over whether the child's attachment to the mother is qualitatively different than that to the father—in many families, it appears to be so, but of course this could be the outcome of the different familial roles customarily taken by the two parents, not of any differential predisposition.

I believe that the case of parenting, like a number of other behaviors that are sexually dimorphic to some degree, is an example of the futility of trying to assign the behavior to either biological or social causes. The two are inextricably linked. Societies are not completely arbitrary in assigning parenting functions mainly to women. The female role in lactation created the historical necessity for women to take the major role in infancy, and a number of customary divisions of labor between men and woman must naturally have grown up in consequence. The biological element may go further than lactation and give females an advantage in the form of a lower threshold for responding to infants, as Money suggests. Some early observations by Harlow (1962) with rhesus monkeys are consistent with this view. He reported that when a bottle-fed infant was caged with an adult male, the male initially rejected the infant's demands for contact comfort, but eventually yielded and began to hole the infant in ventral-ventral contact as a female would do. It was also reported that when an infant rhesus monkey was introduced into a cage with a male-female pair of adolescents, the female directed four times as much nurturant behavior toward the infant as did the male (Chamove, Harlow, & Mitchell, 1967). These reports are consistent with the differential threshold concept. Goldfoot and Neff (Chapter 12) present a somewhat different picture. They report that adult troop-reared male rhesus monkeys, when alone, will respond to the presentation of a stranded infant with parental behavior, doing so with equal latency to that of a female in a similar situation. Only when an adult female is present does the male not respond. This does not fit the picture of a threshold phenomenon. Neither does the study by Frodi and Lamb (1978), which found that men and women showed equivalent physiological arousal when exposed to the cry of a distressed infant, while women reported a stronger impulse to go to the infant and give comfort. If we want to think in terms of threshold, we must postulate different components of parental behavior, some having equivalent thresholds for the two sexes and others having different thresholds. The alternative would be to go to a more complex model, one in which a moderating stimulus determines whether or not a dimorphic threshold phenomenon will manifest itself.

If the threshold dimorphisms that do exist are affected to some degree by perinatal or prenatal hormonal sensitization, social customs could hardly be entirely indifferent to the existence of such predisposition. Societies can differ greatly, however, in the degree to which they function either to exaggerate or to modify the differential sensitivities.

I would like to add another aspect of gender-differentiated behavior to Money's list of possibly sex-derivative or sex-adjunctive behaviors: the tendency for the sexes to segregate themselves and to respond differentially to members of the same or opposite sex in ways other than specifically sexual ones. Gender segregation is important in that it provides the conditions in which two different "cultures of childhood" can emerge and be maintained. When human children have a choice of playmates, they tend to gravitate toward same-sex partners (Maccoby

& Jacklin, in press), and when a child as young as 2-1/2 to 3 years of age is introduced to an unfamiliar age-mate, the play is more active, and more *inter*active, among same-sex than among opposite-sex pairs (Jacklin & Maccoby, 1978). In observations of free play in nursery schools, same-sex clusterings or pairings of children are more common than opposite-sex ones. Although a deliberately engineered program of reinforcing children differentially for cross-sex play can reduce segregation temporarily, the playmate choices return quickly toward a segregated pattern once the behavior-modification program is discontinued (Serbin, Tonick, & Sternberg, 1977). During the early grade school years, segregation is widespread except in activities where adults take charge and direct children toward mixed partnerships (Schofield, 1981; Maccoby, 1986, for a review of studies).

A number of explanations have been offered for sex segregation in childhood. One is that children develop activity preferences that are somewhat different for the two sexes, and that mutual interest in an activity is what draws playmates together. Another explanation is a cognitive one: that children are able quite early to recognize the gender of other children as well as their own. The assumption is that children are drawn to other children who are perceived as the same as the self with respect to a number of attributes (e.g., age or race) and that gender is a salient member of the list of attributes. Another cognitive explanation is that children have been taught some gender stereotypes, and once they can recognize the sex of other children, these stereotypes come into play. Thus, if girls have been taught that boys may be expected to be rough and noisy, they will avoid them. We do not have space here to discuss the merits of these different explanations. My own guess is that there is some truth to each of them, but that same-gender play preferences emerge quite early—earlier than most children are capable of coding other children's gender as "same as me." Furthermore, the segregation in play appears in primates lower than humans—among whom the cultural transmission of cognitive gender stereotypes is surely minimal (Ruppenthal, Harlow, Eisele, Harlow, & Suomi, 1974). I suspect that some signal and response threshold dimorphisms contribute to children's finding some playmates more behaviorally compatible than others. The compatibility may have something to do with dominance relations. In our own work, we found that in our 33-month-old pairs, girls had difficulty influencing the interaction when playing with boys, whereas in girl-girl and boy-boy pairs, influence was mutual. Goldfoot reports (personal communication) that among young monkeys males exert dominance over females at a very high rate if adults are not present in their rearing environments, but that in troop-rearing environments that include both adults and juveniles, females are able to approach (but not fully achieve) equal dominance with males. This suggests a reason why, in nursery school situations, girls stay somewhat closer to the teacher and select female playmates more often; these conditions enable them to establish better control over their interpersonal environment than is possible in unmonitored play with males. Possibly, too, the greater male readiness for rough-and-tumble play is a factor that helps to draw male playmates together.

All this implies that children have a bias toward same-sex play that any society, or any pair of parents, would have to exert considerable pressure to counteract if they were to attempt to get children to choose playmates without regard to gender. Nevertheless, adults clearly do influence the degree to which play is segregated, by establishing the conditions under which play normally occurs.

The nature and degree of parental influence in the etiology of sex differences have been the subject of considerable debate. We have no doubt, as noted earlier, that parents do treat children of the two sexes differently in some respects, although the differences, I would contend, are not so great as many have supposed, particularly in infancy and early childhood. For example, unlike Lewis (Chapter 14), we do not find, in our three cohorts of children studied longitudinally, that mothers are more responsive to children of one sex than the other during the infancy and toddler periods (nor do we find differential rates of crying). Parents do provide sex-differentiated clothing and toys, however, and decorate their children's rooms in ways that give them a distinctive masculine or feminine appearance (pastel colors, ruffles, and floral patterns for girls; stronger colors and stripes or plaids for boys; (Rheingold & Cook, 1975). Fathers engage in more rough, physical play with sons than with daughters. When parents treat the two sexes differently, it is not always easy to determine the direction of effects. Do little boys like rough-and-tumble play because their fathers have trained them to enjoy it, or because they, as well as their fathers, have a low threshold for initiation of this male-male pattern? The same kinds of questions arise with respect to the higher rates of punishment and other coercive treatment directed to boys by their parents. Is this a form of differential pressure, initiated by parents, that will produce some kind of distinctively male behavior in boys, or is it a consequence of something boys are doing in interacting with their parents that elicits this kind of parental behavior? In our own work, we found in the two samples tested that fathers issued more prohibitions to 12-month-old sons than they did to daughters of the same age; but analysis of the conditions preceding paternal prohibitions revealed that sons of this age were getting into mischief more frequently than daughters (Snow, Jacklin, & Maccoby, 1983). When the frequency of child mischievous initiations was partialed out, no difference remained in paternal prohibitions directed toward children of the two sexes.

An even more difficult issue arises in understanding the role of socialization pressures on the two sexes when we consider that the "same" treatment may have differential effects on different children. An illustration comes from an early study by Patterson, Littman, and Bricker (1967), who reported higher rates of aggression in boys than in girls. Observation of nursery school children during free play revealed that for many children, if their attack on another child was followed by that child's yielding (crying, cringing, running away), the likelihood was increased that the aggressor would attack the same victim again on subsequent occasions. Thus, these responses by a victim were, on the whole, reinforcing to aggressors. But girls' attacks, when they did occur, were as likely as boys' to be "successful," so the sex difference in aggression could not be explained by a differential likelihood of reinforcement. The authors noted that some children did not appear to find the cringing or flight of a victim to be forcing. They may have felt sorry for the victim, but for whatever reason, they did not increase their attacks following "reinforcement." The analysis of the effects of reinforcement was not done separately by sex, so we can only surmise that there may have been more girls among the children who were not responsive to reinforcement from the yielding of a victim. The literature now has a number of indications that the sexes differ with respect to the predictions that can be made from parent behavior at one time to child behavior at a later time. Clearly, if boys and girls differ in the way they are affected by a given form of parental pressure, sex differences in the children's behavior cannot be attributed to the parental pressure alone, but must reflect an interaction between what the parent is doing and what the child has brought to the parent-child interchange.

We all give lip service to an interactionist position. We are now beginning to see, however, that this means different things to different people. For some it means that certain environmental conditions are necessary to make manifest preprogrammed dispositions that would otherwise remain latent. For others it means that the same environment will have different effects depending on the predispositions of the individual on whom the environment impinges. I have been suggesting an additional meaning: that individuals shape their environments, including their interpersonal environments, as well as vice versa. Individual male and female children, or sex-segregated groups of children, take an active role in forging environments that are compatible with their dispositions. It is also true that children of the two sexes have their dispositions shaped by socialization pressure from adults to act in sex-appropriate ways. Circular processes of influence and counterinfluence unfold over time; throughout childhood, individuals are engaged in active construction of their own version of the acceptably masculine or feminine behavior patterns to which they attempt to adhere.

References

Chamove, A., Harlow, H.F., & Mitchell G.D.. (1967). Sex differences in the infant-directed behavior of preadolescent rhesus monkeys. *Child Development, 38,* 329–336.

Edwards, C.P., & Whiting, B. (1977). *Sex differences in children's social interaction.* Unpublished report to the Ford Foundation.

Feldman, S. S., & Nash, S. C. (1978). Interest in babies during young adulthood. *Child Development, 49,* 617–622.

Feldman, S. S., & Sharon, S. C. (1977). The influence of age and sex on responsiveness to babies. *Developmental Psychology, 16,* 675–676.

Ferguson, L. R., & Maccoby, E. E. (1966). Intrapersonal correlates of differential abilities. *Child Development, 37,* 549–571.

Frodi, A. M., & Lamb, M. (1978). Sex differences in responsiveness to infants: A developmental study of psychophysiological and behavioral responses. *Child Development, 49,* 1182–1188.

Jacklin, C. N., DiPietro, J. A., & Maccoby, E. E. (1984). Sex-typing behavior and sex-typing pressure in parent-child interaction. In H. F. L. Meyer-Bahlburg (Ed.), *Gender development: Social influences and prenatal hormone effects,* a special issue of *Archives of Sexual Behavior, 13* (5), 413–425.

Jacklin, C. N., & Maccoby, E. E. (1978). Social behavior at 33 months in same-sex and mixed-sex dyads. *Child Development, 49,* 557–569.

Kohlberg, L. (1966). A cognitive-developmental analysis of children's sex-role concepts and attitudes. In E. E. Maccoby (Ed.), *The development of sex differences.* Stanford: Stanford University Press.

Langlois, J. H., & Downs, A. C. (1980). Mothers, fathers and peers as socialization agents of sex-typed play behaviors in young children. *Child Development, 51,* 1237–1247.

Luria, A., & Herzog, E. (1985). Gender segregation across and within settings. Paper presented at Biennial meeting of The Society for Research in Child Development. Toronto.

Maccoby, E. E. (1980). *Social development: Psychological growth and the parent-child relationship.* New York: Harcourt Brace Jovanovich.

Maccoby, E. E. (1986). Social groupings in childhood: Their relationship to prosocial and antisocial behavior in boys and girls. In D. Olweus, M. Radke-Yarrow & J. Block (Eds.), *Development of antisocial and prosocial behavior.* New York: Academic Press.

Maccoby, E. E., & Jacklin, C. N. (1974). *The psychology of sex differences.* Stanford: Stanford University Press.

Maccoby, E. E., & Jacklin, C. N. (in press). Gender segregation in childhood. In Hayne Reese (Ed.), *Advances in child development and behavior,* Vol. 20. New York: Academic Press.

Patterson, G. R., Littman, R. A., & Bricker. W. (1967). Assertive behavior in young children: A step toward a theory of aggression. *SRCD Monographs,* Vol. 35, No.5.

Petersen, A. C. (1976). Physical androgyny and cognitive functioning in adolescence. *Developmental Psychology, 12,* 524–533.

Rheingold, H. L, & Cook, K. V. (1975). The contents of boys' and girls' rooms as an index of parents' behavior. *Child Development, 46,* 459–463.

Schofield, J. W. (1981). Complementary and conflicting identities: Images and interaction in an interracial school. In S. A. Asher & J. M. Gottman (Eds.), *The development of children's friendship.* New York: Cambridge University Press.

Serbin, L. A., Tonick, I. J., & Sternglanz, S. (1977). Shaping cooperative cross-sex play. *Child Development, 48,* 924–929.

Snow, M. E., Jacklin, C. N., & Maccoby, E. E. (1983). Sex-of-child differences in father-child interaction at one year of age. *Child Development, 54* 227–232.

Ruppenthal, G. C., Harlow, M. K., Eisele, C. D., Harlow, H. F., & Suomi, S. J. (1974). Development of peer interactions of monkeys reared in a nuclear-family environment. *Child Development, 45,* 670–682.

Whiting, J. W. M., Kluckbohn, R., & Anthony, A. (1958). The function of male initiation ceremonies at puberty. In E. E. Maccoby, T. M. Newcomb, & E. L. Hartley (Eds.), *Readings in social psychology* (pp. 359–370) New York: Holt.

16

Gender Roles and Achievement Patterns: An Expectancy Value Perspective

Jacquelynne S. Eccles

Delineating the pattern of sex differences across various indexes of achievement-related behaviors has been one goal of social scientists interested in gender roles. For some areas of achievement, these differences have been found to be so large that they can best be described as nonoverlapping distributions. For example, no major league baseball players are female, and 96% of all registered nurses are female. In contrast, many measures of achievement-related behaviors yield no significant sex differences. For example, males and females earn equivalent grades in math and science courses throughout primary and secondary school. Other areas of achievement have either yielded no consistent pattern of sex differences or have been so inadequately studied that no conclusions can now be drawn regarding sex differences.

Explaining the origin and the consequences of sex differences in various achievement-related variables has been a second major goal of interested social scientists. Many theories of presumed differences have been offered. Unfortunately, because the theoretical work and the empirical work have not always proceeded hand in hand, theoretical explanations have emerged for presumed differences without solid evidence that the differences being explained actually exist [see Frieze, Parsons, Johnson, Ruble, & Zellman (1978) and Maccoby & Jacklin (1974) for a full discussion]. In addition, theoretical explanations that rely on presumed sex differences in mediating variables have been advanced when, in fact, neither the sex differences on these variables nor their mediating roles have been established.

Like many of my contemporaries, I am interested in sex differences in achievement patterns and in the link between gender roles and achievement behavior. To avoid the pitfalls inherent in the disjunction between empirical and theoretical work, I have structured this chapter around the following specific goals: (1) to review the sex differences in a subset of achievement behaviors that yield consistent and substantial sex effects; (2) to summarize and evaluate a comprehensive social-psychological theory explaining these differences; and (3) to relate the constructs of this theory to gender roles. The first two goals are firmly based on empirical data; the third goal is more speculative and should be treated by the reader as such.

Sex Differences in Achievement Patterns

Achievement can be operationally defined in many ways. In laboratory studies, it is often defined in terms of task choice, persistence in the face of failure, task performance, speed of performance, and scores on tests of motivation, anxiety, cognitive style, achievement, and aptitude. Field researchers and sociologists have defined it in terms of grades in school, scores on standardized tests of achievement and/or aptitude, course enrollment patterns, activity choices, performance in competitive activities such as sports or spelling bees, persistence in the classroom or on the job, motivational style, occupational choice, income, and career advancement. Because sex differences occur in only some of these variables, we must specify precisely the achievement behavior in which we are interested when we discuss sex differences. Furthermore, because achievement is such a complex array of behaviors, involving so many different constructs, and because individual differences in these various indexes of achievement are shaped by different processes, we should not expect simple explanations for sex differences in achievement patterns; many processes will be involved, and the relative importance of these processes may well vary depending on the particular achievement behavior chosen for study.

My discussion is limited to a set of achievement behaviors that either reflect real-life achievement choices or are linked to these achievement choices. These include scores on standardized tests of academic achievement and/or aptitude, grades in school courses, course enrollment patterns, persistence in laboratory tasks, persistence in or single-minded devotion to occupational achievement activities, and college major and occupational choices.

Test Scores and School Grades

Sex differences in tests of quantitative and verbal skills emerge with some regularity among adolescents and older subjects (Eccles, 1984;

Hyde, 1981). For example, among 13- and 17-year-olds, girls scored better than boys on the National Assessment Tests of reading, literature, art, and music; in contrast, boys scored better than girls on the science and math tests (Grant & Eiden, 1982). These differences were not large, however; the boys' and girls' mean scores differed by only 2–4% correct.

The math and science differences (but not the verbal differences) also show up regularly on the Scholastic Aptitude Tests administered nationally by the Educational Testing Service (ETS, 1980; Maehr, 1983). For example, in 1979 females received an average score of 443 on the SAT math test compared to an average score of 491 for males; in addition, 5% percent of the males scored 650 or higher on the math test, but only 1% of the females scored that high. Similar differences have emerged consistently over the last 18 years on the SAT Advanced Placement Tests for biology, chemistry, and physics, with males outscoring females by 30–50 points (Maehr, 1983) Note, however, that these math and science differences do not emerge with great consistency during the elementary school years. Furthermore, these differences are not very large [accounting for less than 4% of the variance (Hyde, 1981)], are not found universally even in advanced high school populations, and are not evident in course grades at any level including college (Eccles, 1984; Fennema & Sherman, 1977; Sherman & Fennema, 1977).

We know of one exception to the developmental conclusion reached earlier: The quantitative differences (but not the verbal differences) are evident at an earlier age among gifted populations (Benbow & Stanley, 1980, 1983). In addition, the magnitude of this difference remains fairly constant as the children pass through high school (Benbow & Stanley, 1982). As is true for the general population, the mean difference in this population is not large, averaging about 30 points, but the ratio of males to females at the upper end of the distribution is quite striking. For example, boys outnumber girls 4 to 1 in the group of children scoring 600 or better (Benbow & Stanley, 1983). Furthermore, no sex differences exist in these samples on tests of verbal skills; also, girls achieve better grades than boys in their math and science, as well as their language, courses (Benbow & Stanley, 1982).

Spatial Skills

As was the case with tests of mathematics achievement, a fairly consistent pattern of sex differences occurs on tests of spatial skills among adolescents and adults. These differences, however, do not emerge with great consistency among younger children (see Meece, Eccles, Kaczala, Goffs, & Futterman, 1982). In addition, the magnitude of the sex difference is affected by several other factors, such as maturational timing (Waber, 1979), body type (Petersen, 1979), personality characteristics associated with masculinity and femininity (Nash, 1979), previous experience with spatial activities (Burnett & Lane, 1980; Connor, Servin, & Schackman, 1977), and the particular test given (Connor & Serbin, 1980). The relation of these differences to either mathematical achievement or educational and occupational choices has yet to be demonstrated unequivocally.

Persistence

A widespread belief pervades psychology that girls are less persistent in the face of failure on laboratory tasks than boys [see Eccles (Parsons) (1983)]. Both Crandall and Crandall (personal communication, 1983) and I have reviewed the developmental literature related to this hypothesis. None of us find consistent support for it. Although the nature of girls' responses to failure is affected by the sex and age of the evaluator (Dweck & Bush, 1976), girls' behavioral responses in terms of persistence and accuracy following failure on laboratory tasks are, by and large, similar to those of boys (Beck, 1977–1978; Crandall, 1969; Dweck & Reppucci, 1973; Dweck, 1975; Diener & Dweck, 1978; Eccles, 1983a; Eccles (Parsons), Adler, & Meece, 1984; Eccles (Parsons), 1983; Nicholls, 1975; Rholes, Blackwell, Jordan & Walters, 1980; Veroff, 1969). This is not to say that there are no gender effects on the behavioral measures used in these studies. Indeed, under some conditions, boys and girls respond differently to both performance feedback and task manipulations. But, in my opinion, we have little evidence that girls are more likely than boys to give up following academic failures or to exhibit what might be labeled a learned-helplessness response to challenge and/or failure on laboratory tasks.

What about persistence in everyday achievement settings? To define or measure persistence in these achievement settings is difficult primarily because defining real-life achievement is difficult. It is even more difficult to assess sex differences in persistence in everyday achievement activities primarily because males and females engage in different types of achievement activities. Consequently, selecting a criterion activity is also difficult without biasing the results in favor of males or females depending on the activity chosen. For example, defining persistence in terms of occupational status and comparing males and females on this variable clearly biases our conclusion in favor of males. While acknowledging this value bias, we can still compare males and females on a set of variables assumed to be indicators of achievement persistence by the culture at large. The reader is forewarned, however, that these indicators do favor males in part because they represent typical male achievement activities.

One such indicator is advancement through the educational system toward higher degrees. Table 16-1 depicts the proportion of advanced

Table 16-1
Percent of Degrees Awarded to Females by Level of Degree and Selected Discipline Division: 1982

Discipline Division	Level of Degree			
	Bachelor's	*Master's*	*Doctorate*	*Professional*
Business & management	39.0	27.8	17.7	
Computer & informational science	34.8	26.5	8.4	
Education	75.9	72.3	48.6	
Engineering	12.3	9.0	5.3	
Health professions	83.9	75.8	45.3	
Letters	65.7	64.9	52.2	
Mathematics	43.2	33.2	13.8	
Physical sciences	25.7	21.7	13.7	
Social sciences	44.6	37.7	26.9	
Law	50.8	20.2	9.1	33.4
Dentistry (D.D.S. or D.M.D.)				15.4
Medicine (M.D.)				25.0
Veterinary medicine (D.V.M.)				36.2
Pharmacy (D.Pharm.)				36.2
Total	50.3	50.3	32.1	27.5

Source: National Center for Education Statistics, *The Condition of Education: 1984*. Washington, DC: U.S. Department of Health, Education, and Welfare.

degrees awarded to females in 1982. Although males and females received approximately equal numbers of bachelor's degrees, the number of males going on to obtain advanced degrees, even in traditionally female-stereotyped fields, generally exceeded the number of females. Furthermore, this discrepancy increases with the level of the degree being considered (National Center on Educational Statistics, 1984). This pattern has characterized the American educational scene for at least the last 30 years.

A comparable discrepancy exists among gifted populations. Fully 40% of the male college graduates in the Terman gifted sample (Terman & Oden, 1947) obtained a professional degree beyond the master's; in contrast, only 11% of the female college graduates in this sample had received a comparable degree.

Another such indicator is advancement through the occupational system toward ever-higher levels of responsibility and authority. Although institutional barriers undoubtedly contribute to the sex difference on this indicator, females are less likely than males to climb these achievement ladders; when they do, they typically climb at a slower rate than males even in traditionally female-stereotyped fields such as education (Frieze et al., 1978; Oden, 1968; Vetter, 1981). For example, in a recent survey of 1970–1974 Ph.D.s in science, social science, and engineering, 51% of the men had advanced into associate or full professor slots; in contrast, only 32% of the women had moved into these positions (Vetter, 1981). Similarly, among the Terman gifted sample, 24% of the working men had moved into executive and managerial positions by middle age; in contrast, only 9% of the working women had moved into such positions; 48% of these women were either primary or secondary school teachers or high-level secretaries or accountants (Oden, 1968).

One final indicator of persistence is single-minded devotion to one's occupational role. This indicator can be assessed in a variety of ways, including the number of hours one puts into one's work, willingness to ask one's family to make sacrifices for one's career advancement, and excessive concern over one's work to the exclusion of other concerns. Although we do not have extensive data on these or similar variables, several studies suggest that males are more likely than females to evidence the single-career-mindedness pattern (Baruch, Barnett, & Rivers, 1983; Bryson, Bryson & Johnson, 1978; Eccles & Hoffman, 1984; Goff-Timmer, Eccles, & O'Brien, 1985; Maines, 1983; Parsons & Goff, 1980). For example, when asked what concerned them most about their graduate training, female mathematicians were more likely than males to discuss the personal and family costs associated with the long hours of training; in contrast, males were more likely to discuss concern over their academic reputation and status among their professors and peers (Maines, 1983). Similarly, when we compare the distribution of time working women and men devote to their work and to their families, women are much more likely to spread their time across these activities, while men are more likely to focus their time on their work (Goff-Timmer et al., 1985). This pattern is especially true among professionals and individuals in other high-achievement jobs. Finally, when we compare the proportion of males and females that even work outside the home, similar discrepancies emerges. Women, even highly gifted women and those with advanced professional degrees, are less likely to work than are men, and, if they work, they are more likely to work part time and to move in and out of the labor market (Eccles & Hoffman, 1984; Goff-Timmer et al. 1985; National Center on Educational Statistics, 1980, 1984; Oden, 1968; Terman & Oden, 1947; U.S. Department of Labor, 1980). For example, among Terman's gifted sample, only 42% of the women, compared to 96% of the men, were employed in 1960; at which time they were middle-aged.

Course and Occupational Choice

The most marked sex difference in achievement behavior occurs with the achievement activities in which males and females decide to engage. From early childhood, boys and girls select different achievement

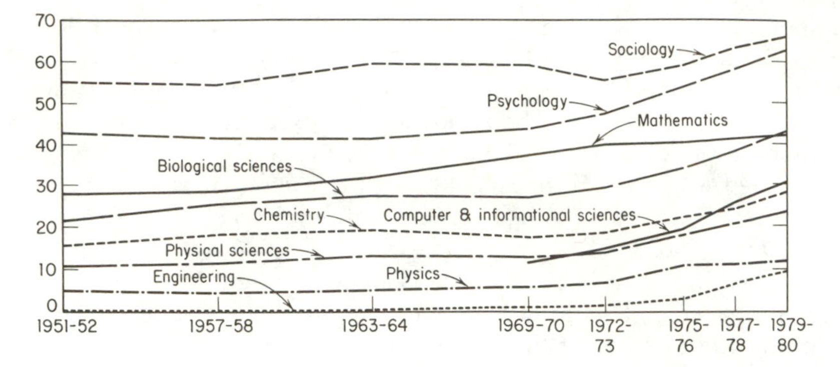

Figure 16–1. Percentage of bachelor's degrees awarded to females by field of study.

activities whenever they are given the choice (Huston, 1983). In spite of some recent changes, the differences are still dramatic; boys still play football and baseball, whereas girls do gymnastics and cheerleading. When they get to high school and have some choice over their courses, males and females still make predominantly sex-stereotyped selections (National Center for Educational Statistics, 1980, 1984), especially on career or vocationally relevant courses. For example, 92% of the students enrolled in traditionally male vocational education courses are male; similarly, 89% of the students enrolled in traditionally female vocational education courses are female (Eccles & Hoffman, 1984).

Even more extreme differences emerge for college majors. Figure 16-1 displays the percentages of bachelor's degrees granted to women in the sciences over the past 30 years. What is most striking is the constancy of the pattern. Although more degrees were awarded to women in 1980 than in 1950, the distribution of these degrees across the sciences has remained fairly stable over this period of time. Projections of the anticipated college majors of the freshman entering college in fall of 1980 suggest that certain majors will continue to have skewed populations. For example, 76% of the students planning to major in foreign languages are females, as are 76% of the students planning to major in education and 67% of the students planning to major in social science. In contrast, only 14% of the students planning to major in engineering and 33% of the students planning to major in physical science are female. Three significant changes have occurred: women are less likely to aspire to a teacher's credential, are more likely to aspire to a professional degree in business, law, or medicine, and are more likely to enroll in engineering and physics than they were 10 years ago (Magarell, 1980). Despite these changes, however, the percentage of advanced degrees in math, physical science, and engineering remains quite low, averaging between 1 and 15% (National Center on Educational Statistics, 1980, 1984).

Fairly similar differences in college major also characterize gifted populations, despite the fact that these males and females all have sufficient math and linguistic skills to major in whatever they choose. For example, among the Terman gifted population, 15% of the males obtained a bachelor's degree in engineering in comparison to less than 1% of the females; similarly, 36% of the females obtained a bachelor's degree in letters in comparison to only 9% of the males (Terman & Oden, 1947).

As is apparent in Tables 16-2 and 16-3, this pattern of sex differences also characterizes the occupational world (see Eccles & Hoffman, 1984). Table 16-2 depicts the percentage of female workers within each occupation (as well as for the total work force) for the period from 1960 to 1979. Although there have been some increases in the percentage of women in certain professions, by and large the labor market is still quite sex segregated, and women are still concentrated in relatively unskilled, low-paid work. In fact, the continued sex segregation of the labor market has been cited as one important cause of the persistence of sex differences in adult earnings (U.S. Department of labor, 1980). One additional fact is apparent in Table 16-2: any movement within an occupation toward more eqalitarian representation of the sexes that has taken place has occurred because females have entered typically male occupations. With few exceptions, female occupations have either retained a stable proportion of females or have increased in their proportion of females over the time period represented.

Once again, this discrepancy is also apparent in gifted populations. Table 16-3 depicts the occupations of Terman's gifted sample as of 1960 when the sample was predominantly middle-aged. Two things are apparent on this table. First, as noted earlier, the women are only half as likely as the men to be employed at all. Second, the distribution of employment differs markedly among these men and women and is consistent with gender-role stereotypes.

We should note at this point that there is nothing sacred about this occupational distribution. There is no biological reason why men should be less interested in teaching than women or that women should be less interested than men in medicine. In fact, in other cultures and at other historical periods, teachers have predominantly been men, and the medical profession has included a higher proportion of women (Newland, 1980). The distribution of men and women across the various occupations in the United States probably has more to do with the cultural definitions of appropriate roles for men and women and with cultural definitions of the occupational characteristics (e.g., status, flex-

Table 16-2
Employment of Women in Selected Occupations: 1960, 1970, and 1979

Occupation	Women as Percent of all Workers in Occupation		
	1960	*1970*	*1979*
Total work force	33.3	37.7	41.1
Professional-technical	38.0	40.0	43.3
Engineers	0.9	1.6	2.9
Lawyers, judges	3.3	4.7	12.4
Physicians, osteopaths	6.8	8.9	10.7
Registered nurses	97.6	97.4	96.8
Teachers, except college and university	71.6	70.4	70.8
Teachers, college and university	21.3	28.3	31.6
Technicians, excluding medical-dental	12.8	14.5	16.1
Writers, artists, entertainers	34.2	30.1	37.8
Managerial-administrative, except farm	14.4	16.6	24.6
Bank officials, financial managers	12.2	17.6	31.6
Buyers, purchasing agents	17.7	30.8	30.2
Sales managers, department heads; retail trade	28.2	24.1	39.8
Sales	36.6	39.4	45.1
Sales representatives, including wholesale	7.3	7.2	12.4
Sales clerks, retail	53.7	64.8	70.7
Clerical	67.5	73.6	80.3
Bank tellers	69.3	86.1	92.9
Bookkeepers	83.4	82.1	91.1
Cashiers	78.4	84.0	87.9
Secretaries, typists	96.7	96.6	98.6
Shipping-receiving clerks	8.6	14.3	21.3
Craft	2.9	4.9	5.7
Carpenters	0.4	1.3	1.3
Mechanics, including automotive	1.1	2.0	1.4
Bakers	15.9	29.4	43.6
Decorators and window dressers	46.2	58.3	72.9
Tailors	20.0	31.4	34.3
Operatives, except transport	34.3	38.4	39.9
Clothing ironers and pressers	N.A.	74.9	76.7
Dressmakers	96.7	94.8	95.4
Laundry and dry cleaning operatives	71.3	62.9	65.9
Transport equipment operatives	1.7	4.5	8.1
Bus drivers	9.8	28.5	45.5
Service	62.8	60.5	62.4
Private household	96.6	96.6	97.6
Food service	70.0	68.8	68.4
Health service	81.5	88.0	90.4
Protective service	4.1	6.2	8.8

Source: U.S. Department of Labor, *Perspectives on Working Women: A Databook.* Washington, DC: U.S. Printing Office, 1980.

Table 16-3
Occupations of Gifted Men and Women Employed Full-Time

	Men		Women	
Profession	*N*	*Percent*	*N*	*Percent*
Professional and Semiprofessional				
Lawyer, judge	77	10	2	<1
College or university faculty	54	7	21	8
Teaching and administration below four-year college level	32	4	68	27
Scientist, engineer, architect	107	15	2	<1
Physician, clinical psychologist	42	6	8	2
Author, journalist	17	2	11	4
Nurse, pharmacist, lab technician	0	—	6	2
Librarian	0	—	15	6
Government work (military and federal agencies)	18	2	0	—
Social work, welfare personnel	0		14	6
Arts and entertainment	21	3	4	2
Other professional	29	4	8	2
Business				
Executive and managerial positions	179	24	23	9
High-level clerical and accountant	61	8	54	21
Real estate, insurance, investments, small business	40	5	7	3
Public relations, promotions, advertising	15	2	5	2
Skilled trades and agriculture	27	4	0	—
Miscellaneous	14	2	5	2

Note: Percentages based on number of full-time employed individuals within each sex. Data are based on Oden (1968) and reflect employment status in 1960. Total population in 1960: males 759; females 597. Total number and percent employed: males 738 (96%); females 253 (42%).

ibility of training and performance, and support for training) than with any "natural" variations in men's and women's interest patterns.

Before leaving this general discussion of college major and occupational choice, I want to focus attention on one particular occupation—engineering—and on two college majors—engineering and physics. More than other occupations and/or majors, these have consistently had very small proportions of females. Given recent concerns of the low levels of participation by women in science and math-related fields, I find it quite interesting that engineering and physics are the major culprits at the bachelor's degree level. Women, in fact, have received a fair share of the degrees in mathematics, biology, and computer science, especially in recent years. To evaluate the low proportions of women in engineering and physics, one needs to consider the question: What is low? How many women should be majoring in engineering and physics, and what criterion can we use in answering this question? The answer depends on your explanation for the difference. If you believe that males inherently have more mathematical or more spatial skills than females, then you would predict that more males would major in fields requiring these skills, and you might believe that this discrepancy was inevitable or at least highly probable. But even if you adopted this conservative perspective (which I do not), how many females ought to be majoring in engineering and physics? To answer this question, we need to estimate the proportion of the pool for these majors that is female. To do this, I used the proportion of students scoring over 550 on the SAT math test because 550 is the approximate mean score of students aspiring to a major in engineering. Table 16-4 displays the comparison of this proportion with the female proportions of each of the following categories in 1979: (1) students who aspire to major in engineering and physics, (2) students who received bachelor's degrees in these fields, and (3) students who received doctoral degrees in these fields. Clearly, females are underrepresented in these categories using even this conservative means to estimate the expected proportion of females for each of these criteria measures. More stringent criteria yield a similar conclusion; for example, even if we use the proportion of students scoring 650 or greater on the SAT math test as our criterion, we would predict that 28% of all engineers and physicists would be female. This underrepresentation of females deserves explanation, as does the overrepresentation of females in fields such as education and letters.

Table 16-4
Proportion of Females in Each Category Within Engineering and the Physical Sciences

	Percentages of Females	
Categories	*Within Engineering*	*Within Physical Sciences*
Math SAT over 550, 1978	37	37
Freshman planning to major, 1978	15	34
B.S. degrees, 1982	12	26
Ph.D. degrees, 1982	5	14

Sources: Educational Testing Service, *National College-Bound seniors, 1979.* (1980). Princeton, N.J.: College Entrance Examination Board. Magarrell (1980). National Center for Educational Statistics, *The Conditions of Education: 1984.* U.S. Department of Health, Education, and Welfare.

Summary

In summary, although no consistent sex differences are reported for course grades and indexes of persistence on laboratory tasks, small but consistent differences are shown on tests of mathematical reasoning, spatial skills, and scientific knowledge favoring males among older children, adolescents, and adults. The differences on tests of language skills and on tests of knowledge in literature, music, and art are less consistent but favor females when found. Finally, fairly consistent differences emerge on indicators of the following: (1) persistence and single-minded pursuit of high levels of adult occupational achievement, (2) achievement-related activity choices in childhood and adulthood, (3) high school course enrollment patterns, (4) college majors, and (5) occupational choice. Each indicator involves an element of personal choice, even if that choice, as well as the options considered, is heavily influenced by gender-role socialization. Furthermore, the largest and most consistent sex differences in achievement tend to occur on behaviors that, in fact, reflect some degree of phenomenological, if not actual, personal choice.

Although very important, institutional barriers and discrimination are not entirely responsible for these differences. Ample evidence shows that psychological factors are also important, and many psychological explanations have been proposed to explain sex differences in achievement patterns. For example, the underrepresentation of females in the professions has been attributed to low self-confidence (Barnett & Baruch, 1978; Crandall, 1969; Nicholls, 1975; Parsons, Ruble, Hodges, & Small, 1976), fear of success (Horner, 1972), fear of loss of feminity (Tangri, 1972), nonconscious sex role idealogy (Lipman-Blumen & Tickameyer, 1972), differential values and orientation (Parsons & Goff, 1980; Stein & Bailey, 1973); Tittle, 1981), and low independence (Hoffman 1972; Stein & Bailey, 1973). Reviewing and evaluating each of these theories is beyond the scope of this chapter [see Frieze, Francis, & Hanusa, (1978); Parsons & Goff (1980); and Eccles (Parsons) (1983) for recent reviews]. Instead, I will focus on a more comprehensive, integrative theory my colleagues and I have been working on for the past 10 years [see Eccles (Parsons), Adler, Futterman, Goff, Kaczala, Meece, & Midgley (1983)]. Although originally proposed as a model of general achievement choices, it is applicable to the issue of sex differences in achievement patterns, especially in terms of the relationship of gender roles to achievement patterns. This model and its relation to gender roles are discussed in the next section.

A Model of Achievement Choices

Over the past several years my colleagues and I have been interested in the motivational factors influencing long-range achievement goals such as career or occupational choice, major selection in college, and the integration of work and family roles. Our interest in this area initially grew out of our concern with the underrepresentation of women in professional careers. Like many of our contemporaries, we set out to explain why bright, capable women were not achieving at the same levels as their male peers. We tried to identify the factors constraining women's efforts to attain these nontraditional, high-level achievement goals. But, troubled by the assumption that choosing a nontraditional career reflects maturity and enlightenment while choosing a traditional career reflects immaturity and sex role rigidity, we have redirected our focus. This assumption inevitably leads the researcher to ask the question, "Why aren't women more like men?" We have concluded that the question, "Why do men and women make the choices they do?" is both more appropriate and less biased.

To answer the latter question, we have elaborated an expectancy/value model of achievement choice based on the theoretical work of Lewin (1938) and Atkinson (1964). This model, depicted in Figure 16-2, links achievement choices to expectancies for success and to the importance and/or value an individual attaches to the available achievement options. It also specifies the relation of these constructs to cultural norms, gender roles, experience, aptitude, and a set of personal beliefs and attitudes associated with achievement activities. The influence of experience on achievement beliefs, goals, and outcomes is assumed to be mediated by one's interpretation of these experiences, by the input of primary socializers, by one's needs and values, by one's self-schemata, and by one's perceptions of the various choices themselves. Each of these factors is assumed to contribute both to the expectations one holds for future success at the options available and to the subjective value one attaches to these options. Expectations and subjective value, in turn, are assumed to influence achievement-related behaviors, including the decision to engage in particular activities, the intensity of effort expended, and one's actual performance. Finally, because gender roles and gender-role socialization affect each of the mediating variables (e.g., the behavior of socializers, one's self-schemata, and one's perceptions of the available options), gender roles should impact on both the expectations one holds for success and the value one attaches to various options.

The model assumes that achievement decisions, such as the decision to enroll in an advanced mathematics class or the decision to major in education instead of engineering, are made in the context of a variety of choices. Furthermore, it assumes that these choices, whether made consciously or nonconsciously, are guided by one's expectations for success on the various options, by such core personal values as achievement needs, competency need, and gender-role schema, by more

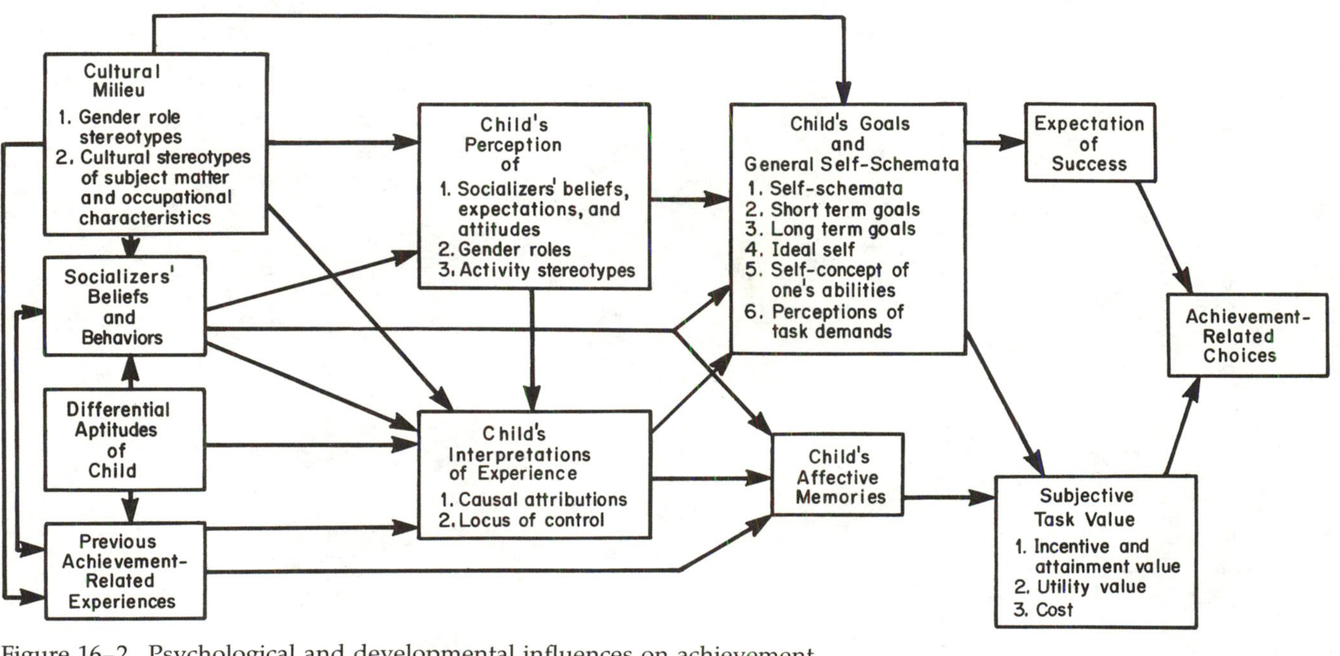

Figure 16–2. Psychological and developmental influences on achievement.

utilitarian values such as the importance of a particular course for one's future goals, and by the potential cost of investing time in one activity instead of another.

For example, consider course enrollment decisions. This model predicts that people will be most likely to enroll in courses in which they think they will do well and that have high value for them. Expectations for success depend on the confidence individuals have in their intellectual abilities and on their estimates of the difficulty of various courses. These beliefs have been shaped over time by experiences with the subject matter and by individuals' interpretations of these experiences (e.g., does the person think that her successes are a consequence of high ability or lots of hard work?).

The value of any particular course for a particular individual is also influenced by several factors. For example, does the person like doing the subject material? Is the course required? Is the course seen as instrumental in meeting long- or short-range goals? Have the individual's parents or counselors insisted that the course be taken or, conversely, have they discouraged the individual from taking the course? Is the person afraid of the material to be covered in the course?

Three features of our model are particularly important for understanding sex differences in educational and vocational decisions. The first is our focus on choice as the outcome of interest. We believe that individuals continually make choices, both consciously and nonconsciously, regarding how they will spend their time and their efforts. Many of the most significant sex differences occur on achievement-related behaviors that involve the element of choice, even if the outcome of that choice is heavily influenced by socialization pressures and cultural norms. Conceptualizing sex differences in achievement patterns in terms of choice takes us beyond the question, "Why aren't women more like men?" to the question, "Why do women and men make the choices they do?" Asking the latter question, in turn, legitimizes the choices of both men and women, allowing us to look at sex differences from a choice rather than a deficit perspective.

Conceptualizing achievement-related sex differences in terms of choice highlights a second important component of our perspective: the issue of what becomes a part of an individual's field of possible choices. Although individuals do choose from among several options, they do not actively, or consciously, consider the full range of objectively available options in making their selections. Many options are never considered because the individual is unaware of their existence. Others are not seriously considered because the individual has inaccurate information regarding either the option itself or the probability of achieving the option. Still others may not be considered seriously because they do not fit in well with the individual's gender-role schema. Assimilation of the culturally defined gender-role schema can have such a powerful effect on one's view of the world that activities classified as part of the opposite sex's role are rejected, often nonconsciously, without any serious evaluation or consideration.

Understanding the processes shaping individuals' perceptions of their field of viable options is essential to our understanding of the dynamics leading women and men to make such different achievement decisions. We have very little evidence regarding these processes and their link to important achievement choices. Socialization theory provides a rich source of hypotheses. For example, nontraditional role models may legimatize novel and/or sex-role-deviant options. Parents, teachers, and school counselors can also influence students' perceptions of their field of options through the information and experiences they provide regarding various options. Parents can affect the options actually available to their children by providing or withholding funds for certain training and educational experiences. They can also affect the options seriously considered by mandating, encouraging, ignoring, and discouraging various options. Finally, peers can affect the options seriously considered by either providing or withholding support for various alternatives. These peer effects can be quite direct (e.g., laughing at a girl when she says she is considering becoming a nuclear physicist), or indirect (e.g., anticipation of one's future spouse's support for one's occupational commitments).

These examples show that social agents can either encourage or discourage students from considering gender-role stereotypic choices. Unfortunately, they typically operate in such a way that students are most likely to consider those options that are consistent with gender-role stereotypes (see Eccles & Hoffman, 1984).

The third important feature of our perspective is the explicit assumption that achievement decisions, such as the decision to enroll in an accelerated math program or to major in education instead of law or engineering, are made within the context of a complex social reality that presents each individual with a wide variety of choices, each of which has both long-range and immediate consequences. Furthermore, the choice is often between two or more positive options or between two or more options that each have both positive and negative components. For example, the decision to enroll in an advanced math course is typically made in the context of other important decisions such as whether to take advanced English or a second foreign language, whether to take a course with one's best friend or not, whether it's more important to spend one's senior year working hard or having fun, etc. Too often theorists have focused on the reasons capable women do not select high-status achievement options and have failed to ask why they select the options they do. This approach implicitly assumes that complex choices, such as career and course selection, are made in isolation. For example, it assumes that the decision to take advanced math is based primarily on variables related to math. My colleagues and I explicitly reject this approach, arguing instead that it is essential to understand the psychological meaning of the roads taken as well as the roads not taken if we are to understand the dynamics leading to the differences in men's and women's achievement-related choices.

Consider, for example, two high school students Mary and Jane, both of whom enjoy mathematics. They reach the end of their junior year in high school and must decide whether to enroll in the advanced math course offered by their school. Both girls want at least a B in all of their courses and believe that getting an A or a B in the advanced math class will take at least 1 hour of homework a night. Although Mary likes math, she hopes to get the lead part in the school play and she plans on majoring in drama in college. In contrast, Jane hopes to become a psychologist and has been told that math is good preparation for a research career. Which young woman will enroll in math? Most likely Jane will and Mary will not, unless Mary gets additional information that makes the cost of taking the course more worthwhile to her.

In summary, we assume that educational and vocational choices, whether made consciously or not, are guided by the following: (1) one's expectations for success on the various options perceived as being available, (2) the relation of these options both to one's short- and long-range goals and to one's core self-identity and basic psychological needs, (3) the individual's gender-role schema, and (4) the potential cost of investing time in one activity instead of another. We believe that each of these psychological variables is shaped by experiences, cultural norms, and the behaviors and goals of one's parents, teachers, role models, and peers. Let me now turn to a more explicit discussion of the link between gender roles and the two primary psychological components of this model.

Gender Roles and Achievement Choices

Both cognitive-developmental and social-cognition theorists suggest that children's self-concepts and goals are derived in part from their interpretations of the attitudes and behaviors of those around them. Gender-role structure and gender stereotypes are two particularly salient components of our social world. Consequently, gender-role beliefs may influence the development of our self-concepts and of our perceptions of the value of various activities as well as our expectations for success on various gender-typed activities. Research has provided support of this hypothesis. By 5 years of age, children have acquired well-defined gender-role stereotypes regarding appropriate behaviors, traits,, and even expectations of success (Huston, 1983; Stein & Smithells, 1969;

Williams, Bennett, & Best, 1975). In addition, both children and adults monitor their behaviors and aspirations in terms of these stereotypes (e.g., Montemayor, 1974; Eccles & Hoffman, 1984; Frieze et al., 1978; Huston, 1983, for recent reviews). Consequently, gender roles probably influence achievement choices through their impact on expectations and subjective task value. Possible mechanisms of this influence are discussed in the next sections.

Expectations for Success

Expectations for success have long been recognized by decision and achievement theorists as important mediators of behavioral choice (Atkinson, 1964; Bandura, 1984; Lewin, 1938; Weiner, Frieze, Kukla, Reed, Rest, & Rosenbaum, 1971). Numerous studies have demonstrated the importance of expectations for a variety of achievement behaviors including task persistence and task choice. But are expectations influenced by gender roles, and do males and females differ in their expectations for success at various achievement tasks? The answers to these questions are not yet settled.

Gender Roles and Expectations

Because females are typically stereotyped as less competent than males, incorporation of gender-role stereotypes into one's self-concept could lead girls to have less confidence in their general intellectual abilities than boys. This, in turn, could lead girls to have lower expectations for success at difficult academic and vocational activities. It could also lead girls to expect to have to work harder in order to achieve success at these activities than boys expect to have to work. Evidence from my own work (see final section) supports these predictions. Similar findings have been reported by others (e.g., Nicholls, 1975; Parsons et al., 1976). Either of these beliefs could deter girls from selecting demanding educational or vocational options, especially if these options are not perceived as especially important or interesting.

Gender roles could also produce differential expectations of success depending on the gender stereotyping of the activity. Both educational programs and vocational options are gender stereotyped in this culture. Many high-level professions and both math-related and scientific technical courses and vocational fields are thought to be male activities; in contrast, teaching below the college level, working in clerical and related support jobs, and excelling in language-related courses are thought to be female activities by both children and adults (see Eccles & Hoffman, 1984; Huston, 1983, for reviews). Incorporating these beliefs into one's self-concept could lead girls to have lower expectations for success in male-typed activities and higher expectations for success in female-typed activities. This pattern of differential expectations could lead girls to select female-typed activities over male-typed activities; the converse argument holds for boys. Some support for this perspective is provided by Eccles (Parsons) et al., (1984), and Huston (1983).

Despite many years of research on sex differences in self-confidence and expectations for success, most of the hypotheses implicit in this analysis have either not been tested or cannot be evaluated because of the equivocal nature of the findings (see Eccles (Parsons) et al., 1984; Eccles, 1984; Meece et al., 1982; Dweck & Elliot, 1983, for opposing opinion). The results are especially equivocal among high-ability students. For example, Fox (1982) found that highly motivated gifted girls have lower self-confidence than equally highly motivated gifted boys; Terman (1926) found that gifted girls were more likely to underestimate their intellectual skills and knowledge, while gifted boys were more likely to overestimate theirs. In contrast, Tidwell (1980) found no sex differences on measures of self-concept among the gifted. Both Tidwell (1980) and Tomlinson-Keasey and Smith-Winberry (1983) found no sex differences on measure of locus of control (a construct often linked to self-confidence and personal efficacy beliefs—see Bandura, 1977). Finally, Schunk and Lilly (1982) found no sex difference in expectations for success among the gifted on a laboratory math task. Similar discrepancies hold for studies based on college students and more normative populations of both children and adolescents (see Eccles (Parsons) et al., 1983; Meece et al., 1982). Note, however, that when differences emerge, males are more self-confident than females, especially in male-stereotyped activities.

The picture is further complicated by the fact that the links assumed to exist in this analysis have rarely been tested. In many of the relevant studies, boys' and girls' expectations were measured, but the link of these expectations to sex differences in achievement choices was not assessed. Expectations do appear to influence choice, but sex differences in choice are most likely the consequence of a variety of factors, some of which may be more important than expectations for success. For example, my own data (gathered on college-bound high school students) suggest that subjective task value is a more critical mediator of sex differences in course enrollment decisions (Eccles (Parsons) et al., 1984). Consequently, at present, the role of differentiated expectations of success as determinants of sex-differentiated educational and vocational behavior needs further study.

Causal Attributions and Expectations

Casual attributions are often linked to self-confidence, expectations, and achievement behaviors (e.g., Dweck & Licht, 1980; Eccles (Parson) et al., 1983; Weiner et al., 1977). The possibility that sex differences in causal attributions might mediate sex differences in achievement behaviors, especially in the motivation to persist despite difficulty and failure, has been suggested by several psychologists (e.g. Bar-Tal, 1978; Dweck & Licht, 1980; Nicholls, 1975; Parsons et al., 1976). The pattern of sex differences on measures of causal attributions is even more equivocal than the findings associated with expectancy and intellectual confidence, and the hypothesized mediating effect of these sex differences on achievement choices is not clear (Cooper, Burger, & Good, 1981; Eccles (Parsons), 1983; Eccles (Parsons) et al., 1984; Frieze, Whitley, Hanusa, & McHugh, 1982).

One interesting sex effect has emerged for mathematical tests and course grades. Girls, especially intellectually bright girls, rank skill, diligence, and/or effort as more important causes of their math success than do boys of equal talent. In contrast, the boys rate high ability as a more important cause of their math success than the girls (Eccles (Parsons) et al., 1983; Eccles (Parsons) et al., 1984; Wolleat, Pedro, Becker, & Fennema, 1980). This pattern of differences may have important consequences for students' decisions regarding future involvement with mathematics consistent with the pattern of sex differences in such involvement. People who view consistent effort (or skill and knowledge acquired through consistent effort) as the important determinant of their success in mathematics may avoid future courses if they think future courses will be more difficult, demanding even more effort for continued success. The amount of effort a student can or is willing to expend on any one activity has limits, and if a female student already believes she is working very hard to do well in math, she may conclude either of the following: (1) her performance will deteriorate in the more difficult future math courses because she is trying as hard as she can at present, or (2) the amount of effort necessary to continue performing well is just not worth it. For some students, especially students who do not place a high subjective task value on math, either of these beliefs would be sufficient justification to avoid both future math courses and math-related careers. The same limits would not apply to students who view ability as a relatively more important determinant of success in math than effort. High levels of math ability should guarantee continued success with little or no increment in one's efforts. If this analysis is correct, then girls should be less likely to enroll in advanced math courses and to aspire to math-related technical fields. This is, in fact, the case among both general and gifted populations.

Summary

Although there are clear theoretical reasons for predicting a link between gender roles and sex differences in confidence and expectations, the empirical work has yielded somewhat equivocal findings. When differences emerge, however, they do support this link. In addition, although both Hollinger (1983) and Eccles (Parsons) et al., (1984) have demonstrated a link between confidence in one's math abilities and students' math-related choices (such as vocational aspirations and enrollment patterns), the mediating role of expectations and self-confidence in fostering sex differences in educational and occupational choices needs further study. Furthermore, consistent with the argument developed earlier, Terman (1926) found a positive relationship between students' subject-matter preferences and their ratings of the ease of the subject for themselves. The boys and girls in his study, however, did not differ in their perceptions of the ease of mathematics. Clearly, more data are needed, especially regarding the putative impact of self-concept-related variables on achievement-related choices and the link between self-concept-related variables and gender roles. Quite possibly confidence is a necessary but not sufficient factor, such that a certain level of confidence in one's ability to succeed is necessary for the inclusion of specific options in one's field of choices, while the actual choice among these options is more closely tied to subjective task value.

It is possible that researchers have been assessing the wrong expectancies. Typically, individuals are asked to report on their confidence about succeeding on an upcoming task or course. They are not asked how confident they are that they could succeed in particular professions or in particular advanced training programs. They are also not asked how much effort they think it will take to succeed in various professions or advanced training programs. Females may be less confident than males of their prospects for success in these more abstract, distant activities. Possibly females are as confident as males are in their ability to succeed but assume that it will take a lot more work, time, and/or effort to succeed than their male peers assume it will take. Either of these beliefs could mediate a sex difference in educational and vocational decisions, especially given the gender stereotyping of most high-status occupations.

Alternatively, critical expectancy beliefs may neither be the expectation one has for success in a particular field nor be the perception one has of the amount of effort it will take to succeed in a particular field; instead, the critical beliefs may be the relative expectations one has for success across several fields and the perceptions one has of the relative amounts of effort it will to take to succeed in various fields. If females think a lot more effort will be necessary to succeed as an engineer or a doctor than to succeed as an elementary school teacher or a journalist or a nurse, they may opt for the more female-typed occupations, especially if they place high importance on having a career that is compatible with their anticipated family roles.

Values as Mediators of Achievement Choices

Value is the second major component of our model of achievement choices. Until recently, value has received less systematic attention in achievement theory, so I would like to elaborate on our interpretation of value and its link to both achievement choices and gender roles. Less empirical work has been devoted to the issue of sex differences, gender roles, and values; therefore, this section is more theoretical than the previous one.

Like others (e.g., Crandall, 1969; Crandall, Katkovsky, & Preston, 1962; Reynor, 1974; Spenner & Featherman, 1978; Stein & Bailey, 1973), we assume that task value is a quality of the task that contributes to the increasing or declining probability that an individual will select it. We have defined this quality in terms of three components: (1) the utility value of the task in facilitating the achievement of one's long-range goals, (2) the incentive value of engaging in the task in terms of more immediate rewards such as the pleasure one gets from doing the activity, and (3) the cost of engaging in the activity.

Incentive and attainment values

Incentive value can be conceptualized in several ways, two of which are of particular interest to me. On the one hand, incentive value can be conceptualized in terms of the immediate anticipated rewards, intrinsic or extrinsic, that performance of a task will provide an individual. For example, playing tennis could be intrinsically rewarding because it makes one feel healthy, or extrinsically rewarding because one is paid for the performance.

Incentive value can also be conceptualized in terms of the needs and personal values an activity fulfills. Individuals develop an image of who they are as they grow up. This image is made up of many component parts including: (1) conceptions of one's personality and capabilities, (2) long-range goals and plans, (3) schemata regarding the proper roles of men and women, (4) instrumental and terminal values (Rokeach, 1973), (5) motivational sets, (6) self-schemata, and (7) social scripts regarding proper behavior in a variety of situations. Some parts of this image are more central to one's self-definition than others; these parts are believed to exert the most influence on one's behavior (Markus, 1980; Parsons & Goff, 1980), perhaps through their impact on the value individuals attach to various activities. We believe that individuals perceive tasks in terms of certain characteristics that can be related to the central components of their self-images. For example, a difficult task requiring great effort to master may be perceived as an achievement task; if it also involves pitting one's performance against others, it may be perceived as a competitive task. Other tasks may be perceived in terms of nurturance, power, aesthetic pleasure, etc. Engaging in a particular task requires the demonstration or exercise of those characteristics associated with the task. Whether this requirement is seen as an opportunity or a burden will depend on the individual's needs, motives, and personal values and on the individual's desire to demonstrate these characteristics to both the self and others.

Essentially, I am arguing: (1) that individuals seek to confirm their possession of those characteristics central to their self-image, and (2) that various tasks provide the opportunity to do exactly this. If one values the characteristics assumed to be inherent in a task, one will regard task involvement as an opportunity to confirm one's self-image and will be more likely to engage in the task than someone who does not value the characteristics associated with the task.

This analysis implies that the incentive value of any particular task will be influenced by three sets of beliefs. First, it will depend on the individual's perception of the characteristics of the task or, more specifically, on the needs and characteristics the individual believes the task will either fulfill or demonstrate. Second, it will depend on the individual's self-schemata and hierarchy of values, needs, and motives. Finally, it will depend on the extent to which the individual believes that participation in the task will fulfill his or her central needs or will affirm his or her self-image. We have labeled this third belief attainment value. Because gender-role socialization influences both our perception of the appropriateness of the activities available to us and our perception of ideal characteristics, gender roles should affect the attainment value we attach to various options and, as a consequence, our achievement choices. In support of this view, Rokeach (1973) has found that variations in core personal values do predict career selection.

Perceived Cost

The value of a task will also depend on a set of beliefs that are best characterized as the cost of participating in the activity. The cost of participating in an activity is influenced by many factors, such as anticipated anxiety, anticipated negative responses from one's peers, friends, parents, colleagues, neighbors, etc., fear of failure, and the negative affective memories one has associated with similar activities in the past. Gender-role socialization can impact on each of these negative affective variables (see Eccles, 1984, for full discussion).

The cost of any given activity or life-defining achievement choice can also be conceptualized in terms of the loss of time and energy for other activities and/or life-defining roles. People have limited time and energy. They cannot do everything they would like. They must choose among activities. To the extent that one loses time for activity B by engaging in activity A and to the extent that activity B is high in one's hierarchy of importance, the subjective cost of engaging in A increases. Alternatively, even if the attainment value of A is high, the value of engaging in A will be reduced to the extent that the attainment value of B is higher and to the extent that engaging in A jeopardizes the probability of successfully engaging in B.

Gender Roles and Task Value

The implications of this analysis for our understanding of sex differences in achievement choices are clear. Because socialization shapes individuals' goals, values, self-schemata, and ideal self, men and women undoubtedly acquire different values, self-schemata, and goals through the process of gender-role socialization. In fact, the very essence of one's sense of femininity and/or masculinity is probably embedded in these self-perceptions and self-ideals.

In terms of task value, gender differences in value structure can manifest themselves in several ways. For one, gender-role socialization could create a gender-differentiated hierarchy of core personal values (such as their terminal and instrumental values, Rokeach, 1973). In support of this suggestion, both Rokeach (1973) and Tittle (1981) have found marked differences in the values of males and females. In addition, differences in values are the most consistent attitudinal differences among the gifted (see Eccles, 1985b). In both cases, the differences reflect gender-stereotyped patterns. For example, on the Allport-Vernon-Lindsey Scale of Values, females are especially likely to hold social, aesthetic, and investigative values; in contrast, males are especially likely to hold investigative and scientific values (McGinn, 1976).

To the extent that males and females differ in their hierarchy of core personal values, tasks embodying various characteristics should have different attainment values for men and women. For example, men may be more likely to engage in athletic activities because they place more importance on demonstrating their athletic competence than do women. Differences in career choice could also reflect this aspect of differential attainment values. For example, Dunteman, Wisenbaker, and Taylor (1978) found that being thing-oriented instead of person-oriented predicted becoming a math or a science major. Similarly, Fox and Denham (1974) found that mathematically talented children are relatively low on social value and high on theoretical, political, and economic values. In both of these studies, the females were less likely to hold the math- and science-related values than were males. Not surprisingly, then, the females were also less likely than the males to aspire to math- and science-related careers.

Alternatively, the structure of men's and women's hierarchies of values might differ. If so, then women ought to rank the importance of various activities differently than men do. For example, if women see the parenting role as more important than, or as important as, a professional career role, while men perceive their career role as more important than their parenting and spousal roles, then women should be more likely than men to resolve life's decisions in favor of the parenting role. This differential would be especially marked if women saw the career options as not only of lower importance but also detrimental to the successful completion of their parenting goals. In support of this suggestion, McGinn (1976) had gifted boys and girls rate various occupations including homemaker on a semantic differential; the girls gave positive ratings to several different professions and to the homemaker role; in contrast, the boys gave positive ratings only to traditional male occupations.

Using a slightly different methodology, I obtained similar results (Eccles, 1983b). I had 48 female college students rate 10 different occupations including mother in terms of difficulty, importance, and the extent of pride and shame one would feel if one succeeded or failed at the occupation. The rank order of their ratings for the various occupations (chosen to represent an array of male- and female-dominated occupations requiring college training) is displayed in Table 16-5. These women rated motherhood as being as difficult, challenging, and potentially rewarding as any of the careers; in addition, they rated success at motherhood as more important and potentially more rewarding than success at any of the other careers, and they rated the potential shame they would feel if they failed as a mother higher than the shame they would feel if they failed at any of the other occupations. Is it any wonder, then, that high-achieving college-trained women select occupations that they believe will not interfere with their role of mother?

Men and women could also differ in the density of their goals and values. As noted earlier, males seem more likely than females to exhibit a single-minded devotion to one particular goal. In contrast, females seem more likely than males to be involved in several activities simultaneously. This difference could reflect differing density patterns for the hierarchy of goals and personal values. That is, females may place high attainment value on several goals and activities, while males may differentiate more among the options open to them. In support of this suggestion, McGinn (1976) found that gifted boys evidenced a more unidimensional value structure than gifted girls on the Strong-Campbell Vocational Interest Test. Additional support is provided by Tittle (1981), who found that high school females are more likely to integrate work and family values in their value hierarchy; in contrast, the males were more likely to keep these values compartmentalized. To the extent that males have a more differentiated hierarchy of values and to the extent that the density of their hierarchy is lower than females', the cost of

Table 16-5
Rank Order of Occupations on Each Dependent Measure in Study

	DEPENDENT MEASURE					
OCCUPATION	*Difficulty*[a]	*Probability of Success*[b]	*Importance of Success*[c]	*Effort*[c,d]	*Positive Affect*[c,e]	*Negative Affect*[c,f]
Mother	3	10	1	1	1	1
Pediatrician	2	4	2	3	3	2
Surgeon	1	3	3	2	2	3
Psychologist	5	7	4	4	4	5
High school teacher	8	9	5	5	5	4
Elementary school teacher	9	8	6	6	6	6
Nurse	7	6	7	7	8	7
Artist	6	2	8	8	7	8
Interior decorator	10	5	9	9	10	9
Mechanical engineer	4	1	10	10	9	10

[a]1 = Most difficult.
[b]1 = Lowest probability of success.
[c]1 = Most importance, most effort, most positive affect for success, most negative affect for failure.
[d]*rho* (with difficulty) = .60, $p > .05$; *rho* (with probability) = .50, $p > .05$; *rho* (with importance) = .89, $p < .001$.
[e]*rho* (with difficulty) = .67, $p < .05$; *rho* (with probability) = −.40, $p > .05$; *rho* (with importance) = .97, $p < .001$.
[f]*rho* (with difficulty) = .55, $p > .05$; *rho* (with probability) = .52, $p > .05$; *rho* (with importance) = .97, $p < .001$.

engaging in their primary goal in terms of other important goals should be less for males than for females.

Finally, a gender-differentiated hierarchy of task values could result from gender differences in people's perceptions of various tasks and in the very definition of success and failure on these tasks (see also Frieze, Frances, & Hanusa, 1983). One of the primary characteristics of gender roles is that gender roles define the activities that are central to one's occupancy of the role. In essence, gender roles define what one should do with one's life in order to be successful in that role. To the extent that one holds success in one's gender role as a central component of one's identity, then activities that fulfill this role will have high value, and activities that detract from one's successful fulfillment of this role will have lower, and perhaps even negative, subjective value. If staying home with one's children and being psychologically available to them most of the time is a central component of one's gender-role schema, than involvement in a demanding, high-level career will have reduced value because it conflicts with a more central component of one's identity.

Adherence to one's gender role may be so central to an individual that merely knowing, even at a subconscious level, that a particular activity is stereotypically part of the opposite gender's role will be sufficient to prevent further consideration of the possibility of engaging in that activity. Consequently, as discussed earlier, gender-role schemata (beliefs regarding the composition of both male and female gender roles) can effectively limit the range of options one even considers as well as affecting the subjective value one attaches to the various options one does consider.

Gender roles can also influence the very definition one has of successful performance of those activities considered to be central to one's identity. Consequently, men and women may differ in their conceptualization of the requirements for successful task participation and completion. If so, men and women should approach and structure their task involvement differently. The parenting role provides an excellent example of this process. If males define success in the parenting role as an extension of their occupational role, they may respond to parenthood with increased commitment to their career goals and with emphasis on encouraging competitive drive in their children. In contrast, if women define success in the parenting role as high levels of involvement in their children's lives, they may respond to parenthood with decreased commitment to their career goals.

Differences in approach to various careers can be interpreted similarly. For example, academic women are found to publish less than academic men. One possible explanation for this finding relies on the reasoning outlined here. Females may define the faculty role more in terms of teaching and service than in terms of publication; in contrast, males may define the faculty role more in terms of research and publication. If so, then male and female faculty members should approach their professional role quite differently and, as a consequence, females should have weaker publication records than men.

Summary

In summary, the model depicted in Figure 16-2 builds on the theoretical base of expectancy/value models of task choice. In addition, by elaborating on the construct of value, it has provided a link between expectancy/value models and the growing literature on the self and on the link between gender roles and self-schemata.

What distinguishes this model from other models of achievement behavior is its attention to the issue of *choice and perceived options.* Whether done consciously or not, individuals make choices among a variety of activities all of the time. For example, they decide whether to work hard at school or just to get by; they decide which intellectual skills to develop or whether to develop any at all; they decide how much time to spend doing homework; they decide whether to take difficult courses or to spend their extra time with their friends; and they decide how to integrate work and family roles, etc. We have tried to address the issue of choice directly. Furthermore, we have tried to specify the kinds of socialization experiences that shape individual differences on the mediators of these choices and on perceived options, especially in the academic achievement domain (see Eccles, 1985b; Eccles (Parsons) et al., 1983; Eccles & Hoffman, 1984; Parsons & Bryan, 1980).

Furthermore, because we have focused on choice instead of avoidance, we believe this model provides a more positive perspective on women's achievement behavior than is common in many popular psychological explanations for sex differences in achievement patterns. Beginning with the work associated with need achievement and continuing to current work in attribution theory, a variety of scholars have considered the origin of sex differences in achievement. Many of these scholars have looked for the origin in either motivational differences or expectancy/attributional differences. It has been argued, for example, that women fear success, have lower expectations for success, are less confident in their achievement-related abilities, are more likely to attribute their failures to lack of ability, are less likely to attribute their success to ability, and are more likely to exhibit a learned helpless response to failure. Furthermore, it has been argued that these differences mediate the sex differences we observe in achievement patterns.

We find several problems in this body of work. First, because they assumed a deficit model of female achievement, researchers have focused their attention on the question of "How are women different from men?"—not on the question of "What influences men's and women's achievement behavior?" Second, the assumption that the differences uncovered in most studies actually mediate sex differences in achievement behavior has rarely been tested. Instead, many studies simply demonstrate a statistically significant difference between males and females and conclude that this difference accounts for sex differences in achievement behavior. Third, the deficit perspective has limited the range of variables studied. Researchers have focused most of their attention on a set of variables that are linked to self-confidence and expectancies because high self-confidence is one of those "good" things that facilitates men's competitive achievement.

Our model provides a very different perspective. By assigning a central role to the construct of subjective task value, we have offered an alternative explanation for sex differences in achievement patterns. This alternative explanation makes salient the hypothesis that differences in male and female achievement patterns result from the fact that males and females have different but equally important and valuable goals for their lives. It also suggests a number of new researchable issues such as the socialization of interest and perceived options. Thus, instead of characterizing females as deficient males, this perspective legitimizes females' choices as valuable on their own terms, and not as a reflection or distortion of male choices and male values. Gilligan (1982) has made a similar point regarding males' and females' moral judgments.

How well does this model do in generating important research questions and in explaining sex differences in achievement choices? To answer this question, we sought out an everyday achievement activity and studied the origin of sex differences in it. Some major components of this research program are described in the next section.

Sex Differences in Course Enrollment Patterns

As noted earlier, two areas of cognitive functioning reveal fairly consistent patterns of sex differences (see Eccles (Parsons), 1984; Wittig & Petersen, 1979). Girls typically perform better than boys on verbal tasks, and boys perform better than girls on quantitative tasks. Sex differences in high school course enrollment, college majors, and adult careers reflect a similar pattern. Applying our model to this pattern of sex differences provided an ideal opportunity to test its utility for explaining sex differences in "real-life" achievement behavior.

To test the utility of our model for explaining sex differences in achievement choices, we conducted two large-scale, cross-sectional/longitudinal studies on the ontogeny of students' achievement beliefs, attitudes, and behaviors regarding math and English. Several forms of data were collected including: student records, student and parent attitude questionnaires, and classroom observations. Information taken from each student's school record included: standardized achievement test scores; final grades in mathematics and English for the 2 years (1975–1977) prior to the study, during the 2 years of the study, and each year following the study until the students graduated from high school; and math and English enrollment patterns. The data summarized here represent effects that are consistent across both studies. Together the two studies include more than 1000 junior and senior high school students and their parents and math teachers. These results are presented in more detail in Eccles (1985a); Eccles (Parsons) et al., (1983); Eccles (Parsons), et al., (1984); Jayaratne (1983); Parsons, Adler, and Kaczala (1982); Parsons, Kaczala, and Meece (1982); and Wigfield (1984).

Student Attitudes and Beliefs: Overview

According to our model, general beliefs influence task-specific beliefs, which, in turn, influence achievement behaviors. To operationalize this model, we created variables to coincide with each of these three levels of the psychological variables. The general beliefs included gender-role schemata, stereotyping of math as a male domain, and perceptions of encouragement to continue taking math by parents, teachers, and peers. The specific beliefs included expectancies for success, perceived ability, perceived task difficulty, perceived amount of effort necesssary to succeed, perceived importance of the subject, perceived cost of success, perceived worth of the amount of effort necessary to succeed, perceived utility value of the subject, causal attributions for success and failure in math, and the reasons why one would take advanced-level math courses. For achievement outcome measures, we asked the students whether they planned to continue taking math and English, and if so, how much; we also collected their grades in their math and English courses and their actual course enrollment patterns.

Factor analysis of the specific student attitude measures yielded three identical factors for both the math and English items: Self-Concept of Ability, Perceived Task Difficulty, and Subjective Task Value. The Self-Concept factor included all items tapping perceived ability, perceived performance, and expectations for success in current and future courses. The Task Difficulty factor included items tapping perceived task difficulty, perceived effort needed to do well, and estimates of actual level of effort. The Subjective Task Value factor included all items related to perceived utility value, enjoyment of the subject, and perceived importance of doing well.

Student Attitudes and Beliefs: Sex Differences

Relatively few sex differences emerged, but those that did formed a fairly consistent pattern. Boys, compared to girls, rated their math ability higher, felt they had to exert less effort to do well in math, and held higher expectancies for future successes in math, even though there were no sex differences on any of the objective measure of math performance. In addition, boys rated math as more useful than the girls did. Finally, both males and females rated math as more useful for males than for females, although the girls endorsed this stereotype to a lesser degree than did the boys. Thus, to the extent that there are sex differences on these math-related self and task perception variables, boys had a more positive view both of themselves as math learners and of math itself. In contrast, girls, rated their English ability higher, had higher expectations for future success in English, and felt that english was both more useful and easier.

These differences were even more dramatic when compared from a developmental perspective. In general, the girls became more positive toward English and more negative toward math as they got older. In contrast, boys' attitudes toward both subjects remained fairly stable over time and across grade levels. For example, female students' estimates of their math ability declined linearly with age. In addition, by eighth grade the females rated their English ability higher than their math ability, and by tenth grade they rated their math ability lower than the boys rated theirs. By eighth grade, the girls also rated English as more important and interesting than math.

These changes in the females' attitudes toward both math and English are especially interesting given the nature of our samples. First, we have no indication that a difference is measurable in math performance between the males and females in these samples on either their course grades or their scores on standardized achievement tests. Second, the male and female students also had comparable test scores on English standardized achievement tests at each grade level. Third, the female students earned higher grades than the male students in their English courses beginning at about the eighth grade. Fourth, when one compares the students' standardized test scores across years, the older females in this sample had higher scores than the younger females. Consequently, the older female students, if anything, had higher math and English ability, on the average, than the younger female students. Nevertheless, even though the older female population was more select than the younger female population in terms of both English and math achievement scores, and there were no apparent sex differences on math performance measures, the attitudes of the female students toward math declined with age while their grades did not. In contrast, both their attitudes and their actual performance in English increased with age.

Given our perspective that choice is the critical mediator of achievement differences, these results certainly lead to the prediction that female students will elect less math than English while males students will continue to take courses in both subject areas. This is, in fact, what has happened in this sample. The females were less likely to take twelfth grade advanced math than the males were, while their English enrollment patterns did not differ. There were no sex differences in math enrollment prior to the twelfth grade.

Attitudes and Behavior

The analyses described thus far suggest several important sex differences in students' attitudes that could mediate sex differences in achievement patterns. The mere existence of these differences does not support their importance as variables mediating sex differences in achievement patterns. The critical question is whether or not these differences, in fact, make a difference. To answer this question, we ran a series of correlational and multivariate regression analyses. Several important results emerged.

As predicted, for both males and females, Self-Concept of Ability and Subjective Task Value correlated positively with students' plans to continue taking math and English, with the students' grades in both math and English 1 year later, and with the students' actual course enrollment decisions in math measured 1 to 3 years later. These results provide initial support for the predicted influence of attitudinal variables on achievement behaviors. But these attitudinal variables are significantly intercorrelated with each other and with past grades. Before we can understand the impact of attitudes on achievement, we need to answer two additional questions: (1) which of these attitudes are

Figure 16–3. Path analysis of influences on enrollment decisions. Path coefficients are standarized betas. Coefficients significant at greater than .05 are shown. All possible paths were tested.

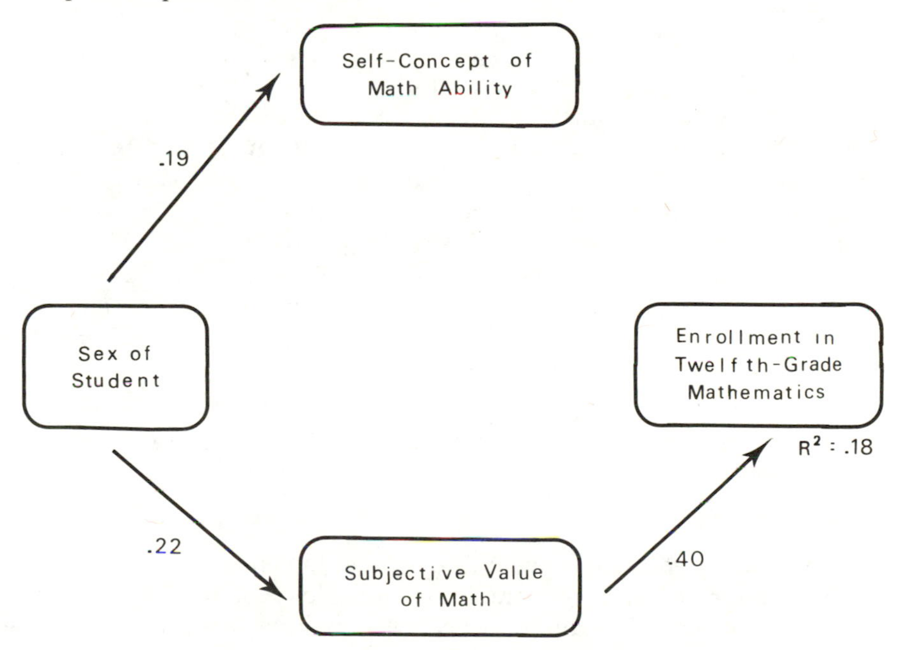

most critical, and (2) are any of the attitudes as critical as past performance in shaping subsequent achievement behaviors. To answer these questions, we used hierarchical multiple regression procedures, regressing students' attitudes toward math and English on each outcome measure (subsequent grades, enrollment plans, and actual enrollment decisions for math) while controlling for the effect of previous grades on both the attitude and outcome measures. As predicted, attitudes had a significant impact on the outcome measures independent of their relation to previous grades. Self-Concept of Ability was a significant independent predictor of subsequent grades; even more interestingly, Subjective Task Value was the most powerful predictor of educational plans.

In light of the sex differences in Subjective Task Value, these results suggest that Subjective Task Value may be the attitude that mediates sex differences in achievement choice patterns. To test this hypothesis, we tested the mediating role of Subjective Task Value in explaining the sex differences in course enrollment patterns for mathematics. We were unable to run a comparable test for English because English is required for all 3 years of high school in the school districts we sampled. These results, depicted in Figure 16-3, are consistent with the hypothesis that sex differences in math course enrollment are mediated by the sex difference in Subjective Task Value, and not by the sex difference in Self Concept of Math Ability.

Summary

As predicted, Subjective Task Value emerged as the most powerful predictor of students' Subjective Educational Plans. Furthermore, the significant sex $\times$ age $\times$ subject area area interaction yielded results consistent with the developmental predictions of our model. High school females had a more positive attitude toward English and less positive attitude toward math than did the junior high school females, especially in terms of Subjective Task Value. Projecting these developmental patterns into the late adolescent years should produce a marked sex difference in attitudes toward the value of math and English and in actual course enrollment decisions. In fact, this happened; the females were more likely to drop math prior to high school graduation than were the males. Finally, our data suggest that Subjective Task Value, not Confidence in One's Math Ability, mediates this sex difference in course enrollment patterns.

Males versus Females

The data discussed thus far were drawn from the entire sample, based on the assumption that comparable relationships would hold for both males and females. The zero-order correlations calculated for each sex separately support this assumption for the variables we have discussed. But this is not case for the correlations of these attitudes with past performance. An important sex difference emerged when we compared the correlations of students' attitudes to their past grades and to a composite score reflecting their relative position within their grade level on their course grades and standardized achievement test scores. The males' attitudes were more strongly related to their performance history than were the females' attitudes, for both math and English. This difference is especially marked for Subjective Task Value; the value males place on both math and English appears to be more highly influenced by their performance in the subject than is true for females.

Considered with the results summarized earlier, these results suggest that sex differences in achievement choice patterns may come from two processes. First, they are a function of the sex difference in Subjective Task Value; second, they are a function of the fact that academic achievement values seem to be shaped somewhat differently in males than in females. The value males attach to various subjects appears to be influenced largely by their performance history. In contrast, the value females attach to various subjects is more independent of their performance history. These results raise two important questions: (1) What factors influence the value individuals attach to various achievement options? (2) Why does the impact of performance history on Subjective Task Value differ for males and females? We are now exploring the variables that shape the value males and females attach to various achievement activities in an effort to broaden our understanding of the ontogeny of sex differences in achievement choice patterns. We are focusing on two sets of variables: gender-role schema and socialization.

Gender-Role Schema

Several investigators (e.g., Nash, 1979) have suggested that the interaction of gender-role identity with gender stereotypes regarding the nature of the task influences students' attitudes toward a subject or an occupation. In addition, we believe that gender roles influence achievement patterns primarily through their impact on the value individuals attach to the many achievement options available to them. To test these hypotheses, we are evaluating the relation between gender-role salience, gender-role stereotypes of various activities, and achievement beliefs and choices.

Our initial attempts to assess these hypotheses were rather unsuccessful (see Eccles (Parsons) et al., 1983). To maintain comparability with the existing literature, we used a personality invertory (the PAQ) to assess gender-role identity. We found that gender-role classification, either as an independent effect or in interaction with gender stereotyping of math, had little effect on students' attitudes toward math beyond the well-established link between scores on the instrumentality scale and self-confidence. These results suggest that one's conception of oneself in terms of gender-stereotyped traits is not related to the subjective value one attaches to academic activities. These findings do not, however, invalidate the significance of one's gender-role identity as an influence in course or career selection. Several critics of personality-based measures of gender-role identity, including Spence and Helmreich (1978), have argued that scales such as the PAQ and the BSRI are not good measures of gender-role identity. We concur with this judgment and are now trying alternative measures.

To better estimate an individual's gender-role identity, we have developed an alternative measure based on activity preferences. We had students rate how important it is for boys and girls to engage in a variety of gender-typed activities (e.g., taking care of an infant, shoveling snow, being interested in makeup, etc.) Kaczala (1983) scored a child as (1) *androgynous* if she or he felt it was important for the same-sex peers to engage in both male- and female-typed activities, (2) *feminine* if she or he felt it only important for the same-sex peers to engage in female-typed activities, (3) *masculine* if she or he felt it was only important for the same-sex peers to engage in male-typed tasks, and (4) *undifferentiated* if she or he felt it was not important whether a same-sex peer engaged in either type of task. Girls whose ideal female was androgynous and boys whose ideal male was androgynous had the most positive attitudes toward math, especially with regard to the subjective task value they attached to math.

We next had the children estimate the frequency with which they themselves engaged in these same activities. Using a similar scoring procedure, Kaczala (1983) classified the children as androgynous, feminine, masculine, and undifferentiated. Once again, androgynous girls placed the highest value on math. In addition, the androgynous girls were the least likely to stereotype math as a masculine activity.

While quite preliminary, these results suggest that gender-role identity, when defined in terms of activity ideals and activity preferences, does affect the value on attaches to academic activities. Whether these effects will also hold up for occupational orientation is yet to be determined.

Socialization

The second set of variables we are exploring relate to the socialization of achievement values. Parental beliefs and attitudes appear to be particularly important (see also Fox, Tobin, & Brody, 1979). Parents, more so than teachers, have sex-differentiated perceptions of their children's math aptitude, despite the similarity in the actual performance of their sons and daughters. Parents also believe that advanced math is more important for boys than for girls and advanced English is more important for girls than for boys. Finally, our initial work suggests that parents' beliefs regarding their children's aptitude are stronger predictors of the students' attitudes toward various subjects than are indicators of the students' actual performance in these subjects (see Parsons, Adler, & Kaczala, 1982). We are now exploring how parents convey their attitudes to their children.

Conclusion

In this chapter, I have summarized a comprehensive model explaining achievement choices, have applied this model to the question of sex differences in achievement choices, and have summarized the results of a developmental study of sex differences in achievement choices generated by this model. The model differs from other explanations of sex differences in achievement behavior primarily in its focus on choice, in its stress on the importance of task value as a critical mediator of sex differences in patterns of achievement choices, and in its explicit linking of gender roles to subjective task value. The results reported support this perspective. Sex differences in the decision to enroll in twelfth grade advanced math courses were mediated by the sex differences in the perceived value of advanced math courses. Furthermore, subjective task value was the most influential attitudinal variable in the course decisions of both boys and girls. Because the subjective value of math was related positively to performance history in math for boys, enrollment in twelfth grade math was predicted primarily by performance history. Subjective task value played a larger predictive role in girls' enrollment decisions. We are now exploring the factors that influence the value both boys and girls attach to various achievement options. We believe gender roles and gender-role socialization are two critical influences but are still gathering the data necessary to evaluate this predication.

Acknowledgments

The research reported herein was supported by grants from the National Institutes of Education, Mental Health, and Child Health and Development. Grateful acknowledgment goes to all of my colleagues who have worked with me on the project described in this chapter. Special thanks go to Susan Goff-Timmer, Carol Kaczala, Judith Meece, and Carol Midgley, without whose collaboration the ideas and the work reported here might never have come to be.

References

Atkinson, J.W. (1964). *An introduction to motivation*. Princeton: Van Nostrand.

Bandura, A. (1984). Self-efficacy: Toward a unifying theory of behavioral change. *Psychological Review, 84*, 191–215.

Barnett, R. C., & Baruch, G. K. (1978). *The competent woman.* New York: Irvington Publishers.

Bar-Tal, D. (1978). Attributional analysis of achievement-related behavior. *Review of Educational Research, 48,* 259–279.

Baruch, G., Barnett, R., & Rivers, C. (1983). *Life prints.* New York: McGraw-Hill.

Beck, J. A. L. (1977–1978). Locus of control, task expectancies, and children's performance following failure. *Journal of Educational Psychology, 71,* 207–210.

Benbow, C. P., & Stanley, J. C. (1980). Sex differences in mathematical ability: Fact or artifact? *Science, 210,* 1262–1264.

Benbow, C. P., & Stanley, J. C. (1982). Consequences in high school and college of sex differences in mathematical reasoning ability: A longitudinal perspective. *American Educational Research Journal, 19,* 598–622.

Benbow, C. P., & Stanley, J. C. (1983). Sex differences in mathematical reasoning ability: More facts. *Science, 222,* 1029–1031.

Boswell, S. (1979). *Nice girls don't study mathematics: The perspective from elementary school.* Paper presented at the annual meeting of the American Educational Research Association, San Francisco.

Bryson, R., Bryson, J. B., & Johnson, M. F. (1978). Family size, satisfaction, and productivity in dual-career couples. *Psychology of Women Quarterly, 3,* 67–77.

Burnett, S. A., & Lane, D. A. (1980). Effects of academic instruction on spatial visualization. *Intelligence, 4,* 233–242.

Connor, J. M., & Serbin, L. (1980). *Mathematics, visual-spatial ability and sex-roles.* Final report to the National Institute of Education, Washington, D.C.

Connor, J. M., Serbin, L. A., & Schackman, M. (1977). Sex differences in children's response to training on a visual-spatial test. *Developmental Psychology, 3,* 293–294.

Cooper, J. M., Burger, J. M., & Good, T. L. (1981). Gender differences in the academic locus of control beliefs of young children. *Journal of Personality and Social Psychology, 40,* 562–572.

Crandall, V. C. (1969). Sex differences in expectancy of intellectual and academic reinforcement. In C. P. Smith (Ed.), *Achievement-related motives in children.* New York: Russel Sage Foundation.

Crandall, V. J., Katkovsky, W., & Preston, A. (1962). Motivational and ability determinants of young children's intellectual achievement behavior. *Child Development, 33,* 643–661.

Diener, C., & Dweck, C. S. (1978). An analysis of learned helplessness: Continuous change in performance, strategy, and achievement cognitions following failure. *Journal of Personality and Social Psychology, 36,* 451–462.

Dunteman, G. H., Wisenbaker, J., & Taylor M. E. (1978). *Race and sex differences in college science program participation.* Report to the National Science Foundation. Research Triangle Park; NC.

Dweck, C. S. (1975). The role of expectation and attributions in the alleviation of learned helplessness. *Journal of Personality and Social psychology, 31,* 674–685.

Dweck, C. S., & Bush, E. (1976). Sex differences in learned helplessness: I. Differential debilitation with peer and adult evaluations. *Developmental Psychology, 12,* 147–156.

Dweck, C. S. & Elliot, E. S. (1983). Achievement motivation. In E. M. Hetherington (Ed.), *Handbook of Child Psychology: Social and Personality Vol. 4* (pp. 643–691). New York: Wiley.

Dweck, C. S., & Licht, B. G. (1980). Learned helplessness and intellectual achievement. In J. Garber & M. E. P. Seligman (Eds.), *Human helplessness: Theory and application*, New York: Academic Press.

Dweck, C. S., & Reppucci, N. D. (1973). Learned helplessness and reinforcement responsibility in children. *Journal of Personality and Social Psychology, 25*, 109–116.

Eccles, J. (1983a). The development of attributions, self and task perceptions, expectations, and persistence. Unpublished manuscript, University of Michigan.

Eccles, J. (1983b). Female achievement patterns: Attributions, expectancies, values, and choice. Unpublished manuscript, University of Michigan.

Eccles, J. (1985a). Sex differences in achievement patterns. In T. B. Sonderegger (Ed.), *Psychology and gender: Nebraska symposium on Motivation, 1984*. Lincoln: University of Nebraska Press.

Eccles, J. (1985b). Why doesn't Jane run? Sex differences in educational occupational patterns. In F. D. Horowitz & M. O'Brien (Eds.), *The gifted and talented: A developmental perspective*. Washington, DC: American Psychological Association.

Eccles (Parsons), J. (1983). Attributional processes as mediators of sex differences in achievement. *Journal of Educational Equity and Leadership, 3*, 19–27.

Eccles (Parsons), J. (1984). Sex differences in mathematics participation. In M. Steinkamp & M. Maehr (Eds.), *Women in Science*. Greenwich, CT: JAI Press.

Eccles (Parsons), J., Adler, T. F., Futterman, R., Goff, S. B., Kaczala, C. M., Meece, J. L., & Midgley, C. (1983). Expectations, values and academic behaviors. In J. T. Spence (Ed.), *Perspective on achievement and achievement motivation*. San Francisco: W. H. Freeman.

Eccles (Parsons), J., Adler, T., & Meece, J. L. (1984). Sex differences in achievement: A test of alternate theories. *Journal of Personality and Social Psychology, 46*, 26–43.

Eccles, J., & Hoffman, L. W. (1984). Sex roles, socialization, and occupational behavior. In H. W. Stevenson & A. E. Siegel (Eds.), *Research in child development and social police (Vol. 1)*. Chicago: University of Chicago Press.

Educational Testing Service. (1980). *National college-bound seniors, 1979*. Princeton, NJ: College Entrance Examination Board.

Fennema, E., & Sherman, J. (1977). Sex-related differences in mathematics achievement, spatial visualization and affective factors. *American Educational Research Journal, 14*, 51–71.

Fox, L. H. (1982). *The study of social processes that inhibit or enhance the development of competence and interest in mathematics among highly able young women*. Final report to the National Institute of Education: Washington, DC.

Fox, L. H., & Denham, S. A. (1974). Values and career interests of mathematically and scientifically precocious youth. In J. C. Stanley, D. P. Keating & L. H. Fox (Eds.), *Mathematical talent: Discovery, description, and development* (pp. 140–175). Baltimore: Johns Hopkins University Press.

Fox, L. H., Tobin, D., & Brody, L. (1979). Sex-role socialization and achievement in mathematics. In M. A. Wittig & A. C. Petersen (Eds.), *Sex-related difference in cognitive functioning: Developmental issues*, New York: Academic Press.

Frieze, I. H., Francis, W. D., & Hanusa, B. H. (1983). Defining success in classroom settings. In J. M. Levine & M. C. Wang (Eds.), *Teacher and student perceptions: Implications for learning*. Hillsdale, NJ: Lawrence Erlbaum.

Frieze, I. H., Parsons, J. E., Johnson, P., Ruble, D. N., & Zellman, G. (1978). *Women and sex roles: A social psychological perspective.* New York: Norton.
Frieze, I. H., Whitley, B. E., Hanusa, B. H., & McHugh, M. C. (1982). Assessing the theoretical models for sex differences in causal attributions for success and failure. *Sex Roles, 8,* 333–343.
Gilligan, C. (1982). *In a different voice.* Cambridge: Harvard University Press.
Goff-Timmer, S., Eccles, J., & O'Brien, K. (1985). How children use time. In F. T. Juster & F. Stafford (Eds.), *Time, goods, and well-being.* Ann Arbor, MI: Institute for Social Research Press.
Grant, W. F., & Eiden, L. J. (1982). *Digest of educational statistics.* Washington, DC: National Center for Educational Statistics, U.S. Department of Education.
Hoffman, L. W. (1972). Early childhood experiences and women's achievement motives. *Journal of Social Issues, 28,* 129–156.
Hollinger, C. L. (1983). Self-perception and the career aspirations of mathematically talented female adolescents. *Journal of Vocational Behavior, 22,* 49–62.
Horner, M. (1972). Toward an understanding of achievement-related conflicts in women. *Journal of Social Issues, 28,* 157–175.
Huston, A. C. (1983). Sex-typing. In P. Mussen & E. M. Hetherington (Eds.), *Handbook of child psychology,* Vol. IV. New York: Wiley.
Hyde, J. S. (1981). How large are cognitive gender differences? A meta-analysis. *American Psychologist, 36,* 892–901.
Jayaratne, T. E. (1983). *Sex differences in children's math achievement: Parental attitudes.* Paper presented at the biennial meeting of the Society for Research in Child Development, Detroit.
Kaczala, C. M. (1983). *Sex role identity and its effect on achievement attitudes and behaviors.* Unpublished doctoral dissertation, University of Michigan, Ann Arbor.
Lewin, K. (1938). *The conceptual representation and the measurement of psychological forces.* Durham, NC: Duke University press.
Lipman-Blumen, J., & Tickamyer, A. R. (1975). Sex roles in transition: A ten year perspective. *Annual Review of Sociology, 1,* 297–337.
Maccoby, E. E., & Jacklin, C. N. (1974). *Psychology of sex differences.* Stanford: Stanford University Press.
Maehr, M. (1983). On doing well in science: Why Johnny no longer excels: Why Sarah never did. In S. Paris, G. Olsen, & H. Stevenson (Eds.), *Learning and motivation in the classroom.* Hillsdale, NJ: Lawrence Erlbaum.
Magarrell, J. (1980, Jan. 28). Today's new students, especially women, more materialistic. *Chronicle of Higher Education* (pp. 3–4).
Maines, D. R. (1983). *A theory of informal barriers for women in mathematics.* Paper presented at the annual meeting of the American Educational Research Association, Montreal.
Markus, H. (1980). The self in thought and memory. In D. M. Wegner & R. R. Vallacher (Eds.), *The self in social psychology.* New York: Oxford University Press.
McGinn, P. V. (1976). Verbally gifted youth: Selection and description. In D. P. Keating (Ed.), *Intellectual talent: Research and development.* Baltimore: Johns Hopkins University Press.
Meece, J. L., Eccles (Parsons), J., Kaczala, C. M., Goff, S. B., & Futterman, R. (1982). Sex differences in math achievement: Toward a model of academic choice. *Psychological Bulletin, 91,* 324–348.

Montemayor, R. (1974). Children's performances in a game and their attraction to it as a function of sex-typed labels. *Child Development, 45,* 152–156.

Nash, S. C. (1979). Sex role as a mediator of intellectual functioning. In M. A. Wittig & A. C. Petersen (Eds.), *Sex-related differences in cognitive functioning: Developmental issues.* New York: Academic Press.

National Center for Educational Statistics. (1979, Sept. 19). Proportion of degrees awarded to women. Reported in *Chronicle of Higher Education.*

National Center for Educational Statistics. (1980, Jan. 28). Degrees awarded in 1978. Reported in *Chronicle of Higher Education.*

National Center for Educational Statistics. (1984). *The condition of education: 1984.* Washington, DC: U.S. Department of Health, Education, and Welfare.

Newland, K. (1980). *Sisterhood of man.* New York: Norton.

Nicholls, J. G. (1975). Causal attributions and other achievement-related cognitions: Effects of task outcomes, attainment value, and sex. *Journal of Personality and Social Psychology, 31,* 379–389.

Oden, M. H. (1968). The fulfillment of promise: 40 year follow-up of the Terman gifted group. *Genetic Psychology Monographs, 77,* 3–93.

Parsons, J. E., Adler, T. F., & Kaczala, C. M. (1982). Socialization of achievement attitudes and beliefs: Parental influences. *Child Development, 53,* 310–321.

Parsons, J. E., & Bryan, J. (1980). *Adolescence: Gateway to androgeny.* Ann Arbor, MI: Women's Studies Occasional Paper Series.

Parsons, J. E., & Goff, S. G. (1980). Achievement motivation: A dual modality. In L. J. Fyans (Ed)., *Recent trends in achievement motivation: Theory and research.* Englewood Cliffs, N.J.: Plenum.

Parsons, J. E., Kaczala, C., & Meece, J. (1982). Socialization of achievement attitudes and beliefs: Classroom influences, *Child Development, 53,* 322–339.

Parsons, J. E., Ruble, D. N., Hodges, K. L., & Small, A. W. (1976). Cognitive-developmental factors in emerging sex differences in achievement-related expectancies. *Journal of Social Issues, 32,* 47–61.

Petersen, A. G. (1979). Hormones and cognitive functioning in normal development. In M. A. Wittig & A. C. Petersen (Eds.), *Sex-related differences in cognitive functioning.* New York: Academic Press.

Raynor, J. O. (1974). Future orientation in the study of achievement motivation. In J. W. Atkinson & J. O. Raynor (Eds.), *Motivation and achievement.* Washington, DC: Winston Press.

Rholes, W. S., Blackwell, J., Jordan, C., & Walters, C. (1980). A developmental study of learned helplessness. *Developmental Psychology, 16,* 616–624.

Rokeach, M. (1973). *The nature of human values.* New York: The Free Press.

Schunk, D. H., & Lilly, M. W. (1982). *Attributional and expectancy change in gifted adolescents.* Paper presented at the annual meeting of the American Educational Research Association, New York.

Sells, L. W. (1980). The mathematical filter and the education of women and minorities. In L. H. Fox, L. Brody, & D. Tobin (Eds.), *Women and the mathematical mystique* (pp. 66–75). Baltimore: Johns Hopkins University Press.

Sherman, J., & Fennema, E. (1977). The study of mathematics by high school girls and boys: Related variables. *American Educational Research Journal, 14,* 159–168.

Spence, J. T., & Helmreich, R. L. (1978). *Masculinity and femininity: Their psychological dimensions, correlates, and antecedents.* Austin: University of Texas Press.

Spence, J. T., Helmreich, R. L., & Stapp, J. (1975). Ratings of self and peers on sex role attributes and their relation to self-esteem and conception of masculinity and femininity. *Journal of Personality and Social Psychology, 32,* 29–39.

Spenner, K., & Featherman, D. L. (1978). Achievement ambitions. *Annual Review of Sociology, 4,* 373–420.

Stein, A. H., & Bailey, M. M. (1973). The socialization of achievement orientation in females. *Psychological Bulletin, 80,* 345–366.

Stein, A. H., & Smithells, T. (1969). Age and sex differences in children's sex-role standards about achievement. *Developmental Psychology, 1,* 252–259.

Tangri, S. S. (1972). Determinants of occupational role innovation among college women. *Journal of Social Issues, 28,* 195–207.

Terman, L. M. (1926). *Genetic studies of genius (Vol. 1).* Stanford, CA: Stanford University Press.

Terman, L. M., & Oden, M. H. (1947). *Genetic studies of genius, Volume IV: The gifted child grows up.* Stanford: Stanford University Press.

Tidwell, R. (1980). Gifted students' self-images as a function of identification procedure, race, and sex. *Journal of Pediatric Psychology, 5,* 57–69.

Tittle, C. K. (1981). *Career and family: Sex roles and adolescent life plans.* Beverly Hills: Sage.

Tomlinson-Keasey, C., & Smith-Winberry, C. (1983). Educational strategies and personality outcomes of gifted and nongifted college students. *Gifted Child Quarterly, 27,* 35–41.

U.S. Department of Labor. (1980). *The employment of women: General diagnosis of developments and issues.* Washington, DC: Report of the USDL, Women's Bureau.

U.S. Department of Labor. (1980). *Perspectives on working women: A databook.* Washington, DC: U.S. Government Printing Office.

Veroff, J. (1969). Social comparison and the development of achievement motivation. In C. P. Smith (Ed.), *Achievement-related motives in children.* New York: Russell Sage Foundation.

Vetter, B. M. (1981). Women scientists and engineers: Trends in participation. *Science, 214,* 1313–1321.

Waber, D. P. (1979). Cognitive abilities and sex-related variations in the maturation of cerebral cortical functions. In M. A. Wittig & A. C. Petersen (Eds.), *Sex-related differences in cognitive functioning: Developmental issues.* New York: Academic Press.

Weiner, B. (1972). *Theories of motivation: From mechanism to cognition.* Chicago: Markham Publishing.

Weiner, B., Frieze, I., Kukla, A., Reed, L., Rest, S., & Rosenbaum, R. M. (1971). *Perceiving the causes of success and failure.* New York: General Learning Press.

Wigfield, A. (1984) *Relationships between ability perceptions, other achievement-related beliefs, and school performance.* Paper presented at the annual meeting of the American Educational Research Association, New Orleans.

Wittig, M. A., & Petersen, A. C. (Eds.). (1979). *Sex-related differences in cognitive functioning: Developmental issues.* New York: Academic Press.

Williams, J. E., Bennett, S., & Best, D. (1975). Awareness and expression of sex stereotypes in young children. *Developmental Psychology, 11,* 635–642.

Wolleat, P. L., Redro, J. D., Becker, A. D., & Fennema, E. (1980). Sex differences in high school students' causal attributions of performance in mathematics. *Journal of Research in Mathematics Education, 11,* 356–366.

17

A Transactional Perspective on the Development of Gender Differences

Anke A. Ehrhardt

In Chapter 16 Eccles presented a new model aimed at enhancing our understanding of achievement choices, particularly in respect to gender differences. In her model various aspects of school and career achievements are valued differently by boys and girls. Gender-related choices result in different tests scores and ultimately influence career goals.

Eccles's approach is in contrast to those of other investigators who attribute differences in abilities to sex-dimorphic performance and achievement patterns instead of choice and preference. This is particularly true for the finding that boys outnumber girls in the high range of math achievement (Benbow & Stanley, 1980, 1983).

These two different approaches are reminiscent of the old controversy between nature versus nurture, innate versus learned. In the past, this dichotomous thinking was applied to many aspects of gender-related behavior. Although most scientists would nowadays agree that positing biology versus environment implies a false bipolarity because all behavior is rooted in the central nervous system, too often unifactorial thinking is still put forth. The distinction between biology versus environment is often drawn in order to divide the different classes of influence into immutable versus modifiable. This presents another false dichotomy, because all effects are more or less modifiable.

In order to emphasize the importance of a particular set of influences (as, for instance, in Eccles's chapter the choice aspect for a particular career), scientists typically focus on specific variables and follow what can best be described as a main-effect model. This type of model postulates that *one* factor or one set of similar factors determines or predominantly influences a particular behavioral outcome. Usually, the one factor is either prenatal or postnatal, or a class of influences that originate in either a person's physiology or the environment. When researchers focus on the one event that is taken to be the most important determinant of the behavior, they are often ready to discard all knowledge of other relevant factors. The problem with the main-effect model is that many cases do not fit such a one-factor model. Therefore, instead of trying to integrate more aspects into one model, we are prone to develop new models as soon as observations are made that cannot be explained. A more adequate explanation of behavior is provided by a more complex model that considers a variety of factors, both of constitutional and of social-environmental origins.

This type of model postulates an interaction of the different factors and is most appropriate if it includes a dynamic concept of development that posits a continual and progressive interplay between the organism and its environment. The transactional model does not assume directionality of one set of factors, be it social-environmental or constitutional, but a relatively greater effect of one set of influences for a particular behavior outcome (for more details, see Ehrhardt, 1985).

Eccles's model stresses that achievement choices are embedded in gender-related behavior. She argues that career choices, for instance, are affected by the value a young woman may put on combining a parenting role at home with her occupational role outside of the home. Nobody would argue that point. Eccles seems to suggest, however, that all aspects of gender-related behavior are solely determined by social-environmental effects. I would argue that she has to integrate prenatal and postnatal constitutional factors into her proposed model. Even if these effects do not directly affect career choice they still have to be seen as links in a long chain of events that determine outcome.

In the area of gender-related behavior, of all the constitutional variables to consider, sex hormones are of particular importance. The field of behavioral endocrinology describing the complex interplay between hormones, pathways to the brain, receptors, and different actions at the cellular level is reviewed elsewhere in this volume (Chapters 2–7). Most of the evidence so far has been accumulated in animal experiments. Up to now, the information on hormonal effects on the human central nervous system is much more speculative, and our knowledge so far is at best fragmentary. Most of the evidence is based on clinical groups of girls and boys who have a documented history of abnormal levels of sex hormones during prenatal development and whose gender-related behavior was studied at different age levels and compared to appropriate controls. The main findings of that research can be stated as follows (see reviews by Ehrhardt & Meyer-Bahlburg, 1981; Money & Ehrhardt, 1972; Reinisch, 1983): Girls who were exposed to unusually high levels of androgen during their prenatal development were found to exhibit increased physically energetic outdoor play behavior and decreased nurturant behavior in terms of parenting rehearsal. They were significantly different in these respects from matched normal controls and, in a separate study, from their endocrinologically normal sisters. The behavior was long-term and could not be solely explained by the various social and environmental factors assessed.

In a number of separate studies, prenatal exposure to pharmacological doses of estrogen and progesterone was assessed, and those sex hormones were found to be associated with the expected opposite effect—namely, relatively less physically energetic play behavior and an increase in more nurturant behavior as exhibited in doll play and infant care in girls and in less aggressive play behavior in boys. This finding could be interpreted as an antiandrogenic effect of some of the estrogen/progesterone compounds, analogous to some of the actions of these hormones demonstrated in animal experiments (Ehrhardt, Meyer-Bahlburg, Feldman, & Ince, 1984). It is important to apply the transactional model to the interpretation of the behavior findings of those clinical samples of boys and girls with prenatal hormonal abnormalities. The behavior differences between experimentals and controls cannot be attributed to any *one* factor, such as a specific prenatal homonal abnormality. Instead, prenatal hormone levels may present one of the links in a long chain of events leading to the expression of a specific trait. Such a developmental sequence needs to be analyzed taking the interplay of constitutional and social-environmental factors into account. More than one developmental sequence may exist for a similar behavior pattern. For instance, although prenatal androgen may predispose a child to rough-and-tumble play, special social-environmental reinforcement is required for the behavior expression to occur. The same behavior pattern may be developed in an individual with a relatively low level of prenatal androgen with a different, more strongly reinforcing social environment. The pliability of the human organism is such that many pathways during the life course may lead to similar behavior sets.

Besides the prenatal phase, puberty is another crucial time of hormonal differentiation. From puberty on, males and females differ greatly in secretion of all sex hormones. Both men and women secrete all major sex hormones (i.e., androgens, estrogens, and progestogens), but the ratios of these sex hormones are vastly different. Sex hormone levels in adulthood contribute to behavior differences in mood, aggression, and sexuality.

At the present time, we still do not know precisely how prenatal hormonal levels relate to adult hormonal levels—that is, does a relatively high level of prenatal androgen predict a relatively high level of pubertal androgen in a man or a woman? Although we know that normal males compared to normal females have a much higher level of androgen both before birth and from puberty onward, the relationship between levels of hormones before birth and after puberty within one sex is less well understood. Therefore, if a correlation between hormones and adult behavior is found, we do not know whether this relationship had hormonal precursors early in the development of that individual. Equally, we do not know whether the behavior affects the hormone levels or whether the hormones preceded the behavior, because the interaction between hormones and behavior is reciprocal.

A good example of this point is the recent observation that testosterone is correlated with the occupational status of women. In a study of 55 normal females, Purifoy and Koopmans (1980) assessed serum androstenedione, testosterone, testosterone-binding globulin, and free testosterone. The authors found that women, independent of age, in professional, managerial, and technical occupations had higher levels of all androgens than clerical workers and housewives. One might interpret these results within a simple cause-and-effect model stating that high levels of androgen in women determine their job status. Within a more sophisticated biosocial framework, one might instead hypothesize that the observed relationship is an indication of a complex interplay of hormones and behavior over time.

For some of these women, it may reflect a developmental sequence that started prenatally with a relatively high level of androgen that predisposed them to a high level of physical energy. Their interest in athletics and active outdoor activity undoubtedly increased their peer contact with boys and therefore may well have included a wider variety of adult roles in their play rehearsal. Their interaction with boys may have given them more exposure to team play and competition with males. This may have led to a wider range of career choices and to more self-confidence and assertiveness enabling them to aspire to competitive careers in male-dominated occupations in adulthood. At puberty, these women might or might not have started to produce relatively high levels of androgen during their menstrual cycle. Within a larger societal context, such women grew up at a time when full-time professional careers for women were permissible and socially reinforced, and thus they might have had a family and school environment that fostered their interests in full-time careers.

On the other hand, some or all of these women might not have had a particular hormonal pattern during prenatal or pubertal development at all, but underwent a developmental sequence that was strongly influenced by social reinforcement for professional careers. Ultimately, fulfilling such an occupational status might have increased their androgen production via a feedback effect of their work on their endocrine system. The point here is that hormonal factors can play different roles in a complex interplay of many variables within a particular developmental sequence. Hence, a correlation between a specific hormone level and a particular behavior pattern has to be seen as a contingency that may signal a whole network of different transactions that may vary from one individual to another.

My plea is obviously for a biosocial approach to the study and interpretation of gender differences. The attempt of Eccles to replace a "deficiency model" in which girls are measured by a male standard with a more sophisticated "choice and preference model" is interesting and presents, in my opinion, progress in our understanding of gender differences in achievement patterns. The inclusion of constitutional variables into her model should be seen as adding an important dimension instead of as invalidating her approach.

References

Benbow, C. P., & Stanley, J. C. (1980). Sex differences in mathematical ability: Fact or artifact? *Science, 210,* 1262–1264.

Benbow, C. P., & Stanley, J. C. (1983). Sex differences in mathematical reasoning ability: More facts. *Science, 222,* 1029–1031.

Ehrhardt, A. A. (1985). The psychobiology of gender. In A. Rossi (Ed.), *Gender and the life course* (pp. 81–96). New York: Aldine.

Ehrhardt, A. A., & Meyer-Bahlburg, H. F. L. (1981). Effects of prenatal sex hormones on gender-related behavior. *Science, 211,* 1312–1318.

Ehrhardt, A. A., Meyer-Bahlburg, H. F. L., Feldman, J. F., & Ince, S. E. (1984). Sex-dimorphic behavior in childhood subsequent to prenatal exposure to exogenous progestogens and estrogens. *Archives of Sexual Behavior, 13,* 457–477.

Money, J., & Ehrhardt, A. A. (1972). *Man and woman, boy and girl.* Baltimore: Johns Hopkins University Press.

Purifoy, F. E., & Koopmans, L. H. (1980). Androstenedione, T and free T concentrations in women of various occupations. *Social Biology, 26,* 179-188.

Reinisch, J. M. (1983). Influence of early exposure to steroid hormones on behavioral development. In W. Everaerd, C. B. Hindley, A. Bot, & J. J. van der Werff ten Bosch (Eds.), *Development in adolescence* (pp. 63–113). Boston: Martinus Nijhoff Publishers.

VI
PSYCHOSOCIAL PERSPECTIVES

18

Psychological Constructions of Masculinity and Femininity

Kay Deaux

The terms *masculinity* and *femininity* are rich in meaning and widespread in use, applied to a diversity of phenomena by both laypersons and scientists. In her recent appraisal of femininity, for example, Susan Brownmiller (1984) pointed to body, hair, clothes, voice, skin, movement, emotion, and ambition as equally important aspects of the concept. I suspect that a similar analysis of masculinity could yield an equally lengthy list.

Given the umbrella-like quality of both of these terms, it is perhaps inevitable that scientific discussions have displayed both confusion and misunderstanding. Different disciplines use different evidential bases, but across the disciplines the terms *masculinity* and *femininity* find continued use. I will not attempt to survey this broad conceptual range; instead, I will restrict my analysis to those uses of the term that are most distinctly psychological—or, perhaps even more narrowly, social psychological.

The terms *masculinity* and *femininity* have been used by investigators in personality and social psychology in at least three areas: (1) stereotypic views of males and females; (2) studies of sex differences and similarities in cognitive abilities, personality traits, and social behaviors; and (3) gender identity and self-reported masculinity and femininity. These three areas are equally important to an understanding of what is meant by masculinity and femininity, and research in each of the areas has been relatively vigorous, particularly in the past 10–15 years. I would like to outline some of the assumptions that underlie each of these approaches, noting representative research and raising some of the questions that each area implies.

Perceptions of Males and Females

A stereotype is, to use the definition offered by Ashmore and Del Boca (1979), "a structured set of beliefs about the personal attributes of a group of people" (p. 222). Originally imbued with both motivational processes and negative connotations, the concept of stereotype has recently been reinterpreted in terms more consistent with the newer work in social cognition. In adopting this theoretical framework, psychologists consider the stereotype in neutral terms, viewing it as simply one type of categorization that shares many features with other cognitive categories.

Stereotypes about women and men represent one particular set of beliefs, wherein males and females are the identifiable groups in question. Gender stereotypes have traditionally been defined in terms of the presence or absence of specific personality traits. Most commonly, a distinction has been made between expressive traits, viewed as more characteristic of women, and instrumental traits, seen as more characteristic of men (e.g., Broverman, Vogel, Broverman, Clarkson, & Rosenkrantz, 1972; Spence, Heimreich, & Stapp, 1974). Thus, women are typically viewed as being warm, gentle, and aware of the feelings of others, while men are described by traits such as independent, dominant, and assertive.

More recently, a number of investigators have begun to recognize that the content of gender stereotypes is not limited to trait adjectives (e.g., Ashmore & Del Boca, 1979; Ashmore, Del Boca, & Wohlers, 1986; Deaux & Lewis, 1983, 1984). Ashmore and Del Boca have suggested three classes of gender-related attributes: defining characteristics (consisting primarily of biological features), identifying characteristics (such as stature, clothing, and other externally visible signs), and ascribed characteristics (exemplified by trait adjectives).

In recent research on gender stereotypes, Laurie Lewis and I have identified four separate components that are reliably used by observers to distinguish males from females (Deaux & Lewis, 1983). These four components are traits, role behaviors, physical characteristics, and occupations. Within each of these component domains is one set of features seen as more characteristic of men than of women and another set seen as more characteristic of women than of men. For example, men are viewed as more likely than women to be strong, to be the financial provider in a household, to be an engineer, and to be competitive. In turn, women are rated more apt to have a soft voice, to take care of children, to be a secretary, and to be able to devote themselves to others. Although the probability attached to a particular characteristic is significantly different for men and women, the features are not viewed in dichotomous terms. The obtained probability of a male being independent, for example, is .77, while the comparable rating for a woman is .58. The pattern for other characteristics is similar, with differences ranging from 15 to 30 points on a 100-point probability scale. Thus, we should most appropriately consider gender stereotypes in terms of relative instead of absolute differences—or as fuzzy sets, in the terminology of cognitive psychologists.

Further complexity is added to the picture when we find that these components are not perfectly correlated with one another. These various components appear to function somewhat independent of one another, at least in the abstract. At the same time, learning that a person possesses a high degree of one component does lead to some convergence in other judgments, particularly when the information is counter to expectations for a particular individual (Deaux & Lewis, 1984). Thus a man who is described as being warm and emotional is also judged to be more likely to have feminine physical characteristics and to engage in more traditionally female role behaviors. Further, such seemingly simple counterstereotypic trait information can also lead to a marked increase in the assessed probability that the man is homosexual.

The Deaux and Lewis research also points to the importance of physical characteristics in the construction of gender stereotypes. Although these obvious indicators have been slighted in most of the earlier research on stereotypy, their influence can not be ignored. Indeed, cues as to physical appearance have been found to carry the greatest weight in subsequent gender-related judgments, influencing inferences of traits, role behaviors, and occupational position (Deaux & Lewis, 1984). Thus, the male who is slight of build or whose voice is relatively soft may be particularly vulnerable to judgments of femininity and of homosexuality. The ready availability of information about physical characteristics thus serves as a point of initial inference—a point beyond which the casual observer may not pass. In other words, the stereotypic inference process may begin as soon as the visual information is available, and observers may not wait to find out whether their inferences are actually based in fact.

To the best of our knowledge, all societies have a set of beliefs (or schema) about gender. Comparison of gender stereotypes within the United States to stereotypic concepts held in other parts of the world reveals both similarities and differences. In perhaps the most extensive recent investigation, one involving data collections in 30 different nations, Williams and Best (1982) find "pancultural generality" in many aspects of gender stereotypes. Specifically, they report that men are typically viewed as stronger, more active, and higher in achievement, autonomy, and aggression. Women are seen as weaker, less active, and more concerned with affiliation, nurturance, and deference. Both patterns are obviously consistent with the instrumental-expressive distinction found in previous single-nation studies. This same pattern can be detected in Block's (1973) cross-cultural studies, wherein specific traits may differ across countries but the underlying instrumental and expressive dimensions remain constant. Other variations across country as a function of religious affiliation have been noted by Williams and Best (1982); in Catholic countries, for example, women are viewed more positively than they are in Protestant countries. Dwyer (1978) notes variations in Morocco based on age and social interaction, with a heavy element of sexuality contained in the female stereotype. Strathern (1976), basing her observations primarily on New Guinea cultures, notes apparent contradictions in some elements of gender stereotypes (for example, she points to views of women as both hard-headed and soft-brained). She also finds evidence for a multiple-component view of stereotypes, as do Deaux and Lewis (1983, 1984), with components often implicating one another.

Thus, the evidence suggests that male and female, or masculinity and femininity, are not simple unidimensional concepts. Instead, they are loosely constructed categories that contain, with varying degrees of probability, a variety of characteristics and associations. From the cognitive perspective, we could view these attributes as varying in their degree of prototypicality to the central concepts—concepts that are, at best, fuzzy sets instead of clearly delineated categories.

At the same time that the concepts of masculinity and femininity, considered separately, show evidence of multidimensionality, the two concepts themselves are typically viewed as opposite ends of a single dimension. Thus, a common perception exists that what is masculine is not feminine, and vice versa, a finding that has been demonstrated in a number of investigations (Deaux & Lewis, 1984; Foushee, Helmreich, & Spence, 1979; Major, Carnevale, & Deaux, 1981). The commonly used term *the opposite sex* captures this mode of thought, positing opposition to explain differences. Such a dialectic assumption is evident not only in lay conceptions, but also in the postulates of scientific investigators.

To summarize our knowledge of gender stereotypes, we can safely say that all known societies have constructed a set of beliefs that differentiate males and females along some dimensions. Most often, the general concepts of masculinity and femininity are viewed in terms of opposition, with the presence of one set of characteristics implying the absence of the other. In addition is a set of elements central to the gender stereotypes held in most cultures. One can speculate that certain defining or identifying characteristics of males and females support these distinctions; for example, the typically greater strength and stature of the male and the female's stronger link to childbearing may provide the seeds from which attributes such as instrumentality and nurturance develop. Considerable diversity, however, is in other ascribed attributes that are associated with masculinity and femininity. These attributes are formed in the context of the existent social system, and undoubtedly reflect both cause and consequence of the distribution of women and men into specific social roles (cf. Eagly, 1983; Eagly & Steffen, 1984). Such variations suggest caution in assuming a simple or universal gender stereotype; conceptions of male and female are elaborated in different ways depending on the cultural context in which they emerge.

Sex Differences and Similarities

Research on sex differences and similarities can be examined to determine whether perceived stereotypes are reflected in reality—that is, do gender stereotypes possess a kernel (or a larger portion) of truth? One possible hypothesis is that gender stereotypes develop as widely used cognitive categories because they accurately mirror the reality of male and female behavior. Such a hypothesis would be supported by evidence of reliable sex differences in specifically designated behaviors and would be strengthened by evidence of universality of beliefs. Alternatively, one might hypothesize that differences in the behavior of women and men are far less evident than the stereotypes would suggest. Cognitive processes such as biased scanning and preferential recall of category-consistent information might then be invoked to explain the persistence of gender stereotypes when they are not totally justified by experience. This perspective would be more compatible with evidence of variation in stereotypes across culture and changes in stereotypes across time. Suggestive of this approach is the statement of Strathern (1976): "Myths about masculinity and femininity (gender constructs) endure and change, as language endures and changes, because of their usefulness as symbols in the society at large" (p. 68).

The answers to these questions are less easily obtained than one might suppose. Leaving aside for the moment the relative contributions of biological and experiential determinants, one still finds a great deal of uncertainty as to what sex differences are reliable and stable. The uncertainty does not result from any lack of effort. Since before the turn of the century, investigators have explored various behavioral areas for evidence of sex differences, arriving at different conclusions throughout. Summaries of such findings have often been challenged, for both methodological and political reasons.

The most recent attempt to provide a broad-scale assessment of possible sex differences in behavior is that of Maccoby and Jacklin (1974). Their ambitious review of the literature pointed to only four areas in which reliable sex differences could be established—verbal, mathematical, and spatial abilities, and aggressive behavior. In many other areas, they stated that they had simply no evidence for sex differences in behavior, including sociability, suggestibility, self-esteem, and achievement. In still other areas, Maccoby and Jacklin concluded that the questions are still open, with insufficient data or ambiguous findings precluding conclusions in either direction. These unsettled areas included competitiveness, dominance, compliance, and nurturance (all areas that are closely tied to pervasive gender stereotypes).

Research in the past decade has probed these issues further, supporting some of Maccoby and Jacklin's conclusions and questioning others (see Block, 1976, for a general critique of the Maccoby and Jacklin analysis). Typically, investigators have focused on a narrower set of behaviors and applied meta-analytic procedures to determine the existence and extent of differences. Eagly, for example, has found evidence of sex differences in social influence and in prosocial behavior (Eagly & Carli, 1981; Eagly & Crowley, 1985). Hyde (1982) reports reliable sex differences in aggression. Yet these and other differences have also been found to be small, generally accounting for less than 5% of the variance. Within-sex variation is substantially greater than between-sex variance in these analyses. Furthermore, most of the behavioral differences are strongly affected by situational factors, cautioning against any sweeping statements about invariant sex differences.

These patterns of small but significant differences recommend caution but not dismissal. In constrained situations—what one might call the maximal performance setting of the psychological laboratory—sex differences appear to be rather limited. On many of the measures that psychologists have devised, males and females are capable of comparable performance. Furthermore, experimental studies indicate that most, if not all, of the behavioral differences observed in the laboratory are highly susceptible to variations in the situation and in experience. Thus, immutability can not be assumed.

One exception to this pattern of small sex differences in laboratory investigations concerns the case of self-report. Self-assessed tendencies by males and females often show stronger sex differences than do actual performance measures. Such a pattern has been reported both for analyses of aggression (Frodi, Macaulay, & Thome, 1977) and for empathy (Eisenberg & Lennon, 1983). These differences are difficult to interpret. On the one hand, sex differences of this type may reflect the operation of demand characteristics, with both males and females reflecting an awareness of prevailing gender stereotypes instead of accurately assessing their own behavior. Alternatively, one could hypothesize that in self-report, people are summarizing a broader range of experience than that tapped by the psychologist's repertoire of tasks—experience that may evidence differences in choices, if not in capability.

From a somewhat different perspective, Carol Gilligan (1982) has postulated fundamental sex differences on the basis of self-reported reactions to moral dilemmas. Men and women, Gilligan asserts, speak in "different voices"—men concerned with separation and women with attachment. These different voices have implications for developmental sequences, perceptions of self and other, language, thought, and moral choices. Although Gilligan acknowledges the existence of some overlap, her basic message speaks to the differences between the sexes. Her thesis thus affirms that there is a "masculine" and a "feminine" mode, quite different from each other, with implications for a wide range of human behavior. Researchers may question Gilligan's conclusions, and in particular the evidence on which her contentions are based (see Benton et al., 1983). Nonetheless, her thesis is important if for no other reason than it moves beyond the piecemeal investigations so characteristic of sex-difference research.

The constrained environment of the laboratory, although important in assessing differences between women and men on specific tasks, cannot remain the sole arbitrator of questions regarding sex differences. Observations in less constrained contexts provide more grist for the mill of sex differences. A simple glance at the social world shows considerable variation in the behavior of women and men. Occupational distributions, child-care practices, and other realms of social life suggest distinctly different male and female patterns (see Eagly, 1983; Eagly & Steffen, 1984). These situations, not the experimental laboratory, form the experiential base for the development of gender belief systems; here the evidence of behavioral differences is far more pervasive.

Thus, in assessing sex differences, a distinction must be made between what men and women *can* do, when put to the test by psychological investigators, and what they *do* do in situations that provide greater choice among options. The two settings pose different questions, and researchers may arrive at different answers. Concepts of masculinity and femininity must take both types of evidence into account—inherent capabilities and behavioral choices—but the distinction between the two needs to be acknowledged in any conceptualization.

Questions of Gender Identity

Acknowledging the element of choice in the conceptualization of masculinity and femininity relates to the issue of self-definition by women and men. Questions of gender identity have been approached from two slightly different perspectives: developmentalists have focused on the initial acquisition of a sense of one's maleness or femaleness (e.g., Stoller, 1968; Money & Ehrhardt, 1972), while social and personality psychologists have emphasized the concomitants of the adult's self-defined masculinity and femininity. Within the tradition of developmental psychology, three major models have been postulated over the years: the psychoanalytic model, the social-learning model, and the cognitive-developmental model. Each model posits that the child acquires a relatively stable sense of gender identity at a fairly early age, but the models differ in the hypothesized processes leading to this acquisition.

For the psychoanalytic theorist, masculinity and femininity are the inevitable outcomes of biological givens. Events occurring during the well-known Oedipal period determine sex role identity, an identity that is believed to remain stable throughout life. Subsequent theorists in the psychoanalytic tradition, such as Karen Horney and Erik Erikson, suggested modifications of the original Freudian position, but they maintained adherence to the biological determinants and the stability of gender identity.

Social-learning theorists point to the environment instead of to biological factors as determinants of gender identity, but they are similar in their acceptance of a stable gender identity formed relatively early in life. The child's identification with the same-sex parent is the principal mechanism by which gender identity is established, and both direct reinforcement and observational learning are invoked to explain acquisition of gender-appropriate behaviors.

Cognitive-development theory, as proposed by Lawrence Kohlberg, is similar to psychoanalytic theory in its postulation of a universal and invariant sequence. Also similar is its assumption that a stable gender identity is formed by the time a child is 5 or 6 years of age. This theory differs, however, in its suggestion that the child first categorizes himself or herself as a boy or girl, and then structures experience to be consistent with this categorization. Thus, identification with same-sex parents is a consequence, not a cause, of gender identity.

Each of these models, described only briefly here, postulates fundamentally different patterns of male and female behavior, reflecting differences in masculine and feminine identity. Specific examples of gender-related behaviors are provided by each theory. In general, however, sufficient attention has not been paid to delineating the precise set of behaviors that should be considered ramifications of gender identity. Although one may be tempted to conclude, in ad hoc fashion, that any and all observed sex differences in behavior are reflections of basic differences in self-conceived masculinity and femininity, such an easy conclusion runs a heavy risk of circularity. As Spence (1985) has recently suggested, the influence of gender identity in guiding the acquisition of gender-congruent behaviors may diminish with time, becoming less influential to the adult than it is to the developing child.

Such a position does not necessarily imply that a basic sense of gender identity is weakened. It does suggest that the link between gender identity and any specific behavior may be tenuous. Thus, the distinction must be maintained (or perhaps initially recognized) between the global concept of gender identity and the specific referents that are invoked. Individuals may, as Spence (1985) suggests, create their own calculus for assessing maleness or femaleness, "using those gender-appropriate behaviors and characteristics they happen to possess to confirm their gender identity and attempting to dismiss other aspects of their make-up as unimportant" (p. 84).

Similar problems in distinguishing between a global construct and specific attitudinal and behavioral referents can be detected in the measurement of self-described masculinity and femininity by investigators in the personality and psychometric traditions. Over the past 50 years, several instruments have been developed to assess these constructs. Implicit in most has been the assumption that masculinity and femininity are bipolar concepts that can be represented on a single dimension (Constantinople, 1973). Further inherent in many of the measures is the belief that masculinity is what males do and femininity is what females do. Thus, to cite one of my favorite examples, a preference for a shower to a bath is scored for indicating masculinity, while the reverse preference indicates femininity. Because males and females typically endorsed these two alternatives with different frequencies, the item became, by definition, a valid indicator of masculinity and femininity. (The cultural limitations of this definition should be obvious.)

Recent work in this area has abandoned the notion of a single construct. In its place are two orthogonal scales that are proposed to measure masculinity and femininity separately, thus allowing people to vary from high to low on one or both of the scales. The most prominent examples of this approach, among several that have surfaced, are the BSRI developed by Sandra Bem (1974) and the PAQ introduced by Spence, Helmreich, and Stapp (1974). For both scales, males have been found to score significantly higher than females on the masculinity dimension, and females to score higher than males on the femininity dimension. The differences are relatively small in absolute terms, however, and the overlap is considerable. Thus, substantial numbers of both males and females score high on both "masculinity" and "femininity," suggesting for some investigators (e.g., Bem, 1974) the utility of postulating an androgynous personality type that combines both attributes.

Note that the specific terms *masculinity* and *femininity* were included as items in the original version of the BSRI. As Pedhazur and Tetenbaum (1979) noted in their critique of the scale, self-ratings for these two items are typically far more polarized than are ratings by women and men on any of the other items. Further, as Bem acknowledges, these two items "are actually the worst two items on the scale" (1979, p. 1050). They do not correlate with other items presumed to measure concepts of masculinity and femininity, but are more directly linked to simple sex distinctions. Thus, males almost always report that they are highly masculine, and females almost always report that they are highly feminine, independent of their self-reported possession of other traits on the scale. This apparent discrepancy is less interesting as a methodological problem, which has been discussed at length in the literature, than as a conceptual issue relevant to the basic definitions of masculinity and femininity. What does it mean to consider oneself masculine or feminine, contingent on being a male or a female? If this self-designation bears little relationship to behavioral referents or to other hypothesized correlates of masculinity and femininity, exactly what can we conclude? In her penetrating critique of the maculinity-femininity construct, Constantinople observed, "We are dealing with an abstract concept that seems to summarize some dimension of reality important for many people, but we are hard pressed as scientists to come up with any clear definition of the concept or indeed any unexceptionable criteria for its measurement" (1973, p. 390).

The extensive debates as to precisely what is being assessed by the masculinity and femininity scales of the PAQ and BSRI reflect this uncertainty. Bem, as the title of her instrument implies, has conceptualized these dimensions as relevant to a broad range of sex-related role behaviors, including, in her more recent work, gender-belief systems and cognitive schemata for processing gender-related information (Bem, 1981). Spence and Helmreich, in contrast, have advocated a closer link between the traits being measured by the scale and their behavioral consequents. Hence, they are prone to speak of clusters of instrumental and expressive traits, eschewing more extended links to other gender-linked behaviors and attitudes. Evidence for strong relationships among various gender-related behaviors has not thus far been provided (Spence & Helmreich, 1980). Further, we have some evidence that simple biological sex is more predictive of choices of sex-linked tasks than are dimensions of masculinity and femininity (Helmreich, Spence, & Holahan, 1979).

As discussed earlier with respect to the development of gender identity, we should distinguish between the global concepts of masculinity and femininity and more specific referents that may or may not be associated with the self-defined masculine and feminine person (see

Spence, 1985). Knowing that a person describes herself as feminine may offer little predictability for any specific gender-related behavior.

Masculinity and Femininity: Dictum or Diversion?

Masculinity and femininity are indeed fuzzy sets, not only for the lay observer attempting to form meaningful categories, but also for the scientific investigator. The prototypical concepts of masculinity and femininity are comprised of a variety of associated characteristics, related to their prototypes with varying degrees of probability. Lulled into security by terms that everyone seemingly understands, we have tended to ignore the many signs of confusion and lack of clarity inherent in these concepts. Each of the three areas that I have reviewed points to problems in conceptualization. Two problems stand out most clearly: the assumption of unidimensionality and the absence of behavioral referents.

Psychologists in the past assumed that masculinity and femininity were opposite ends of a single dimension, an assumption that is still evident in lay conceptions. In recent years, the two concepts have been separated in experimental investigations, and empirical evidence supports the hypothesized orthogonality between the two dimensions. Although one loaf has become two, however, we have seen insufficient consideration of the homogeneity of each of these loaves. Both are most often assumed to be unidimensional concepts, encorporating a variety of gender-related traits and behaviors. Available evidence does not support this assumption. Recent studies of gender stereotypes demonstrate that a number of separate and identifiable components are only moderately correlated with one another. Assessment of the attitudes, self-descriptions, and behaviors of males and females shows these gender-related characteristics to be only weakly related. Cross-cultural work shows diverging patterns of gender stereotypes and considerable variation in the division of labor between males and females (e.g., Murdock, 1965). Taken together, these lines of evidence pose a strong indictment of the unidimensional assumptions that underpin much of the psychological research.

A second major weakness of most research on masculinity and femininity is the tendency to point to ephemeral concepts in the absence of clearly defined behavioral referents. In post hoc fashion, one can point to many behaviors as indicants of masculinity and femininity. Yet few attempts have been made to chart systematically the occurrence of these behaviors and to determine the relationships among them. Trait psychologists have often been criticized for their proclivity to lump diverse behaviors together without first assessing their co-occurrence, a habit nearly as evident today as in the often-maligned Freudian era. Behavioral ecologists may have much to offer in their display of caution when assuming interrelationships among behavior patterns (e.g., Daly & Wilson, 1983).

Also missing from most of the research dealing with masculinity and femininity is an appreciation of the fluidity of these concepts and the behaviors they presumably describe—a fluidity that is evident in animal behavior as well as human (see Goldfoot and Neff, Chapter 12 in this volume). Considerable variation across situations is exhibited in many behaviors—most certainly in that vast range of behaviors that are caught up in the masculinity-femininity net. Furthermore, an element of choice is in many aspects of an individual's gender display that defies assumptions of constancy and inevitability. Let me speak in just a bit of detail about these more social-psychological considerations.

Biological distinctions undoubtedly exist between male and female, many of which are addressed in this volume. Apart from the less visible differences that may be detected in neurological and hormonal analysis, we see more apparent signs of sex differences. Two features that may be most critical are the typically greater size and strength of the male, and the female's greater involvement in childbearing. Yet these basic differences cannot realistically account for the great diversity of traits and behaviors associated with masculinity and femininity, much less for the considerable overlap in virtually all behaviors associated with the concepts.

Many investigators have pointed to the range that masculinity and femininity may cover. Birdwhistell (1970), for example, in discussing nonverbal displays of masculinity and femininity, argues that although certain physical distinctions are common to males and females, the nonverbal displays are learned and do differ across cultures: "Let me stress again that these positions, movements and expressions are culturally coded—that what is viewed as masculine in one culture may be regarded as feminine in another" (Birdwhistell, 1970, p. 44). Goffman (1979), analyzing gender depictions in the medium of advertising within U.S. culture, makes an even stronger statement about the role of individual choice: "What the human nature of males and females really consists of, then, is a capacity to learn to provide and to read depictions of masculinity and femininity and a willingness to adhere to a schedule for presenting these pictures. . . . One might just as well say there is no gender identity. There is only a schedule for the portrayal of gender" (p. 8).

At a more micro level, we have growing evidence from social psychological experiments that people consciously choose to portray different aspects of themselves depending on situational influences. To cite just a few examples, potential reinforcement can affect the degree to which achievement behavior is enacted (Jellison, Jackson-White, Bruder, & Martyna, 1975), aggression is displayed (Bandura, Ross, & Ross, 1961) and makeup and feminine clothes are worn (von Baeyer, Sherk & Zanna, 1981). Thus, masculinity and femininity can be consciously selected, displayed in some situations and absent in others. These findings do not deny that the concepts of masculinity and femininity are in some sense understood and acted on by the individual. Yet at the same time, they caution against the more prevalent assumption that the concepts are static and uninfluenced by the environmental circumstances.

To summarize the arguments presented here, I contend that traditional conceptions of masculinity and femininity have oversimplified that which is not simple, have unidimensionalized that which is multidimensional, and have conveyed a sense of stability and permanence to that which is inherently flexible. Perhaps the concepts of masculinity and femininity can be maintained in our repertoire. Yet if they are to remain, important distinctions need to be made between the theoretical concepts of masculinity and femininity and specific gender-related behaviors and attitudes. We need to specify the exact features that the concepts are believed to incorporate, we need to pay much more attention to situations and environments that encourage or discourage gender displays, and we need to give more thought to the functions that such behaviors may serve. Analyses of possible differences between males and females have not been well served by the easy invocation of the concepts of masculinity and femininity, and recognition that these terms are not explanations but simply labels is a necessary step for further understanding.

Acknowledgments

This chapter was prepared while the author was a Fellow at the Center for Advanced Study in the Behavioral Sciences, supported in part by the John D. and Catherine T. MacArthur Foundation. A grant to the author from the National Science Foundation (BNS-8217313) is also gratefully acknowledged. Thanks are extended to Janet T. Spence for her helpful comments on an earlier version of this manuscript.

References

Ashmore, R. D., & Del Boca, F. K. (1979). Sex stereotypes and implicit personality theory: Toward a cognitive-social psychological conceptualization. *Sex Roles, 5,* 219–248.

Ashmore, R. D., Del Boca, F. K., & Wohlers, A. J. (1986). Gender stereotypes. In R. D. Ashmore & F. K. Del Boca (Eds.), *The social psychology of female-male relations: A critical analysis of central concepts.* New York: Academic Press.

Bandura, A., Ross, D., & Ross, S. (1961). Transmission of aggression through imitation of aggressive models. *Journal of Abnormal and Social Psychology, 63,* 575–582.

Bem, S. L. (1974). The measurement of psychological androgyny. *Journal of Consulting and Clinical Psychology, 42,* 155–162.

Bem, S. L. (1979). Theory and measurement of androgyny: A reply to the Pedhazur-Tetenbaum and Locksley-Colten critiques. *Journal of Personality and Social Psychology, 37,* 1047–1054.

Bem, S. L. (1981). Gender schema theory: A cognitive account of sex typing. *Psychological Review, 88,* 354–364.

Benton, C. J., Hernandez, A. C. R., Schmidt, A., Schmitz, M. D., Stone, A. J., & Weiner, B. (1983). Is hostility linked with affiliation among males and with achievement among females? A critique of Pollak and Gilligan. *Journal of Personality and Social Psychology, 45,* 1167–1171.

Birdwhistell, R. L. (1970) *Kinesics and context: Essays on body motion communication.* Philadelphia: University of Pennsylvania Press.

Block, J. H. (1973). Conceptions of sex roles: Some cross-cultural and longitudinal perspectives. *American Psychologist, 28,* 512–526.

Block, J. H. (1976). Issues, problems, and pitfalls in assessing sex differences: A critical review of "The Psychology of Sex Differences." *Merrill-Palmer Quarterly, 22,* 283–308.

Broverman, I. K., Vogel, S. R., Broverman, D. M., Clarkson, F. E., & Rosenkrantz, P. S. (1972). Sex-role stereotypes: A current appraisal. *Journal of Social Issues, 28*(2), 59–78.

Brownmiller, S. (1984). *Femininity.* New York: Linden Press/ Simon & Schuster.

Constantinople, A. (1973). Masculinity-femininity: An exception to a famous dictum? *Psychological Bulletin, 80,* 389–407.

Daly, M., & Wilson, M. (1983). *Sex, evolution, and behavior* (2nd ed.). Boston: Willard Grant Press.

Deaux, K., & Lewis, L. L. (1983). Assessment of gender stereotypes: Methodology and components. *Psychological Documents, 13,* 25.

Deaux, K., & Lewis, L. L. (1984). The structure of gender stereotypes: Interrelationships among components and gender label. *Journal of Personality and Social Psychology, 46,* 991–1004.

Dwyer, D. H. (1978). *Images and self-images: Male and female in Morocco.* New York: Columbia University Press.

Eagly, A. H. (1983). Gender and social influence: A social psychological analysis. *American Psychologist, 38,* 971–981.

Eagly, A. H., & Carli, L. L. (1981). Sex of researchers and sex-typed communications as determinants of sex differences in influenceability: A meta-analysis of social influence studies. *Psychological Bulletin, 90,* 1–20.

Eagly, A. H., & Crowley, M. (in press). Gender and helping behavior: A meta-analytic review of the social psychological literature. *Psychological Bulletin.*

Eagly, A. H., & Steffen, V. (1984). Gender stereotypes stem from the distribution of women and men into social roles. *Journal of Personality and Social Psychology, 46,* 735–754.

Eisenberg, N., & Lennon, R. (1983). Sex differences in empathy and related capacities. *Psychological Bulletin, 94,* 100–131.

Foushee, H. C., Heimreich, R. L., & Spence, J. T. (1979). Implicit theories of masculinity and femininity: Dualistic or bipolar? *Psychology of Women Quarterly, 3,* 259–269.

Frodi, A., Macaulay, J., & Thome, P. R. (1977). Are women always less aggressive than men? A review of the experimental literature. *Psychological Bulletin, 84,* 634–660.

Gilligan, C. (1982). *In a different voice: Psychological theory and women's development.* Cambridge: Harvard University Press.

Goffman, E. (1979). *Gender advertisements.* New York: Harper & Row.

Helmreich, R. L., Spence, J. T., & Holahan, C. K. (1979) Psychological androgyny and sex-role flexibility: A test of two hypotheses. *Journal of Personality and Social Psychology, 37,* 1631–1644.

Hyde, J. S. (1984). How large are gender differences in aggression? A developmental meta-analysis. *Developmental Psychology, 20,* 722–736.

Jellison, J. M., Jackson-White, R., Bruder, R. A., & Martyna, W. (1975). Achievement behavior: A situational interpretation. *Sex Roles, 1,* 369–384.

Maccoby, E. E., & Jacklin, C. N. (1974). *The psychology of sex differences.* Stanford: Stanford University Press.

Major, B., Carnevale, P. J. D., & Deaux, K. (1981). A different perspective on androgyny: Evaluations of masculine and feminine personality characteristics. *Journal of Personality and Social Psychology, 41,* 988–1001.

Money, J., & Ehrhardt, A. A. (1972). *Man & woman, boy & girl.* Baltimore: Johns Hopkins University Press.

Murdock, G. P. (1965). *Culture and society.* Pittsburgh: University of Pittsburgh Press.

Pedhazur, E. J., & Tetenbaum, T. J. (1979). Bem Sex Role Inventory: A theoretical and methodological critique. *Journal of Personality and Social Psychology, 37,* 996–1016.

Spence, J. T. (1985). Gender identity and its implications for concepts of masculinity and femininity. *Nebraska symposium on motivation, 32,* 59–95. Lincoln: University of Nebraska Press.

Spence, J. T., & Helmreich, R. L. (1980). Masculine instrumentality and feminine expressiveness: Their relationship with sex role attitudes and behaviors. *Psychology of Women Quarterly, 5,* 147–163.

Spence, J. T., Helmreich, R., & Stapp, J. (1974). The personal attributes questionnaire: A measure of sex-role stereotypes and masculinity-femininity. *JSAS Catalog of Selected Documents in Psychology, 4,* 43. (MS No. 617).

Stoller, R. (1968). *Sex and gender: On the development of masculinity and femininity.* New York: Science House.

Strathern, M. (1976). An anthropological perspective. In B. Lloyd & J. Archer (Eds.), *Exploring sex differences* (pp. 49–70). New York: Academic Press.

von Baeyer, C. L., Sherk, D. L., & Zanna, M. P. (1981). Impression management in the job interview: When the female applicant meets the male (chauvinist) interviewer. *Personality and Social Psychology Bulletin, 7,* 45–51.

Williams, J. E., & Best, D. L. (1982). *Measuring sex stereotypes. A thirty-nation study.* Beverly Hills: Sage Publications.

19

Masculinity and Femininity Exist Only in the Mind of the Perceiver

Sandra Lipsitz Bem

Like many of the other participants in this volume, I am going to be self-indulgent and not comment directly on the paper I was asked to discuss. Instead, I am going to make explicit some of the underlying assumptions and hypotheses about male-female differences that I use both in my own thinking and in my own research. I hope that in so doing, I will be able to clarify (or at least to bring into sharper focus) some of the larger issues discussed throughout this volume. I warn readers, however, that I do not have space to present data and that I am going to be extremely speculative. Readers who would like to see the relevant data should note that my remarks represent a highly condensed extract from a lengthy theoretical and empirical retrospective of my work on Gender Schema Theory written for a 1985 Nebraska Symposium on Motivation (Bem, 1985).

As a feminist and a psychologist, I have long wrestled with one central question: How does the culture transform male and female infants into masculine and feminine adults? How does it create the many gender differences in behavior, motivation, and self-concept that transcend the dictates of biology?

Although this question has no single or simple answer, I will focus on what I see as the most psychologically interesting source of gender differentiation and what has also been the central concern of my own research, namely, the psychology of the individual as it pertains to gender. That is, I will focus on the thoughts and feelings about gender that individuals carry around in their heads and how these influence the way in which they perceive, evaluate, and regulate not only their own behavior but also the behavior of others. Before proceeding to that topic, however, I would like to comment briefly on two other important sources of gender differentiation, biology and the contemporaneous environment.

First, biology. In this volume is a great deal of discussion saying that it is not biology alone, but biology in interaction with culture, that produces many gender differences. Another way to say this is that whereas the sex-differentiated aspects of human biology are relatively constant, the cultural context varies a great deal, sometimes exaggerating the influence of biology, sometimes counteracting the influence of biology, and sometimes—in a more neutral fashion—simply letting the influence of biology shine through without either exaggerating or counteracting it.

I said explicitly that I am a feminist. For the record, this does not mean that I don't think there are any biologically based sex differences in behavior. It also does not mean that I think we should try to manipulate the culture so as to eliminate whatever biologically based sex differences there are. What it does mean, however, is that insofar as possible, we should let the distribution of activities and roles across males and females reflect nothing but biology. That is, we should try to arrange our social institutions so that they do not themselves diminish the full range of individual differences that would otherwise exist within each sex. If, under those conditions, it turns out that more men than women become engineers or that more women than men decide to stay at home with their children, I'll live happily with those sex differences as well as with any others that emerge. But I am willing to bet that the sex differences that emerge under those conditions will not be nearly as large or as diverse as the ones that currently exist in our society.

Let me shift now to the contemporaneous here-and-now environment. Clearly, environment impacts differentially on males and females in many ways. For example, some institutional practices explicitly treat males and females differently, such as the greater availability of athletic training for males and of parental leave for females. Similarly, some institutional practices have the effect of treating males and females differently even though the practices themselves are not explicitly based on sex, such as giving bonus points to veterans on civil service examinations, setting age requirements on admissions to graduate and professional schools, and the fact that the schoolday and the workday do not coincide. Finally, some normative expectations distinguish between males and females, such as the fact that women but not men are asked to bake cookies for bake sales and are called home from work when their children get sick at school.

Many if not most of these practices probably derive historically from a time when the sex-differentiated aspects of human biology were more pertinent to the demands of daily living than they are now and thereby dictated a sex-based division of labor. Nevertheless, these practices now establish a contemporaneous environment that is quite sufficient by itself to create and maintain many gender differences without any added contribution from biology. Whatever gender differences the contemporaneous environment can produce, of course, it can also take away. Accordingly, if the many gender-differentiated practices that abound in our culture were ever to be eliminated, we could finally begin to get back to the basics of biology, which is another way of saying that males and females could finally begin to be as *similar* as their biology allows.

Not all nonbiological differences between males and females are maintained by external forces, of course, for individuals bring to the contemporaneous situation preferences, skills, personality attributes, behaviors, and self-concepts that were acquired during childhood and that are consistent with their culture's definitions of masculinity and femininity. Accordingly, we turn now to a discussion of sex typing, the psychological process whereby male and female children become "masculine" and "feminine," respectively.

Sex Typing

As every parent, teacher, and developmental psychologist knows, male and female children become "masculine" and "feminine," respectively, at a very early age. By the time they are 4 or 5 years old, for example, girls and boys have typically come to prefer activities defined by the culture as appropriate for their sex and also to prefer same-sex peers. The acquisition of sex-appropriate preferences, skills, personality attributes, behaviors, and self-concepts is typically referred to within psychology as the process of sex typing.

The universality and importance of this proces is reflected in the prominence it has received in psychological theories of development, which seek to elucidate how the developing child comes to match the template defined as sex appropriate by his or her culture. Three major theories of sex typing have been especially influential: psychoanalytic theory (Bronfenbrenner, 1960; Freud, 1959a, 1959b), social learning theory (Mischel, 1970), and cognitive-developmental theory (Kohlberg, 1966) (for reviews of these theories, see Bem, 1985; Huston, 1983; Maccoby & Jacklin, 1974; Mussen, 1969). More recently, I have introduced a fourth theory of sex typing into the psychological literature—gender schema theory (Bem, 1981, 1985).

Gender Schema Theory

Gender schema theory begins its account of sex typing with the observation that the developing child invariably learns his or her society's cultural definitions of femaleness and maleness. In most societies, these definitions comprise a diverse and sprawling network of sex-linked associations encompassing not only those features directly related to female and male persons, such as anatomy, reproductive function, division of labor, and personality attributes, but also features more remotely or metaphorically related to sex, such as the angularity or roundedness of an abstract shape and the periodicity of the moon. Indeed, no other dichotomy in human experience appears to have as many entities linked to it as does the distinction between female and male.

Gender schema theory proposes that in addition to learning such content-specific information about gender, the child also learns to invoke this heterogeneous network of sex-related associations in order to evaluate and assimilate new information. The child, in short, learns to encode and to organize information in terms of an evolving gender schema.

A schema is a cognitive structure, a network of associations that organizes and guides an individual's perception. A schema functions as an anticipatory structure, a readiness to search for and to assimilate incoming information in schema-relevant terms. Schematic information processing is thus highly selective and enables the individual to impose structure and meaning onto a vast array of incoming stimuli. More specifically, schematic information processing entails a readiness to sort information into categories on the basis of some particular dimension, despite the existence of other dimensions that could serve equally well in this regard. Gender-schematic processing in particular thus involves spontaneously sorting persons, attributes, and behaviors into masculine and feminine categories or "equivalence classes" regardless of their differences on a variety of dimensions unrelated to gender—for example, spontaneously placing items like "tender" and "nightingale" into a feminine category and items like "assertive" and "eagle" into a masculine category. Like schema theories generally (Neisser, 1976; Taylor & Crocker, 1981), gender schema theory thus construes perception as a constructive process in which the interaction between incoming information and an individual's preexisting schema determines what is perceived.

What gender schema theory proposes, then, is that the phenomenon of sex typing derives, in part, from gender-schematic processing, from an individual's generalized readiness to process information on the basis of the sex-linked associations that constitute the gender schema.

Specifically, the theory proposes that sex typing results, in part, from the assimilation of the self-concept to the gender schema. As children learn the contents of their society's gender schemata, they learn which attributes are to be linked with their own sex and, hence, with themselves. This does not simply entail learning the defined relationship between each sex and each dimension or attribute—that boys are to be strong and girls weak, for example—but involves the deeper lesson that the dimensions themselves are differentially applicable to the two sexes. Thus, the strong-weak dimension is absent from the schema to be applied to girls just as the dimension of nurturance is implicitly omitted from the schema to be applied to boys. Adults in the child's world rarely notice or remark upon how strong a little girl is becoming or how nurturant a little boy is becoming, despite their readiness to note precisely these attributes in the "appropriate" sex. The child learns to apply this same schematic selectivity to the self, to choose from among the many possible dimensions of human personality only that subset defined as applicable to his or her own sex and thereby eligible for organizing the diverse contents of the self-concept. Thus do children's self-concepts become sex typed, and thus do the two sexes become, in their own eyes, not only different in degree, but different in kind.

Simultaneously, the child also learns to evaluate his or her adequacy as a person according to the gender schema, to match his or her preferences, attitudes, behaviors, and personal attributes against the prototypes stored within it. The gender schema becomes a prescriptive standard or guide (cf. Kagan, 1964), and self-esteem becomes its hostage. Here, then, enters an internalized motivational factor that prompts an individual to regulate his or her behavior so that it conforms to cultural definitions of femaleness and maleness.

From the perspective of gender schema theory, then, males and females behave differently from one another on the average because, as individuals, they have each come to perceive, evaluate, and regulate both their own behavior and the behavior of others in accordance with cultural definitions of gender appropriateness. Thus, cultural myths become self-fulfilling prophecies, and thus, according to gender schema theory, many gender differences emerge.

As a real-life example of what it means to engage in gender-schematic information processing, consider how a particular college student at Cornell might go about deciding which new hobby to try out from among the many possibilites that are available. Among other things, he or she could ask about each possibility how expensive it is, whether it can be done in cold weather, whether it can be done between and among classes, exams, and term papers, etc. Being gender schematic, however, means having a readiness to look at this decision through the lens of gender and thereby to ask first: What sex is the hobby?

What sex am I? Do they match? If so, the hobby will be considered further. If not, it will be rejected without further consideration.

The student in this example would not necessarily be consciously aware of his or her own gender-schematic processing. Quite the contrary. Just as the fish is unaware that its environment is wet (after all, what else could it be?), so too are most people unaware that their perceptions are (but need not be) organized on the basis of gender. The child learns over time to utilize certain dimensions instead of others as cognitive organizing principles, but the child does not typically become aware that alternative dimensions might have been utilized. The dimensions chosen as cognitive organizing principles thus function as a kind of nonconscious ideology, an underlying or deep cognitive structure influencing one's perceptions without conscious awareness. Such is the nature of schematic processing generally. Such, in particular, is the nature of gender-schematic processing.

One of the stated goals for this volume was to work toward a consensus on a set of definitions of masculinity and femininity acceptable to a wide spectrum of research areas. At the more biological end of things, a stable network of male behaviors and female behaviors may someday be identified that can reasonably be called masculinity or femininity. At the more cultural end of things, however, gender schema theory contends that masculinity and femininity do not exist "out there" in the world as objective realities waiting to be found, but they are merely cognitive constructions produced by gender-schematic processing. Put somewhat differently, masculinity and femininity exist only in the mind of the perceiver. Look through the lens of gender and you perceive the world as falling into masculine and feminine categories. Put on a different pair of lenses, however, and you perceive the world as falling into other categories.

From this perspective, the important feature of individuals is thus not whether they engage in some arbitrary set of behaviors that we, as investigators, happen to call masculine or feminine, but whether individuals spontaneously organize their self-concepts and their behavior on the basis of gender. In this regard, note that my own research (Bem, 1985) looks at individual differences in this domain as a way of demonstrating both that gender-schematic processing exists and that it plays a major role in creating and/or enhancing male-female differences.

The Antecedents of Gender-Schematic Processing

To this point, I have argued that many gender differences derive, at least in part, from the spontaneous readiness of individuals to sort information into equivalence classes, to evaluate their adequacy as persons, and to regulate their behavior on the basis of gender. From where does gender-schematic processing come? That is, how and why do so many males and females develop a readiness to organize information in general—and their self-concepts in particular—on the basis of gender?

Gender schema theory proposes that gender-schematic processing derives in large part from the society's ubiquitous insistence on the functional importance of the gender dichotomy even in situational contexts where sex needn't matter at all, from the culture's insistence that an individual's sex matters in virtually every domain of human experience. The typical American child cannot help observing, for example, that what parents, teachers, and peers consider to be appropriate behavior varies as a function of sex; that toys, clothing, occupations, hobbies, the domestic division of labor—even pronouns—all vary as a function of sex. An implicit but monolithic message is thereby communicated to the child: one's sex matters in all contexts; pay attention to sex and gender.

Gender schema theory thus implies that children would be far less likely to become gender schematic and hence sex typed if the culture were to temper its insistence on the functional importance of the gender dichotomy. Ironically, even though our society has become sensitized to negative sex stereotypes and has begun to expunge them from the media and from children's literature, it remains blind to its gratuitous emphasis on the gender dichotomy itself. In elementary schools, for example, boys and girls line up separately or alternately; they learn songs in which the fingers "are ladies" and the thumbs "are men;" they see boy and girl paper doll silhouettes alternately placed on the days of the month in order to learn about the calendar. Note that children are not lined up separately or alternately as blacks or whites; fingers are not "whites" and thumbs "blacks"; black and white dolls do not alternately mark the days of the calendar. Our society seeks to deemphasize racial distinctions but continues to exaggerate sexual distinctions.

Because of the role that sex plays in reproduction, perhaps no society could ever be as indifferent to sex in its cultural arrangements as it could be to, say, eye color, thereby giving the gender schema a sociologically based priority over many other categories. For the same reason, sex may even have evolved to be a basic category of perception for our species, thereby giving the gender schema a biologically based priority as well. Be that as it may, if the culture were to temper its ubiquitous insistence on the functional importance of the gender dichotomy, individuals would probably be far less likely to encode so much of the world as masculine or feminine, and many gender differences would thereby be diminished or eliminated. (In this regard, I am just beginning some research with children to ask, among other things, whether gender schematicity varies with the extent to which the child's socialization history has emphasized the functional significance of gender over and above its significance in the context of reproduction.)

I will end this discussion by calling attention to a provocative analog between one's gender identity and one's human identity. Both what sex we are and what species we are surely have profound biological influences on both our bodies and our behavior. Note, however, that none of us spend much time pondering our humanness, asserting that it is there, fearing that it might be in jeopardy, or wishing that it were otherwise. I would like to suggest that the same might be true of our sex if only society would stop projecting gender into situations irrelevant to genitalia. The biological sex differences that exist—whatever they are—would still remain and would still be perceived (perhaps they would even be cherished), but they would not function as imperialistic schemata for organizing everything in the world, and the artificial constraints of gender on the individual's unique blend of temperament and behavior would thereby be eliminated.

References

Bem, S. L. (1981). Gender schema theory: A cognitive account of sex typing. *Psychological Review, 88,* 354–364.

Bem, S. L. (1985). Androgyny and gender schema theory: A conceptual and empirical integration. In T. B. Sonderegger (Ed.), *Nebraska Symposium on Motivation 1984: Psychology and Gender* (Vol. 32). Lincoln; University of Nebraska Press.

Bronfenbrenner, U. (1960). Freudian theories of identification with their derivatives. *Child Development, 31,* 15–40.

Freud, S. (1959a). Some psychological consequences of the anatomical distinction between the sexes (1925). In E. Jones (Ed.), *Collected papers of Sigmund Freud,* Vol. 5. New York: Basic Books.

Freud, S. (1959b). The passing of the Oedipus complex (1924). In E. Jones (Ed.), *Collected papers of Sigmund Freud,* Vol. 5. New York: Basic Books.

Huston, A. C. (1983). Sex-typing. In P. H. Mussen & E. M. Hetherington (Eds.), *Handbook of child psychology,* Vol 4. New York: Wiley.

Kagan, J. (1964). Acquisition and significance of sex-typing and sex role identity. In M. L. Hoffman & L. W. Hoffman (Eds.), *Review of child development research,* Vol. 1. New York: Russell Sage Foundation.

Kohlberg, L. (1966). A cognitive—developmental analysis of children's sex-role concepts and attitudes. In E. E. Maccoby (Ed.), *The development of sex differences.* Stanford: Stanford University Press.

Maccoby, E. E., & Jacklin, C. N. (1974). *The psychology of sex differences.* Stanford: Stanford University Press.

Mischel, W. (1970). Sex-typing and socialization. In P. H. Mussen (Ed.), *Carmichael's manual of child psychology,* Vol. 2. New York: Wiley.

Mussen, P. H. (1969). Early sex-role development. In D. A. Goslin (Ed.), *Handbook of socialization theory and research.* Chicago: Rand McNally.

Neisser, U. (1976). *Cognition and reality.* San Francisco: Freeman.

Taylor, S. E., & Crocker, J. (1981). Schematic bases of social information processing. In E. T. Higgins, C. P. Herman, & M. P. Zanna (Eds.), *Social cognition: The Ontario Symposium,* Vol. 1. Hillsdale, NJ: Lawrence Erlbaum Associates.

VII
CULTURAL PERSPECTIVES

20

Masculinity and Femininity in Elaborated Movement Systems

Anya Peterson Royce

I

To elaborate means to work out in detail, to develop, to perfect. In the case of movement, this involves the process of the selection of elements, their stylization, and their arrangement into a system with recognized rules. The latter implies mastery of a technique, technique being the rules for selection and stylization. The kinds of elaborated movement systems I will be concerned with here include dance, mime, theater of movement, female impersonation, and everyday movement and gesture that have become the object of stylization and elaboration for context-specific reasons.

One could approach the topic of masculinity and femininity in these systems with a focus on any of three areas: form, content, and context.

Form refers to the form and structure of movement systems and also to the body in its capacity as an instrument. Form is usefully divided further into body, technique, and style. The body is the physical instrument. One can evaluate bodies in terms of attributes on which everyone will agree for a given period of time. Notions about what makes the best body for a particular form change across time and region. Technique refers to the set of movements, gestures, and steps that are the foundation of the performance genre that every performer must master. The system has some flexibility in that body type and technique can be adjusted to each other. Sometimes less desirable bodies can still produce superb effects on stage. Last, we have style, where performers distinguish themselves from other instruments. Style is what makes Fernando Bujones different from Anthony Dowell, and Marcel Marceau different from Jean-Louis Barrault. Style, in its original meaning, implies choice: choosing ways of moving or interpreting that complement one's body, technique, and mind. With regard to technique, choice is minuscule—rules about a battement tendu or an impulse do not vary from dancer to dancer, mime to mime. It is not in the grammar that one has license to show individuality; it is in the literary style—the choice of words, rhythm, pacing. In sum, a qualitative difference exists between physical type and technique on the one hand and style on the other. We have more or less objective standards for technique and bodies. Performers convey their own meanings to audiences in the realm of style. I will take up each of these subdivisions of form as it impinges on notions of masculinity and femininity.

Content refers to the meanings, narrative, plot, or emotions conveyed by a performance.

Context is the arena in which these forms develop and present themselves, society's notions about the forms and the performers and other cultural features that inform and color an audience's or performer's attitude about the genre or specific performances.

I will deal with all three areas in this paper. Peformance is, by its nature, an interaction between performer and audience. Form and content, therefore, are negotiated in terms of the larger context.

Any kind of conscious performance—indeed, any form of art—generates effect from two sources: the form itself and the ideas associated with the form. This twofold source of effect is especially significant in movement systems such as dance, mime, and theater of movement. Here the substance of the form is the human body and therefore identical to the substance of ordinary movement and gestures. Cultural codes of aesthetic preference transform the basic building material into the stylized, restricted forms we recognize as art (Royce, 1984). The human body carries certain connotations, and however much the body may be changed in a particular performance genre, viewers will respond automatically if subconsciously to those cues. In this paper, I am concerned with cues that deal with masculinity and femininity. Performers and directors use them in three ways: (1) reinforce them, (2) contradict them, (3) neutralize or deemphasize them. Four categories of forms result from these actions.

1. Gender-free forms—neutralize or deemphasize masculine and feminine traits.
2. Cross-gender forms—either contradict or reinforce stereotypic traits.
3. Stereotyped forms—reinforce traits.
4. Unmarked masculine forms—deemphasize masculine traits while maintaining a male identity.

One has always to keep in mind the complexities created by the fact that these three strategies and the resulting forms apply to both form and content. They may apply in one area and not in another. There may be a deliberate attempt to convey conflicting messages. The performance forms we are dealing with here are all multichanneled systems of communication where the potential for ambiguity or outright deception and manipulation is quite high. Some female impersonation may manipulate the message in appearance and movement or gesture so that the audience receives conflicting evidence. A choreographer may deal with a traditional love story in terms of content but set the pas de deux on two males or two females (this occurs far less often) instead of on a male and a female. Bodies may be totally dehumanized by costume and still convey human relationships.

Finally, we must be aware of the complication introduced by the frequent disparity between what performers believe about themselves and their art as it touches on masculinity and femininity and what audiences see and comprehend. Audiences' reactions and to some extent performers' self-images are forged in the context of culture and society, and this is yet another dimension to be kept in mind.

II

Performers working in a form that uses the human body in motion frequently have gender-free images of themselves. I mean by this that in class, in rehearsal, or in performance, the fact that they are biologically male and female is irrelevant. When biological facts do intrude, they are resented, as the following lament by Toni Bentley, a Balanchine dancer, illustrates (1982).

> I can see no way, no way at all, to be a woman and dance. When I was five pounds thinner with no monthly cycles, I was more constant in mood. But this fluctuation and the desires that often come in cycles now distract all my thoughts and weaken my ability to dance. . . . How can I work, how can I stand at the barre every morning and every night, when one day is an ecstasy and the next an agony? (pp. 135–136)

The desired state is very clear—to be free of the complications of sex, indeed, humanness, so that one can be a more perfect instrument. Again, Bentley is instructive: "We must be like trained animals and active instruments, responding to the learned stimulus of music" (p. 142). Out of class and offstage, the images might be quite different, but in the act of performing, certain metaphors occur again and again: the performer as instrument, as athlete, as abstraction, as celibate.

The performer's body *is* the instrument and the product both, so it is not strange that performers should think of themselves in these terms. Murray Louis, one of the most influential of America's modern dancers has a particularly succinct articulation of this notion (1980).

> The vocabulary of abstraction that the dancer and the choreographer use begins with their instrument, the body. The body is, in a sense an orchestra composed of various instruments: arms, legs, head, and torso. They can be played in solo fashion or in various combinations. The dancer works very hard to develop the widest possible range of flexibility and expression in all of these parts. (p. 162)

In this imagery the body and the performer are two different entities with the former being treated as any other instrument. It is starved, stretched, and developed until it becomes the best vehicle for the communicative act of performance. It is frequently spoken of and to as if it were an inanimate instrument. It is subjected to impersonal scrutiny in the mirror of the studio, each part being viewed with an eye to its ability to come through in performance. Bodies are not viewed as human males and females but as designs in space—is this line pleasing, does this gesture convey this particular meaning in the best way? When one looks in the mirror, one sees oneself not as a person but as an aesthetic product. To achieve the best aesthetic product, bodies are punished and mistreated, often pushed to the limits of what they can do and beyond. In forms such as dance the serviceability of these instruments is often very short. Ironically while separating their bodies from themselves in developing and evaluating their status, performers and their bodies are one and the same when it comes to serving art. Many performers think of themselves as being finished not just for performing but for the business of living once their bodies can be longer serve them as artists.

Because performers in movement-based arts must master the physical demands of the techniques and must engage in strenuous physical effort in class, rehearsal, and performance, they often think of themselves as analogous to athletes. They also speak of the respite from other demands of the body given them by the exhaustion brought by the physical labor of their art. Erika Eisinger (1981) echoes these sentiments in her article on Colette's music-hall days.

> [The performance is] a display of muscular control, of discipline, and power. Her beauty on stage . . . is neither fragile nor ornamental: it is athletic. . . . In the world of the stage, men and women experience each other as co-workers, comrades, not as lovers or enemies. The exhausting physical labor of the stage pro-

> vides a reprieve from love. . . . The theatre requires both men and women to earn their livelihood by virtue of their physical attributes. Both sexes put on make-up, exhibit their bodies, receive propositions, and fear the loss of their ability to please. Stage life equalizes or neutralizes sexuality. (p. 98)

Eisinger's description of Colette's attitudes, which were developed while she was a music-hall performer, raises a related notion, that of celibacy. In their work, performers are loathe to invest energy in matters of sexuality because it detracts from performance; It is distracting, irrelevant, and, for some, contamination of the purity of the form. Audiences frequently share these notions, attributing a kind of purity to the performer, partly because of the purity of the form, partly because they know the sacrifice that the performer makes in giving up anything that resembles a normal life. This is especially true of classical mime, where the absence of speech coupled with the use of whiteface and simple white or black costumes reinforces the sense of purity. In the classic film *Les Enfants du Paradis,* the truth about who stole the pocket watch is "told" in mime by Baptiste, a player with the Funambules. His mimed account of the incident is both compelling and convincing. In a larger sense, all artists are like this—we give over to them the task of articulating truths for us. The cost of this gift for them is sacrifice and purity.

Purity in both the physical and spiritual dimensions is described by Bentley, the Balanchine dancer (1982).

> We have a different bodily structure than most humans. Our spirits, our souls, our love reside totally in our bodies, in our toes and knees and hips and vertebrae and necks and elbows and fingertips. Our faces are painted on. We draw black lines for eyes, red circles for cheekbones and ovals for a mouth. Any hint of facial wrinkles, teary eyes, drops of sweat, audible breathing or diminishing energy levels is a sign of imperfection. They are symptoms of mortality. (pp. 16–17)

Finally, performers may view what they do as an abstraction and hence view themselves as a component of that abstraction. Louis uses this metaphor when he speaks about a particular duet (1980).

> I didn't perform the emotion of boy meets girl when I danced the duet from "Kaleidoscope" with Gladys Bailin. I played it with the cool sensuality called for by the austerity of its design. . . . It had an Oriental calm . . . our costumes were full tights. For Gladys and me, the dance was about two people on a long journey, per-

> haps through life. . . . But this was in 1956, when dancing in full-body tights was not common practice in modern dance. Men danced in men's clothes and women in dresses or skirts; otherwise it was difficult to get "emotional" about them. Seeing the human form revealed in tights gave you two choices; you could either think dirty or come to grips with dance as abstraction. (p. 145)

One of the best contemporary dance critics, Arlene Croce, shares Louis's view, referring to the ballerina as an abstraction of a woman, not a woman, and ballet as a world of signs and designs. In response to those who find sexist meanings in ballet, she replies (1979):

> To impute sexist meanings to standard ballet usages . . . is to indulge in fantasy-land explorations at the Disneyland level. Ballet is fantasy, true, but even when it is erotic fantasy its transfigured realism reorders the sensations that flow from physical acts, and our perceptions change accordingly. The arabesque is real, the leg is not. (pp 80–81)

In order to make general statements instead of particular ones, all art forms must abstract from the everyday world, must highlight and select; otherwise, they never rise above the everyday. In forms that use the human body, abstraction is more difficult because the same body is used both as a vehicle of artistic expression and in ordinary activities.

Is it the same body? In dance, mime, and contemporary rock groups, the body is physically altered to conform to whatever prevailing preferences may be. Some of this is accomplished by selecting the desirable body types from the beginning. From there, training shapes the body. A great deal of variation is found from company to company and country to country, but male and female bodies tend toward the gender-free, or the "androgyne" of Colette's music-hall colleagues. Their androgynous character can be further highlighted by common practices such as the removal of hair. Bentley comments: "We are hairless. We have no leg hairs, no pubic hair, no armpit hair, no facial hair, no neck hair and only a solid little lump at the top of our heads. Any sign of stubble must be closely watched out for and removed" (1982, p. 16). Not coincidentally the eunuchs in the ballet *Prodigal Son* are bald. Their choreography also reflects the indeterminate nature of their sex with movements neither masculine nor feminine. Lindsay Kemp, director of his own company, plays female roles as often as male. When not actually donning wigs and female dress, he appears as a hairless creature about whose gender we are never certain.

At this point we may consider the attitudes of the general public and

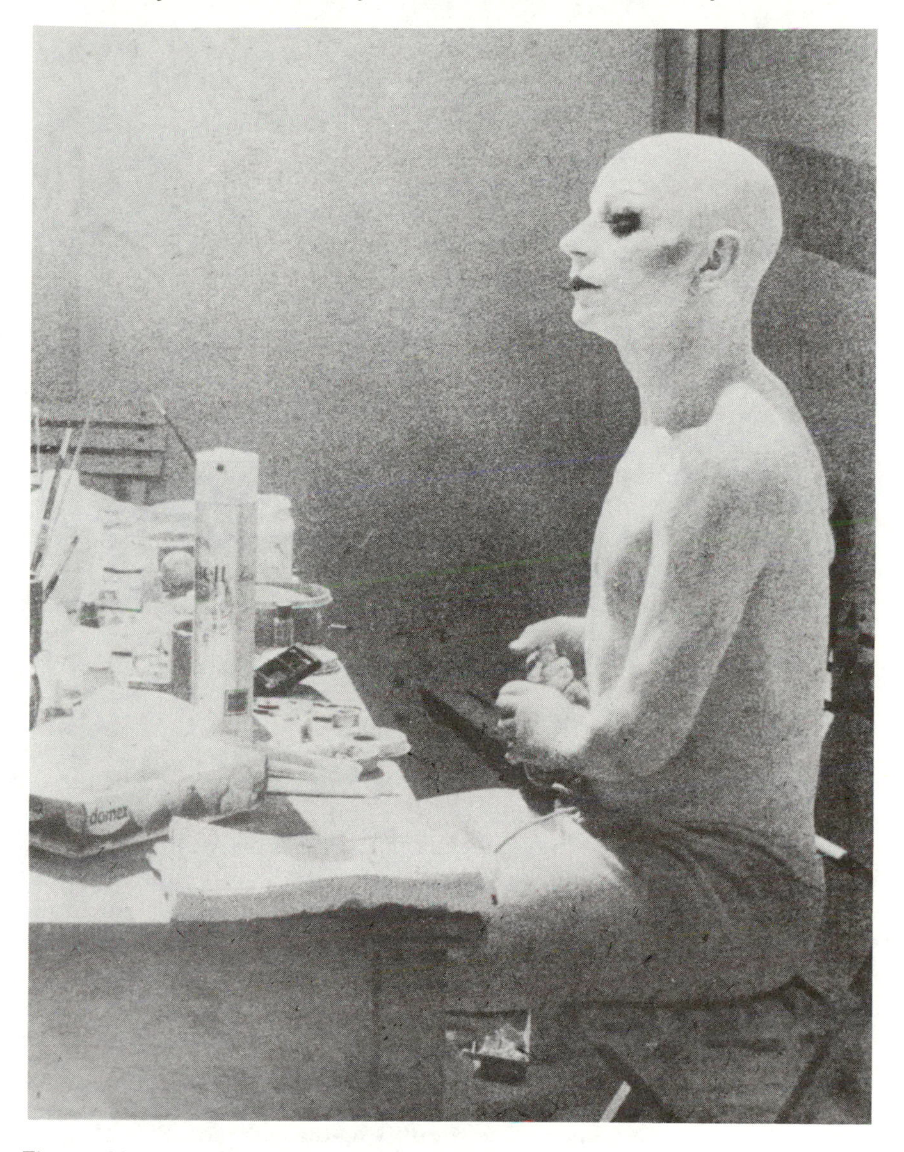

Figure 20–1. Lindsay Kemp before performance. (Photo by Guido Harari, Haughton & Harari, 1982.)

critics about the preferred images of masculinity and femininity in performance. Let us take classical ballet as an example, because with its long history, we can document changes in preference. From the early days of ballet throughout most of the eighteenth century, the male dancer reigned and the technique reflected that dominance. In the first ballets,

Figure 20–2. Lindsay Kemp in performance. (Photo by Guido Harari, Haughton & Harari, 1982.)

men danced both male and female roles. Women began participating as early as the late seventeenth century but heavy, voluminous costumes and heeled shoes prevented them from participating in the displays of technical virtuosity exhibited by the men. When enterprising women such as Marie Camargo, Barberina, and Marie Sallé rid themselves of the constraining costumes, they began competing with men in terms of the essentially male techniques instead of developing a technique that emphasized feminine qualities. They excelled in multiple beats and jumps. Camargo was praised by Voltaire, in fact, for being the first woman to dance like a man (see Philp & Whitney, 1977, p. 34). Until the nineteenth century, the dancers whose names everyone remembers were almost without exception men: Pierre Beauchamp, Louis Dupré, Jean Balon, Gaetan and Auguste Vestris, Jean-Georges Noverre, Jean Dauberval. The nineteenth-century Romantic movement removed the male dancer from his position of dominance. Thanks to the Romantics and dancers such as Fanny Essler and Marie Taglioni, this period became the Age of the Ballerina. Ballet became the realm of the ethereal, of haunted forests and convents, of fairy creatures and sylphs—all these illusions aided by the inventions of gas-lighting and the blocked pointe-shoe. The dance of the spirits of the nuns from *Robert Le Diable* in the eery flickering light of the gas-lamps with the wraithlike Taglioni appearing to hover above the stage was one of the early performances that ensured the cult of the ballerina. Theophile Gautier led the Romantics and eventually the theatergoing public to applaud feminine qualities of movement and appearance and deride or ignore the masculine. Comments such as the following from the age's supreme dance critic set the tone (Gautier in Philp & Whitney, 1977).

> Dance has as its sole object the revelation of beautiful form, in graceful attitudes and the development of lines that are pleasing to the eye. . . . Nothing is more abominable than a man who displays a red neck, great muscular arms, legs with calves like a church beadle, his whole heavy frame shaken with leaps and pirouettes. . . . [Men] affect that false grace, those revolting manners that have sickened the public of male dancing. (p. 34)

Few male dancers could compete successfully in the new Romantic style. Ironically, one of the few, Jules Perrot, was lauded by Gautier as "the male Taglioni." Gautier elaborated: "Jules Perrot displays perfect grace, purity, and lightness: it is music made visible . . . his legs sing very agreeably to the eye. These praises are the less to be doubted coming from us, since there is nothing we like less than to see male dancers" (see Philp & Whitney, 1977, p. 40).

The Romantic movement had the same effect on mime. Early nineteenth-century mime still demonstrated strongly its roots in the commedia dell'arte. It was a rough-and-tumble form characterized by movement, slapstick, acrobatics, and popular themes. Grimaldi's Clown in British pantomime and Jean-Gaspard Deburau's Pierrot in the pantomime of Paris's Funambules epitomized this Italianate form of mime. Both were heroes of the people in acknowledgment of their acrobatic and mimic skills but also because they stood for the underdog who, through cleverness, trickery, and mayhem, fought back. After Gautier, Jules Champfleury, and Charles Nodier "discovered" the Funambules and began to improve the fare presented there, Pierrot was transformed into a simple, lovesick, moonstruck waif, scarcely ever fighting back and never winning. Instead of being a flesh-and-blood man of many appetites, he was a fantastic, melancholy, sexless creature of the Romantic imagination. Unlike the ballet that elevated all the graceful qualities of the ballerina under the influence of Romanticism, pantomime created an androgynous character with qualities of neither sex.

Although the lyricism of the Romantic ballet was replaced by brilliance and technical virtuosity in Russia under Petipa, the male dancer did not fare much better. In ballets such as *Sleeping Beauty* and *Swan Lake,* he was relegated to the role of princely porteur, carrying the bal-

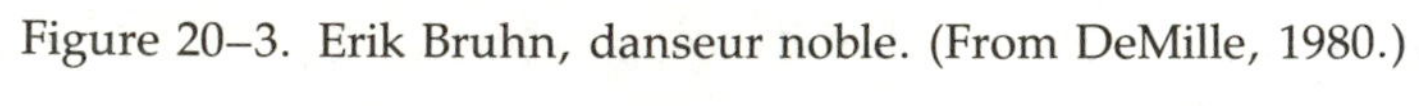

Figure 20–3. Erik Bruhn, danseur noble. (From DeMille, 1980.)

lerina through one pas de deux after another and paying obeisance to his parents the king and queen. This image of the male dancer is embodied in the term *danseur noble.* Then Diaghilev changed it all. He broke away from the massed swans and the row upon row of courtly attendants, his choreographers producing shorter pieces with themes ranging from the abstract as in *Les Sylphides* to the primitive as in *Sacre du Printemps* to the oriental as in *Schéhérazade.*

Although the Diaghilev Ballets Russes included a number of fine female dancers, Karsavina and Danilova among them, the male dancers are of particular interest when one is assessing the relative importance of masculine and feminine qualities. Even while dancing roles with traditional male and female values, most of Diaghilev's male dancers exhibited a notable absence of traditional masculine form and movement. This is true of Adolph Bolm, Anton Dolin, Leonide Massine, Serge Lifar, and, perhaps most strikingly, Vaslav Nijinsky. Nijinsky created the male role in *Chopiniana* (later, *Les Sylphides*) and gave it the androgynous quality that characterized so many of his roles. Cyril Beaumont described the movement style: "the billowing of his white silk sleeve as he curved and extended his arm; then that lovely movement when, on extending his leg in a développé, his hand swept gracefully from thigh to shin in a movement so graceful and so delicate as to suggest a caress" (in Philp & Whitney, 1977, p. 58). In *Schéhérazade,* Nijinsky was described by Benois as "half-cat, half-snake, fiendishly agile, feminine yet wholly terrifying" (Philp & Whitney, P. 62). Nijinsky also danced many roles of a half-human, half-fantasy nature and totally identified with this neither-nor state. *L'Après-midi d'un Faune, Le Spectre de la Rose,* and *Petrushka* belonged to him above all other dancers because he so completely portrayed that not wholly of this world state. We may be tempted to argue that the more one moves away from traditional masculine and feminine form and movement, the better able one is to stand for Everyman. Robert Joffrey, a contemporary choreographer, expresses a similar idea:

> It is my feeling that movement can have tremendous sensuality that is not directed to specific sexual gender. Certainly movement for me is totally asexual—both men and women can do any dance step. Of course some steps have become particularly associated with one sex rather than the other, but . . . this is because of some archaic connotations and training habits that still exist in the dance world. We should not forget that Fokine released the male in ballet and enabled him to become a significant performing artist in his own right. As he pointed out, movement must flow from one's emotional responses and should not be tied to the rigidity of classical technique. (Philp & Whitney, 1977, p. 9)

Figure 20–4. Vaslav Nijinsky as the Golden Slave in *Schéhérazade*. (From Philp & Whitney, 1977.)

If Fokine released the male dancer, Balanchine imprisoned him once more, this time within the constraints of Balanchine's neoclassical technique and his adamant stand against the primacy of emotion in ballet. Balanchine also clearly favored female dancers; his statement "ballet is woman" is often quoted. We can name great male dancers from Balanchine's company over the years—Edward Villela, Arthur Mitchell, and Peter Martins—but the succession of female stars is simply overwhelming. Balanchine was proud that his dancers called him "mother" because, for him it was better than "father." His muse was female—and, in his view, the male was put on earth for the sole purpose of serving and admiring women. They should draw attention to women's beauty (see Taper 1973, p. 342). His female dancers did not necessarily express stereotypes of femininity in movement, his technique in its demands was essentially the same for males and females. Certainly Balanchine's preoccupation with thinness created female bodies with few visible female attributes. But in terms of placement within the choreography, women were highlighted.

If Balanchine felt that ballet was woman, Maurice Béjart expresses the opposite view: ballet is man. Because of this perspective, Jorge Donn, Béjart's principal male dancer, went back to Béjart after being courted by Balanchine. Béjart has choreographed several pas de deux for two males with romantic or erotic themes. He also uses Patrice Touron, a soloist, in roles *en travestie* such as *The Dragonfly*, his homage to Anna Pavlova. Or Touron as Beauty in *Wien, Wien Nur du Allein* whom Joan Pikula (1984, p. 57) describes as "moving with persistent grace, a gentle and gentling figure." Neither Balanchine nor Béjart used women and men in traditional ways. Both, in fact, seemed to use the single sex to stand for both. Donn articulates this in terms of his own qualities in performance which sometimes have been referred to as androgynous. Commenting on Donn, Pikula (1984, p. 64) says: "In Donn, this is not a confusion of sexual roles, nor is it nonsexual. Rather, it is an expression of sexuality on a plane above individual differences, something like a universal sexuality, which casts a potent spell on both the men and women in the audience."

A similar sexuality is conveyed by many contemporary rock groups. Boy George, the lead singer in Britain's Culture Club, while made up as a woman, uses both male and female clothing. His movements tend to be stereotypically feminine, his speaking voice and singing ambiguous. He has a strong sexual appeal to teenagers and young adults of both sexes. A performer of even broader appeal, if we are to judge by the number of awards, is Michael Jackson, who exudes an unmistakable sexuality while his physical appearance is neither typically masculine or feminine. We must not forget one of the first of such performers, Mick Jagger, who made this kind of supersexuality explicit in his film *Performance*.

How are audience and performer images of masculinity and femininity translated into the form and content of performance? I have spoken about this in the contrast between Balanchine and Béjart. Let us turn to some companies that do not consciously select either the male or the

Figure 20–5. Alwin Nikolais Company in performance. (From publicity photo.)

female as the focus. Twyla Tharp begins from the premise of an identical technique for men and women. One of the few exceptions is *Deuce Coupe,* choreographed for American Ballet Theatre but now sometimes danced by Tharp's company. A "ballerina" role in it is danced *en pointe.* Tharp's technique derives in part from her early training with Paul Taylor, exaggerating some of Taylor's flowing qualities. Typical Tharp is loose-spined, syncopated, relaxed of limbs and head. Some pieces are incredibly fast-paced like *Fugue;* others move at a more leisurely pace like most of *Sue's Leg.* Both men and women could be interchangeable except in those pieces where the content calls for clear gender roles. *The Catherine Wheel* is such a piece. Traditional notions of roles are reinforced in its themes. Interestingly, the choreography itself could be danced equally well by men or women, and the bodies of the dancers are almost totally obscured for most of the piece by baggy costumes. The interchangeability of dancers is demonstrated in the changes made in the casting of *Fugue.* This piece was originally set on Tharp, Sara Rudner, and Rose Marie Wright. Subsequently, it has been danced by two men and a woman, and most recently by three men. Some viewers may prefer one or the other of these combinations, preference

Figure 20–6. Nikolais dancers. (From publicity photo.)

sometimes having to do primarily with the personalities of the different dancers. The choreography and technique execution, however, remain identical regardless of which mix of sex is performing. Personalities of the individual dancers are extremely important in Tharp's company. She tries to bring out these personalities instead of using the dancers in stereotyped male or female roles. Like personality, there is no single Tharp body. Her dancers come in all shapes and sizes from the tall and lanky Wright to the petite, compact Tharp herself. Costumes that hide the body are another Tharp trademark. The looseness of some of the

Figure 20–7. Pilobolus. (From Whitaker, 1980.)

silk sweatpant costumes matches the looseness of some of the choreography and the bodies. In addition, Tharp sometimes uses a layered look, with tunics and sarongs on top of baggy pants and leg warmers.

Alwin Nikolais, like Tharp, underscores the same technique for men and women.

> In his abstract expressionist theatre, the men and women wore the same costumes and freely interchanged roles. . . . He often used dynamic Coral Martindale as one of the "men" without giving a second thought to what role she should be playing because she was a "female." With the exception of lifts and other acts requiring certain strengths, he used all his dancers interchangeably. (Louis, 1980, p. 146)

Nikolais also uses costumes that disguise the body or minimize its impact in terms of gender. Dancers' bodies are totally hidden within stretch fabric tubes, and the point of the choreography is the pattern made by these "creatures." Or dancers in leotards and tights manipulate ribbons and wider swathes of material.

Pilobolus is a company that began in a college modern dance class as a project of three men who were dancing for the first time but who all had been involved in gymnastics and other sports. The choreography they evolved was based on gymnastic principles and transference of weight. They built shapes in space out of their bodies metamorphosing into strange animals or geometric shapes. The "theme" of most of their early work was abstract—they were simply manipulating their bodies to make patterns in space. This continued even after they added two women to the group, building their bodies into pyramids and other shapes. The solos for women were not noticeably different than those for the men. Recently, Pilobolus has done pieces that differentiate, even call attention to, the two sexes, but here their sense of the incongruous colors the treatment of traditional masculine and feminine stereotypes. In one piece, (*Untitled*, 1975) the two women appear in demure cotton frocks with long skirts. The skirts hide their male partners on whose shoulders they are. The skirts do not hide, however, the partners' legs and bare feet. So one has the contrast between the small, delicate women from head to skirt and the rather large, hairy male legs. At a later point, a male dancer, nude, emerges from the skirt in a kind of birth. But he remains, while clearly male in physical attributes, presexual in his lack of social development and his relationships with the women.

III

The three actions possible for performers to take with regard to ideas about masculinity and femininity—reinforcing, contradicting, and neutralizing—interact in complicated ways in terms of solo mime, both illusion and corporeal. For the purposes of this paper, illusion mime is best represented by Marcel Marceau and corporeal mime by Etienne Decroux. In these kinds of mime, masculine traits are deemphasized, while a male identity is maintained. The form associated with the action is best described as unmarked masculine; that is, males stand for both male and female as humanity in a way in which females cannot because they are marked as female.

Comments by mimes illustrate the feeling they have about the fundamental sexual identity of mime. Mimes, they say, have to be able to portray both masculinity and femininity. The latter should be a kind of virile femininity. The former should not be sexual but poetic and strong. Men can achieve this balance, they say, but it is difficult for women to capture either quality. They also have interesting ideas about body image. Male bodies are less distracting, more neutral than female bodies. Here, those male bodies that are hypermasculine in terms of cultural stereotypes are not thought to be the perfect stuff of mime. Instead, again, the male is representative of all humankind, the body being a kind of archetype for humanity. Here, we have moved from masculinity and femininity in content to images of these in body shape. Finally, we look at the matter of technique and its relationship to the masculine and the feminine in mime. Mime technique relies heavily on weight and tension in movement and gesture. Women are regarded as not having sufficient strength and endurance to sustain the weight and tension of mime. Further, mime technique requires one to breathe from the stomach. Men do this naturally; women, in contrast, breathe naturally from the chest. They can learn mime breathing but it will always be unnatural. Breathing, in fact, is inextricably bound up with weight. Marceau, for example, spoke of it in the context of a discussion of counterpoint, weight, and volume, and summarized these concepts as "feeling the weight" (interview, April 5, 1981).

Etienne Decroux (1963) is specific about the importance of weight as a defining feature of mime: "Le mime manque légèreté mais le danseur manque de poids" (the mime lacks lightness but the dancer lacks weight) (p. 71). Again, he refers to weight when he remarks that mime is like work or life in that the movements are sudden, irregular, and earthbound (in Royce, 1984). Decroux prefers to focus on the body instead of the face because the body is heavy and has the weight necessary for mime. Another way to contrast mime and the kind of dance epitomized by classical ballet is in the use of effort elements. One extreme use of the effort elements of space, weight, time, and flow is in movements characterized as direct (space), strong (weight), sudden (time), and bound (flow). This characterization parallels what Marceau and Decroux say about the nature of mime.

Effort elements have been used recently in a dissertation that compared expressed psychological masculinity and femininity with qualities of movement in the performance of three common tasks. (Roberts, 1978). The results lend credence to mimes' notions about the centrality of weight and the relationship of that with their perception that males are better suited to mime. The study indicates that the effort qualities most associated with masculinity and femininity are strong and light, those qualities assigned to weight. Those associated with space, time, and flow—namely, flexible, direct, sudden, sustained, bound, or free—are not linked to masculinity and femininity. The psychologically feminine female in the study used more light movements than any other subject, while the psychologically masculine male used more strong movements than anyone else. Although the author notes that raters' decisions were affected when the biological sex of the subject was apparent, the pattern of coded movements for those subjects belonging to opposite biological and psychological sex groups does suggest that we need to consider anatomy, specifically musculature, when we look for masculine and feminine in movement. For example, the psychologically feminine male used the least light movements, while the psychologically masculine female used strong movements that put her only seventh in frequency compared to all subjects. Perhaps weight is a quality with specific connotations for masculinity and femininity and a quality, moveover, that may be difficult to manipulate.

Let me return then to the issue of mime technique and notions of masculinity and femininity. Mime technique, as we know it today, was developed in the 1930s and 1940s. The fundamentals derived from the efforts of Decroux, Barrault, and Marceau, although since then Decroux has focused on corporeal mime and Marceau on illusion mime, with Barrault concentrating his energies in the legitimate theater. It was a conscious elaboration of a technique over a relatively short period of time with strong intellectual underpinnings. This kind of history makes it very different from classical ballet, for example, or even from modern dance. Ballet evolved its fundamental technique over 100 or more years with contributions from many individuals representing many nationalities and no single underlying principle. Twentieth-century mime was developed by men; modern dance, a form very close to mime in the use of weight and other effort qualities, was created and dominated by women. This appears to be an important factor because both mime and modern dance have only one technique for men and for women, one that features many supposedly masculine movement qualities, and yet modern dance is a form in which women have excelled and mime is not.

Classical ballet differs with regard to technique. It presupposes a basic technique for everyone but recognizes the different physical capabilities of men and women in its two subsets of technique. Women learn the whole vocabulary of *pointe* work and supported adagio, while men practice leaps, multiple beats, and *tours*. Dancers then are judged in terms of the appropriate subset of technique, and, theoretically, outstanding dancers have the same prestige and opportunity regardless of sex. In mime, with only one technique perceived as requiring male qualities, women are at a disadvantage because they are judged within the context of that technique and its perceptions.

Alternatives exist to the unmarked masculine mime style. In mime companies such as that of Henryk Tomaszewski, the repertoire consists primarily of dramas with male and female roles that are played by males and females. Tomaszewski also has notions about what good bodies are, and his company tends to have men and women who are physically attractive in conventionally male and female ways.

Mummenschanz, a Swiss and Italian group whose first company is made up of two men and a woman, work through transformations that involve playing with society's tendency to anthropomorphize. They perform in black leotard, tights, and masks. Sometimes the "mask" covers the entire body as in their representation of a tube that plays

Figure 20–8. Danuta Kisiel-Drzewińska (Helena) and Stefan Niedzialkowski (Parys). Photo by Marek Czudowski.

with a large beach ball. The bodies that are the instruments of performance are essentially gender-free. The subjects that they portray are almost always nonhuman in terms of form. The content, however, involves such human characteristics as love, envy, or pride. The result is that the audience attributes human qualities to gender-free bodies dressed in nonhuman costumes.[1] One example will illustrate this process. In a piece for two, the performers appear in leotards, tights, and black hoods covering their heads and faces. Toilet paper rolls are affixed to the hoods and become the features of the face—eyes, nose, mouth, and ears. The plot is a love story with the two participants

1. Elsewhere (Royce, *in press*), I have discussed the implications of the new show of Mummenschanz for the attribution of human qualities. Briefly, the new show has eliminated most human traces. The face and the body have disappeared beneath head objects and whole-body masks. Reinforcing this is the insistence on abstraction with no distracting naturalistic cues.

Figure 20–9. Mummenschanz. (From publicity photo, International Concert Artists Management.)

becoming increasingly involved—a mild flirtation (a sidelong glance represented by tearing paper off the eye roll) becomes a misunderstanding (agitated talk represented by frantic pulling off the mouth roll) with tears (again, the eye roll), then a reconciliation (listening with the ear rolls and kissing with the mouth rolls).

Finally, female mimes now appear with more frequency both as soloists and as heads of troupes. Ella Jaroszewicz, a former member of Tomaszewski's company, creates both solo pieces for herself and mime dramas for the troupe. Her technique is not different from the male solo mimes like Marceau or Daniel Stein, and she is proficient. The dramas use men and women in roles and costumes that usually reinforce cultural notions about masculinity and femininity, although they sometimes contradict them.

Female solo mimes such as Mamako Yoneyama or Lotte Goslar, while using classic mime technique, take approaches that do not use the leotard-clad body nor do they illustrate classic mime themes. Yoneyama, for example, has developed what she calls Zen Mime. Her major work, *The Ten Bulls,* is based on Japanese women's experience, but she clearly uses the latter to stand for the experiences of humanity in general—a case of the unmarked feminine. Goslar is both a solo performer and the head of a company of dancer-clowns called the Pantomime Circus.

Her pieces stem more directly from the tradition of circus clowns in terms of content and costume. One of her more famous solos, *Old Clown,* is typical of her work. She describes the costume thus: "My body was completely covered with an outrageous potato sack; huge shoes and gloves hid my feet and hands; my hair was under a wig; my nose under a big bulb; and makeup covered every bit of my own skin" (Rolfe, p. 191).

Jaroszewicz, Yoneyama, and Goslar all speak to general themes affecting both women and men and do so on the basis of a technique identical to or very close to that of the classic mime developed by men. There are also feminist mimes whose training has been in the classic idiom but whose pieces speak specifically to women and the feminist movement.

We have a number of examples of female mimes who succeed in the unmarked masculine form with its deemphasis on masculine qualities. Can a female mime be convincingly masculine and vice versa? One reporter-critic believes so. Catherine Freeling observed a class taught by a Decroux-trained mime now teaching in Berkeley, Leonard Pitt. She comments on two student attempts to work with masks (1979).

> A pretty, smiling, young woman with long brown hair puts on the mask of an evil-tempered brute. She is no longer herself: the well-proportioned female body begins moving in a heavy, aggressive fashion. The masculine posture is more convincing than the feminine form. The attractive woman is forgotten—we see only the brutish man. When a tall young man wears the mask of a coquettish woman, the same process occurs. Soon we believe, because through his movement he makes us believe, that we see the coquette before us. (p. 24)

Having observed Pitt and his classes, I can concur with Freeling's analysis. There are some important observations to be made by way of explanation and to avoid the temptation to make this phenomenon more broadly applicable. First, one is influenced in this case by the use of masks. Every mime since Decroux who has worked with masks has said that audiences tend to focus on the face or on the masked face. Masks carry more meaning because our attention is focused on them and distracted from the body. Hence, a mask of an evil-tempered brute signals maleness to us more powerfully than the female body wearing it says femaleness. If the body adopts masculine posture and movement—heavy and aggressive—we see a male character. Second, just as the mask is stronger than the naked face, so the movements and postures are stronger than the body moving naturally. They are, in one sense, caricatures, and therefore they speak more loudly to us. It requires a masterful use of the body to deflect attention from the mask or face when the latter clearly portray some emotion. All the more so when the face conveys one emotion and the body another. Marceau's *Maskmaker* is one of the best examples of this. When the "happy" mask will not come off, Marceau convinces us of his despair through his body all the while the face is fixed in a broad grin.

IV

Although society's notions about masculinity and femininity affect any performer whose art is based on the human body, some categories of performers play on these notions more intentionally than others. Contradicting or reinforcing the stereotypes leads to cross-sex forms. Female impersonators and such high-camp troupes as the Ballets Trockadero are the most obvious manipulators. Les Ballets Trockadero de Monte Carlo, an all-male ballet company that does the classics, is a good source of information on stereotpyes of masculinity and femininity because both are caricatured in the company's performances. The company is clearly out to present parody and makes no attempt to disguise the fact that all the ballerinas are really ballerinos. The classic white tutus reveal large expanses of chest hair, and raised arms show armpit hair—just the reverse of the image of the dancer given us by Toni Bentley. A feminine movement—tiny pas de bourrées accompanied by lightly rippling arms—is abruptly terminated with a flat-footed caricature of a tough masculine walk. The dancers take the classical technique and turn it into a parody of itself. The humor is heavy-handed: dancers fall off point, trip over carelessly tied ribbons, bump into each other. One might argue that no matter how proficient the dancers are technically, there is still something fundamentally wrong about recreating a traditional ballet company with its clearly defined ideas about men and women with a group of male dancers. Arlene Croce speaks to the issue of humor: "Watching ballet in drag can be instructive. . . . Drag ballet provides one answer to the question of why men impersonating women are funny, while women impersonating men are not: it has to do with gravity. (A heavy thing trying to become light is automatically funnier than a light thing trying to become heavy)" (1979, p. 80). We have already noted that the effort-shape quality that most distinguishes between masculinity and femininity in movement is that of weight. The Trockadero dancers, when they want to stress the incongruity of men impersonating women, imbue their movements with weight. The traveling arabesques of the wilis in *Giselle* become heavy and earthbound instead of conveying the image of weightless sylphs.

Female impersonation is a different kind of stylized gender-switching. Although some female impersonators perform in the same flamboyantly heavy-handed style of the Trockadero, many work to present an image of femininity that is indistinguishable from that of a biological female. It is in the area of movement that they sometimes destroy the illusion, particularly with ordinary movement patterns—walking, gesturing, standing, sitting. It happens far less frequently when the performer presents a dance that is identified as being feminine. I have seen male hula performances where the illusion of femaleness was perfect because the performer learned a codified movement system well. Anyone who has mastered the hula—that is, the stereotyped version seen in the mainland United States—will appear female because that is what the form itself implies. Kabuki and bugaku are other codified systems in which males dance all the roles, male and female. Because the movements are so stylized, we have no difficulty accepting a male performance of a female role. It is infinitely easier to learn a movement pattern presented as a systematic, codified form with its own internal logic than to learn ordinary movement patterns. Unless one is exceptionally aware of movement and has the ability to translate what one sees into one's own movement, this capturing of ordinary movements presents insurmountable problems (Royce, 1984). This is neatly put by Jean Genet in *Our Lady of Flowers* (1963), where he describes Divine, a transvestite, responding to a delicate lover by becoming more masculine: "She tried for male gestures, which are rarely the gestures of males" (p. 133).

Lindsay Kemp and his company engage in neither the obvious caricature of the Ballets Trockadero nor the approximation of reality of some female impersonators. His kind of theater shocks, outrages, charms, and seduces by a subtle manipulation of notions of masculinity and femininity. There are two women in his company, but they do not necessarily play roles for women. Conversely, the men in the company play both male and female roles. Kemp himself most often plays female roles or at least roles where the gender is ambiguous. In *Flowers*, based on the Genet play, Kemp plays Divine. He makes his entrance 20 minutes into the production after a shocking scene of prisoners masturbating accompanied by ear-splitting percussion and an equally disturbing scene in the Montmartre cemetery. He glides onstage on the balcony of the cafe. The spot picks him up and follows him for the several long minutes it takes for him to cross to center stage and descend the staircase. His gown is long and rose-colored. He wears a fur around his neck and has a beaded cap that fits closely to his head. He carries a fan. The movement is a perfect display of kabuki style for female characters—each step slides forward with the entire foot flat, raising the toe just at the completion and lowering it again. Its extreme deliberateness and the measured delicacy signify the feminine in kabuki. Kemp's use of the fan strengthens the stylistic impact. Following the two previous scenes which assault the audience with sound, light, and images of death, imprisonment, robbery, nakedness, and a seeming woman who is revealed to be a man, the illusion of an elegant, feminine creature is convincing. The cafe scene then becomes a courting and seduction of Divine/Kemp. Kemp switches from kabuki to the waltz, still preserving that quality of shy elegance. The consummation occurs later with Kemp portraying Divine first as shy as her clothes are stripped from her, then lustily passionate. A rival appears at dawn and Divine is left desolate. At this point the audience gradually realizes that Kemp has led them into an almost perfect rendition of the mad scene from *Giselle:* a nice, symbolic association of the two naive and abandoned women. His performance is heartbreaking and a little mad, ending with a frantic grande jeté offstage, wedding veil trailing from his upflung hand. Kemp, while capable of imitating movements that we regard as feminine, chooses to convey Divine's femininity instead by using highly stylized forms that symbolize femininity. In many ways, it is more effective because we respond to the symbol instead of being caught up in seeing whether he is *really* convincing as a female. At other points, Kemp plays with sexuality in a deliberate attempt to shock. A "woman" appears in the cemetery scene, outfitted with flaming red hair and a black cloak. "She" is partner to a sadistic interlude with a man. Then her cloak is ripped off, leaving her naked and clearly male. More possibilities are introduced by the physical appearance of some of the members of the company. Several of the males have delicate, feminine features and slight builds. One of the women, while small, has the back and arm musculature of a male dancer. In *Salome,* Kemp plays on the feminine features of one of the males who is cast as John the Baptist while Kemp takes the role of Salome. In sum, Kemp is a keen observer of masculine and feminine movement and behavior and creates his theater out of contradiction, reinforcement, paradox, and double transformations. In his theater one can neither ignore sexuality nor take anything as it appears on the surface.

V

The Ballets Trockadero and Kemp's company are instances of cross-sex performance where notions of masculinity and femininity are deliberately caricatured or reversed. I want now to present material on biological males who, from a relatively early age, are raised in the company of women subsequently taking on that gender identity.

Since 1971, I have been observing Isthmus Zapotec society in the southern Mexican city of Juchitan. One component of Zapotec society is individuals known as *mušxe* (Zapotec gloss this in Spanish as *efeminado*). From a very early age (for some, 6 months), the reference group

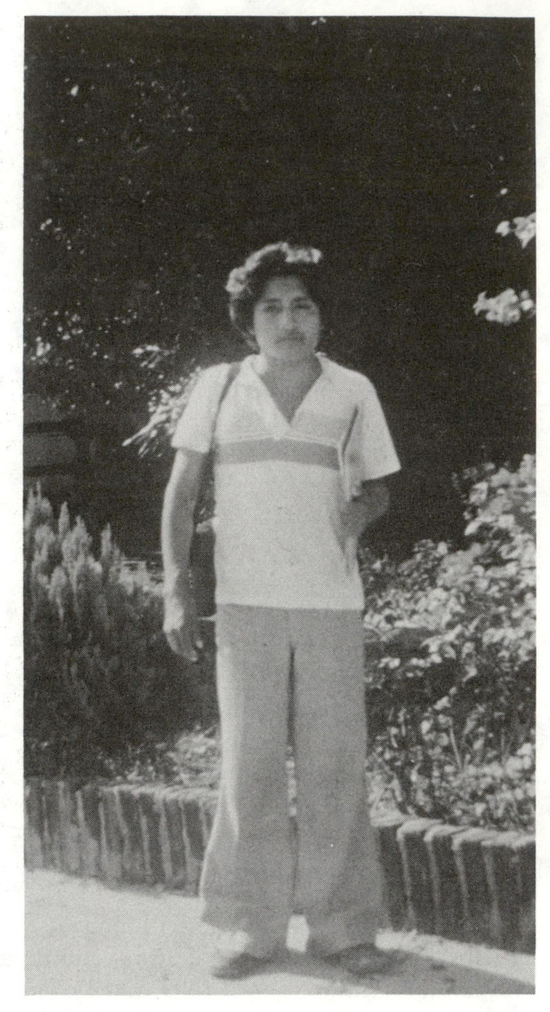

Figure 20–10. Isthmus Zapotec *mŭxe* from Juchitán. (Photo by Anya Peterson Royce.)

of *mŭxe* is female. They spend time in the compound or household helping their mothers, cleaning, taking care of younger siblings. They frequently form play groups with other *mŭxe* or with girls. When pre-adolescent, instead of accompanying their fathers and male relatives to the fields or lagoons, they apprentice themselves to someone in one of the recognized *mŭxe* trades—embroidery, baking bread, making paper flowers, designing and making decorations and floats for *velas* and *vela* parades, or creating the small pottery figurines known as *tanguyu'*. The masters from whom they learn their skills are either older *mŭxe* or, less frequently, women. Those who do not go into these trades find employment as servants, in which case their reference group again is fe-

Figure 20–11. Isthmus Zapotec *mux̌e* from Juchitán. (Photo by Anya Peterson Royce.)

male, or in low-level white-collar jobs where many of their colleagues are women.

Many *mux̌e* are indistinguishable from women in terms of movement—both ordinary and in dance. One of the biggest contrasts between Zapotec males and females in this area is posture and carriage. Zapotec women have an erect carriage, with shoulders held back and chests and hips thrust forward. Standing in conversation, they often place both hands on their hips, knuckles instead of the palmar surface of the fingers resting on the hips. This is a listening posture. When talking, they gesticulate often with wide, expansive gestures, punctuated by sharp, gestural full stops. Walking is also distinctive; the strides are long and usually slow but on occasion brisk. A distinctive kicking forward of the lower leg is seen most clearly among women who wear the traditional costume with its long, gathered or gored skirt and white petticoat. The effect of this walk is to thrust the skirt forward, creating a kind of swirl of petticoat and skirt. In addition to the aesthetic effect created, the movement also frees the leg. Women who have never worn traditional dress seldom show this kind of walk. *Mux̌e,* however do, even though cross-dressing is rare. I should note here that the women with whom *mux̌e* associate most often are those who wear traditional dress and so they are most influenced by traditional patterns of movement. The posture and the walk are influenced, in part, by the heavy burdens that women carry on their heads. *Mux̌e* do this too. Men carry such loads on their shoulders. Everyday movement and gesture are heightened in dance. Some *mux̌e* prefer dancing the woman's part, while others dance the man's. Most retain typical female posture and movement regardless of which part they elect.

Zapotec males stand and move in ways clearly different from women. Carriage is upright or slightly forward with the hips under the line of the shoulders. Watching a conversation among Zapotec men, one is struck by the absence of expansive gesture. If standing, the arms rest at the sides or are crossed in front. If sitting, the palms usually rest on the thighs. Like women, men touch each other's shoulders or arms, although they do not engage in the stroking that women do. The male walk has none of the distinctive characteristics of the women's walk. The man's part in traditional Zapotec dance is characterized by *zapateados* (heel stamps) and a bent-forward posture with one or both hands behind the back.

Women and *mux̌e* seem much more aware of posture, gesture, and movement than do men. Even in ordinary conversation, they seem to be making aesthetic choices and evaluations. They move as if they are conscious of being observed and judged. The only time Zapotec men engage in this kind of stylization is in speechmaking, which is regarded as a kind of art form. Women are also frequent participants in speechmaking and may call on a whole new repertoire of gesture and movement. The general population has clear ideas about what constitutes a good speech and can articulate criteria.

In the previous discussion, I spoke about the difficulty of learning the ordinary movement patterns of the other sex. *Mux̌e* do not experience this difficulty because for the most part their reference group is female from an early age.

In examining elaborated forms that are free of goals and therefore not constrained by anything other than the rules of technique they set for themselves, we see more clearly ideas about masculinity and femininity. Development, changes, and performers' notions of masculine and feminine imagery are influenced by the broader context. Even in reversal or contradiction, society's ideas about gender are present in art forms based on movement of the human body. We must examine them for what they say about themselves and for what they tell us about culture and society.

References

Bentley, Toni. (1982). *Winter season*. New York: Random House.
Croce, Arlene. (1979). *Afterimages*. New York: Vintage Books.
Decroux, Etienne. (1963). *Paroles sur le mime*. Mayenne: Gallimard.
DeMille, Agnes. (1980). *America dances*. New York: Macmillan.
Eisinger, Erika. (1981). The vagabond: A vision of androgyny. In E. Eisinger & M. McCarty, *Colette: The woman, the writer*. University Park, PA: Pennsylvania State University Press.
Freeling, Catherine. (1979). Leonard Pitt. *The Berkeley* (weekly), California.
Genet, Jean. (1963). *Our lady of flowers*. (Bernard Frechtman, Trans.). New York: Grove Press.
Haughton, David, & Harari, Guido. (1982). *Lindsay Kemp*. Milan: Editoriale Domus.
Louis, Murray. (1980). *Inside dance—Essays*. New York: St. Martin's Press.
Philp, Richard, & Whitney, Mary. (1977). *Danseur: The male in ballet*. New York: Rutledge, division of McGraw-Hill.
Pikula, Joan. (1984, January). Sex and death: The lush world of Béjart. *Dance Magazine*.
Roberts, Jean E. (1978). A comparison of expressed psychological masculinity and femininity with movement qualities displayed in the performance of common tasks. Unpublished doctoral dissertation, Temple University, Philadelphia.
Rolfe, Bari (Ed.). (nd.). *Mimes on miming*. Los Angeles: Panjandrum Books.
Royce, Anya Peterson. (1984). *Movement and meaning: Creativity and interpretation in ballet and mime*. Bloomington: Indiana University Press.
Royce, Anya Peterson. (in press). Limits of innovation in dance and mime. *Semiotica*.
Taper, Bernard (1973). *Balanchine*. New York: Harper & Row.
Whitaker, Leslie. (1980). *Pilobolus*. Master's thesis, American University, Washington, DC.

21

Alienated from the Meanings of Reproduction?

Lionel Tiger

Royce's chapter (20) is self-contained and clear on its own terms, and thus it is difficult if not superfluous to comment on it. What struck me as most salient about the material was the constant tension, both in the behaviors described and in her account of them, over the whole question of gender identity. That is, the audience very much seems to want to know: Is it male? Is it female? The choreographer or the mime almost toys with this question, and yet sustains our interest throughout because we don't know for certain whether it is male or female. This stresses one function of gender identity, not only as self-perception but also as a signaling device. We can recall fish that in order to become female actually had to change color. What is the function of this change of color? The signaling function, the communicative function—perhaps much more so than self-conception, gender identity, or other psychological states—is apparently critical, a central function of gender as a source of social transaction. We see this quite brilliantly in the art forms Royce shows us where individuals are seen as certain kinds of people *because* of their gender display, which becomes perhaps more important than their politics or even their economic status.

Although the whole issue of gender identity is scientifically of great importance—and perhaps of paramount importance to individual people and those with whom they interact—we must invoke the law of parsimony in considering it. For example, some persons argue that nonhomosexuals dislike homosexuals, often bitterly, because of unresolved personal conflicts, denial, moral prejudice, and so on. These are self-evidently pervasive features of some social groups—no question about it. Other factors may be at work, also, and I will recite only one case to illustrate what several of these might be. I took part in New York City in a political planning session of a group of people interested in changing the city's legislation, to make it illegal to discriminate against homosexuals in housing, employment, city services, and so on. The meeting was being held at the home of the son of a wealthy and famous woman in the fashion industry—a son who had been fiercely and obviously disowned by his mother because of his sexual preference. I was in a conversation with Harold Brown, the chief medical officer of the city, who indicated that he could not understand why the mother had acted as she had. I asked whether she had any other children. No, he said. Then, I answered, that was the end of Mrs. X. If she had an interest in her own reproductive destiny—one already reflected in having a child in the first place—then presumably her son's effective infertility could scarcely leave her without an opinion on the nature and consequences of his choices in life. Perhaps her antipathy toward her son was unfair, exaggerated, and finally without central compassion. But we do not need to look to the psychology of grievance or even to psychology in general to begin to explain the origin of her strong feeling. The underlying biology of the matter is stark enough to suggest the nature of what is going on.

We needn't do this only with respect to individual cases. I think we can try to use the law and go to large numbers. I regret being scientifically solipsistic by talking principally about my own work, but I will do that nonetheless. I did a study with an Israeli colleague about women in the kibbutz movement in Israel (Tiger & Shepher, 1975), where we had a real population of 34,040 case histories, involving detailed census material we were able to follow over three generations. We also did detailed ethnographic work in two communities and attitude testing in two communities; thus, we had a quite unusual and large data base. Using these very large numbers, perhaps we may make generalizations about what might be sex-linked, gender-identity, or whatever kind of behavior we decide to call it.

In sociology, where I was first trained, the initial theories were generated from the earliest studies, which looked at prisons, unemployed people, or individuals who were homeless or deviant in one sense or another. The theory of social behavior came out of the deviant case. For many years in sociology the central bafflement and technical problem was how there could be society at all, because the theory had been built up on the basis of the tails of the normal curve, not from the norm. I have therefore been principally interested in looking at large numbers, in even crudely normative circumstances, to see if it is possible to make comments about the sexuality of the species.

This position is obviously controversial and perilous intellectually and to some extent politically. Of course, I take John Money's point that the matter has often been highly politicized. I do not accept one feature of the point, however: that people who are interested in sex differences or observe them are necessarily representative of the status quo or even wish to maintain it. That seems to me a completely arbitrary attribution—quite unfair and certainly unnecessary. The contrary argument, in fact, is much stronger: if you want to change a system, you had better understand it. I have had some curious experiences in organizations as various as AT&T and the Pentagon, where individuals were well-meaningly trying to change the positions and demeanor of men and women in the organization. Because they fundamentally misunderstood what they were dealing with, they caused trouble for many people while solving none of the problems that they were willing to spend time and money to solve.

One comment about variation: Eleanor Maccoby once remarked that she was weary of talking with anthropologists who, when she makes a comment about psychology of sex differences, will say, "Yes, but what about the Bonga-Bonga?" The Bonga-Bonga loom large in this kind of problem because there is always somebody in some part of the world who does an extraordinary thing in a strange manner, who has incomprehensible customs or curious costumes or whatever, and this person is therefore held to invalidate the norm. Intellectually, of course, this is not so. But one does have to deal with what the function of variation may be. For my own purposes, I have tried to adapt Romer's rule in biology to this function. Romer's rule, as you may recall, says that animals mutate in order to remain the same; that is, they make minor adjustments in order to maintain the basic structure. So, instead of changing the form of locomotion when faced with a new predator moving into the environment, one changes one's plumage (if one is a bird), skin color, or whatever. I prefer to think of cultural variation in the human context as an application of Romer's rule. That is, cultural variation is a social way of permitting individual cultures, all of whom encounter different circumstances, histories, and so on, to vary small things in order to maintain the basic mammalian, *Homo sapiens sapiens* kind of pattern. If that is the case, our task is to try to discern the broad rules that might constitute a description of *Homo sapiens* in sexual terms. I think we are obligated from the outset to adopt a hypothesis that we are not likely to be random. Just as we are not random physically, as our bodies are essentially predictable, as life cycles are predictable, so it is also likely in behavioral terms that we are not random.

If not random, how can we find out with respect to what, and what is the range of variation of this nonrandomness?

I return to my own study because in a way it is dramatic; and, again, the numbers are so big that it becomes interesting. The kibbutz in Israel is made up of a group of individuals who live collectively. Kibbutzim are largely agricultural, or have been; they were established as long as 80 years ago—it varies. We studied two confederations of kibbutzim, one right-wing and one left-wing. (The differences between them are salient mainly to the kibbutzniks.) We took both these total populations and asked certain questions of the data and of the people.

Perhaps the most innovative things about the kibbutzim have to do with the relationship between the productive and reproductive system. Child rearing, for example, has been communal (or has been until recently; I will return to that). Everybody has the same income, which is no income—everyone gets a small allowance. Everybody has the same access to the means of production, which are totally communally owned. Everybody receives housing and so on from the commune. All food is taken communally, by and large. Laundry, purchasing, and the usual quotidian activities of running a household or a life are done in a professional manner at a central point. Most important, in the customary pattern, after 3 to 6 weeks of age, children are taken from the natal home and raised by professionally trained people in a children's house. Originally this constituted an ideological assault on the nuclear family, from which most of the early Zionists came—the eastern European ghetto family, highly structured, seemingly coercive, and so on. The kibbutz system was an effort to break this down. Similarly, because of the socialistic tradition, the whole question of private property was finessed by not having any. So here we have a real experiment; this is why it is easy to become irritated with fake experiments in the lab, because we have so many real ones sitting there. This kibbutz study was there for 40 years, and nobody did it.

What we found out was really quite dramatic. The initial generation tended to have relatively similar labor-force gender patterns; that is, males and females tended to do more or less the same thing, independent of physical difficulties—hacking rocks to make roads, dragging things here and there, digging latrines, and so on. These tasks, as well as to some extent working in the kitchen and so on, were relatively equally shared—not entirely, but there was a fair symmetry of endeavor between males and females. With each succeeding generation the polarization of work within the labor force increased, so that by the time we did our study, in 1971–72, the polarization of work in the kibbutz was greater than that of Israeli society in general. Note that this is not dependent on income—the customary argument is that women go into lower paying jobs, and that is why they take the work they do. Here income is not a factor, and we had a mysterious case of real polarization based, it appears, on the choices made by individual men and women, and by the community at large, in terms of allocating its scarce resources of labor.

If you take the conventional sociological position about the origin of gender—that it depends on role models, one generation teaches the next, and so on—what you have in every situation in the kibbutz is exactly the reverse. The first generation is the least traditional, and each succeeding generation becomes more traditional in conventional terms. Female groups were more resistant to males joining them than vice versa. We had predicted that male groups would be more resistant, but females tended to resent males more in their work place than the other way around.

The political conduct of the community is, again, totally decentralized and communal; everyone takes part. Here again, we found strong differences between the males and females in terms of what they were interested in. We found increasing with each generation a traditional distinction between women going into management of morale, health, education, nursing, and so on, and males into economic management, security planning, and the business that in our own culture is traditionally associated with male persons.

We wanted to see who spoke more in the assembly, who contributed more to the public activity of making decisions. Careful protocols were taken; we measured the protocols of utterance by males and females, simply by column inches. We discovered that the males talked three times as often as the females; they are endlessly repetitious, they make the same intervention four times on the same point, etc. The only occasion in any meeting in which women spoke more than men was on the question of whether the kibbutz should provide curtains or whether people should be given material and allowed to make their own curtains. We saw a strange assymetry between males and females in terms of the public forum, but we would have predicted it in terms of more conventional patterns of societies with different socioeconomic and political bases.

In terms of education, everyone had an equal option. Boys and girls grew up together, after all; they had roughly the same clothing, they used the same toys, they slept in the same rooms, they were raised under the same circumstances. Their educational choices became highly polarized, however, as they got higher up the system. When they reached the university level, women almost inevitably went to the traditional female choice of social work, psychology, sociology, anthropology—rarely medicine, sometimes nursing. The males went in for engineering, economics, etc. Again, this does not indicate a preference for economic employment, because all the money these individuals received would return to the commune. They made a choice.

The most interesting finding, of course, had to do with the reproductive system. It was here that the most drastic innovation or experiment had occurred. Just before and while we were doing the study, a process began that was inexorable and is still going on—a changing preference against children living in the children's house. Parents want them home. In kibbutz after kibbutz, discussions occurred about repatriating or renatalizing the children. This idea was almost inevitably supported strongly by the females and resisted by the males. In kibbutz after kibbutz, whenever this question came to a vote the females always outvoted the males, and children are now returning to the home.

The males' objection always was: "This is ideologically impure; it's expensive, because all the artifacts you need for raising children you must replicate in each dwelling. Women will withdraw more from the labor force, because they will inevitably end up doing more of the child care, and economically we can't afford this."

There was still a generation gap. The pioneer women would generally vote against the change on the grounds that it was against the intention of the kibbutz. Their daughters and granddaughters said firmly that they wanted the change. When asked why, they came up with answers of remarkable predictability: "We like our kids and we want to be with them. Sometimes I don't trust the nurse. I don't think my kid gets on with the other kids. I think too many kids is not right. I don't like it. I want to be with my kid. Anyway it's more fun." Or something like that.

Bettleheim (1969) has the usual Viennese psychoanalytic interpretation about how these women wanted to be with their babies in order to get back at their mothers, but when we asked them, they said they just wanted to be with their children.

Something else is interesting; it raises a question about humans that we have not really addressed in this volume. The kibbutz women had a very high birthrate. Some researchers claim this is for military and similar reasons, but I don't think so. The women have on the average four children over an 18-year period. They extend the childbearing years almost to the maximum, given the limits of reproductive life. In this situation almost complete contraception is available; nobody need have children if she does not want to, and abortion was always available in the kibbutz movement in years past. The point is their reproductive system is highly productive. Let's contrast that for a second with our own system and indeed the system of most of the industrial countries in which the birthrate is either at or below replacement.

This emerges as a kind of problem. In terms of the world's ecology, perhaps it is fine. If one runs a zoo, the customary view is that if the animals are reproducing it's a good zoo. In our case, we have a very low birth rate. Perhaps signaling ambiguity about maleness and femaleness (that Royce has so elegantly described) has consequences for reproductive capacity. That is, when the women in the kibbutz did not want the men with them at work, were they trying to make a separation between the productive and reproductive systems and somehow maintain, if you will, the breeding tonus of a community, which to produce four children per female took a lot of effort, a lot of money, and so on? It was a choice made by informed and self-conscious women and men.

My point is about the grammar of gender. It has to be seen not simply as a kind of entertainment function or a form of self-expression, as if it were a secretion. This is not just an issue of volition; it has finally to do with what the whole gender business is about, namely reproduction. Today when one looks at Royce's intriguing, tempting question, is it male or is it female?, one wonders whether the enthusiasm for that question has to do with a lack of enthusiasm for the reproductive procedure, revealed after all by very large numbers. We are talking about decisions that have yielded in population terms a new level and pattern of human reproduction. Bear in mind that generally the countries that are richest in terms of the wherewithal to afford children have the fewest. This situation is in a deep sense quite paradoxical and puzzling, unless one considers that communities that move to a gender-generalized position may therefore reduce the salience of the reproductive function. Less grammar, less speech.

References

Bettleheim, Bruno. (1969). *Children of the dream.* London: Collier-Macmillan.
Tiger, Lionel, & Joseph Shepher. (1975). *Women in the kibbutz.* New York: Harcourt, Brace, Jovanovitch.

Author Index

Subject Index